Angle-of-Arrival Estimation Using Radar Interferometry

Angle-of-Arrival Estimation Using Radar Interferometry

Methods and Applications

E. Jeff Holder

SciTech Publishing
an imprint of the IET
Edison, NJ scitechpub.com

SciTech Publishing
an imprint of the IET

Published by SciTech Publishing, an imprint of the IET.
www.scitechpub.com
www.theiet.org

Editor: Dudley R. Kay

10 9 8 7 6 5 4 3 2 1

ISBN 978-1-61353-184-6 (hardback)
ISBN 978-1-61353-185-3 (PDF)

Typeset in India by MPS Limited
Printed in the US by Integrated Books International
Printed in the UK by CPI Group (UK) Ltd, Croydon

Contents

List of Figures

Photo Credits:

Figs 1.1–1.4: Technovative Applications
Fig. 1.5: Trackman A/S (left); XtremeRadar (centre); Flightscope (Pty) Ltd (right)
Fig. 1.6: XtremeRadar (left); Trackman A/S (right)
Fig. 1.7: InfoCenter for Environmental Geology, South Korea (left); EADS Astrium (right)
Fig. 1.8 and 1.9: ERS data © ESA (1995, 1996, 1999), DEM © Eric Fielding, Oxford (1999)
Fig. 1.10: Eric J. Fielding/NASA/JPL
Fig. 1.11: Netherlands Institute for Radio Astronomy (ASTRON)
Fig. 1.12: Commonwealth Scientific and Industrial Research Organisation (CSIRO)
Fig. 1.13: Mullard Radio Astronomy Observatory
Fig. 1.14: Professor Tony Wong, University of Illinois
Fig. 1.15: NRAO/AUI, J. Uson
Fig. 8.6: Propagation Research Associates, Inc

The editor and publisher gratefully acknowledge permission to use copyright material in this book. Every effort has been made to trace and contact copyright holders. If there are any inadvertent omissions we apologise to those concerned, and ask that you contact us at books@theiet.org so that we can correct any oversight.

List of Tables

Preface

The term "radar interferometry" has more than one implication among radar engineers which generally fall along the lines of whether the application is used for imaging or for tracking. In the former case, specific applications include synthetic aperture radar and radio astronomy where radar interferometry facilitates multi-dimensional imaging. For the latter case, radar interferometers provide a cost-effective radar architecture to achieve enhanced angle accuracy for enhanced target tracking. The objective of this book is to quantify interferometer angle estimation accuracy by developing a general understanding of various radar interferometer architectures and presenting a comprehensive understanding of the effects of radar-based measurement errors on angle-of-arrival estimation. As such, this book is primarily directed toward tracking radars but will also discuss imaging applications as well.

Radar interferometers can process either analog or digital signals. The analog interferometer combines energy from two (one-dimensional angle) to three or four (two-dimensional angle) widely separated antennas in free space, whereas the digital interferometer combines received signals in the signal processor usually in the digital domain. As a result, the analog interferometer is transmitting and/or receiving energy with an antenna pattern consisting of multiple lobes referred to as grating lobes, and the digital interferometer processes signals that are ambiguous in angle-of-arrival corresponding to the locations of the grating lobes for the analog interferometer. These angle ambiguities are the result of the interferometer spacing not satisfying the spatial Nyquist rate (one-half wavelength) in order to trade number of antenna elements for angle accuracy. Thus, the radar interferometer can be a cost-effective radar architecture to provide enhanced angle accuracy due to the reduced number of antenna elements but large effective aperture.

In this book we concentrate on the angle-of-arrival estimation performance for various types of radar interferometers that are capable of achieving significantly improved estimation performance over traditional radar antenna architectures. Several types of interferometers are considered, such as a basic digital interferometer, a monopulse interferometer, an orthogonal interferometer, and others. The basic digital interferometer is defined by the use of digitized phase outputs from each antenna of the interferometer array. The phase difference between antennas is then used to derive the angle-of-arrival. In contrast, the monopulse interferometer performs phase differencing in the analog domain and then uses monopulse ratio voltages to determine angle-of-arrival. The orthogonal interferometer is an alternative implementation of digital interferometry through simultaneous transmission of "nearly orthogonal" waveforms.

The synthetic aperture radar (SAR) interferometer is another variety of interferometer that is used to create three-dimensional images for terrain mapping, deformation measurements, earthquake monitoring, and others. Conventional SAR provides two-dimensional imagery (down range and cross-range), and SAR interferometry provides another dimension (up) for imagery. Another type of imaging interferometry is radio astronomy where the motion of the earth is used to create two-dimensional stellar images. We distinguish imaging interferometers from tracking interferometers and focus the book on the latter application but include discussion of the former.

The classical interferometer can be thought of as a radar array that is not fully populated with antenna elements but can achieve angle accuracy comparable to the fully populated array. The interferometer is a special example of a sparse array with specialized angle-of-arrival processing that deals with angle ambiguities in specialized ways. However, there are many examples of sparse arrays that are not interferometers that eliminate angle ambiguities as well as large sidelobes due to the spacing of the elements. In some of these sparse arrays interferometric methods can be used to process angle-of-arrival while in others monopulse techniques are applicable. In Chapter 8 we present a mathematical development for sparse arrays and relate general sparse arrays to interferometry and interferometric processing.

Angle-of-arrival estimation does not come naturally for a radar system. By its very acronym, radio detection and ranging, RADAR, there is no mention of angle estimation primarily because early radar systems were integrating energy to detect targets and determine the time of arrival of pulses for range estimation. To estimate angle-of-arrival, a more complicated antenna is required that essentially differences (differentiates) the signal returns at each antenna, requiring multiple coherent receive antennas or a complex antenna structure. This differencing is performed quite simply for an interferometer with a minimal number of antennas; however, the trade in achieving enhanced angle accuracy is that other effects become significant such as angle ambiguities and sensitivity to phase errors. The primary focus of this book is to (1) define the various interferometer architectures including defining signal processing algorithms required to enable these architectures and (2) identify and quantify the error effects that impact interferometer angle-of-arrival performance.

The book begins with a discussion of applications of radar interferometry in Chapter 1 which include military, sports, radar imaging, satellite tracking, and astronomy. The latter is an application of receive-only analog interferometry, whereas the other applications are radar interferometer architectures. Chapter 2 lays the foundation of probability theory required to formulate the angle accuracy performance for the various interferometer architectures. In particular, the Cramer-Rao Lower Bound is derived for the cases of Gaussian additive noise, and other lower bounds are derived for non-Gaussian errors such as the Bhattacharyya, Bobrovsky-Zakai, Weiss-Weinstein, and Ziv-Zakai bounds, which are tighter bounds for non-linear estimation. In Chapter 3, we summarize some of the fundamentals of basic radar theory and discuss various waveforms that can be used in interferometer architectures. In Chapter 4, we define and discuss the fundamentals of radar angle-of-arrival estimation and derive accuracy estimates for monopulse radar systems. Chapter 5 introduces numerous waveform types that can be used in interferometer arrays including waveform classes that have low cross-correlation. Chapter 6 introduces the various interferometer architectures and derives Cramer-Rao Lower Bound angle accuracy performance for each architecture. Chapter 7 provides a detailed discussion of interferometer signal processing including angle ambiguity resolution and angle estimation algorithms. In Chapter 8 we discuss sparse array geometries and define minimal redundancy arrays. The radar interferometer is an extreme case of sparse array geometry that is not minimum redundancy, which is the reason that angle ambiguity resolution is required for the interferometer. The objective of minimum redundancy sparse arrays is to avoid angle ambiguity resolution with as few antenna elements as possible. Chapter 9 presents an exposition of the effects of certain errors other than additive random noise on angle error performance, and in Chapter 10, we present a detailed discussion of angle errors resulting from tropospheric refraction and turbulence effects.

For the most part this book is an attempt to derive fundamental relationships that quantify the angle-of-arrival performance of a radar interferometer and develop an intuitive understanding of radar angle-of-arrival measurement theory. Most of the results derived in this book can be found to some degree in other sources whose presentations may or may not be obvious and intuitive to the reader. This book attempts to present and justify the fundamental results required to understand the subject matter using a basic mathematical approach that hopefully provides intuition as well. However, some of the material presented is new and attempts to provide alternative derivations and insights into these basic results. As such, the objective is to develop a theory of radar interferometry that provides a fundamental basis for angle-of-arrival performance as affected by noise sources of various types. In some ways, the book explores topics beyond interferometry, primarily because a radar interferometer is, after all, a radar, and most of the basic theory of radar also applies to interferometry. However, the book is not meant to be a radar book per se since there are numerous excellent radar text books that encompass the theory of radar well beyond the scope of this book. Instead this book offers an in-depth look at the derivation of angle error equations for a radar interferometer as affected not only by additive noise but by other error effects such as multipath, glint, and spectral distortion.

Radar angle estimation is explored through antenna architectures other than an interferometer architecture. The basic monopulse antenna is introduced to provide a basis for comparison with conventional radar angle estimation, so-called eigen-based super-resolution techniques are presented to provide a comparison with modern angle estimation techniques, and sparse array architectures are presented to show how interferometry can be generalized to multiple antennas. The common thread throughout the book is the development of an understanding of not only angle-of-arrival estimation but also radar interferometry as it applies to angle estimation.

Finally, a word of caution about terminology is warranted. The term *precision* denotes the second-order statistics about a mean zero process denoted as the variance, whereas *accuracy* denotes both first- and second-order statistics. For Chapters 1–7, the discussion of error assumes that the mean of the error distributions is zero and thus the terms *accuracy* and *precision* are synonymous. The second-order moment (variance) or standard deviation (square root of variance) is what is computed in this book for the various interferometer architectures, and, as such, the book is really a discussion of radar precision. In Chapter 8, we consider error sources that affect the mean of the error distribution where the term *accuracy* is really the appropriate terminology.

Acknowledgments

A colleague, Mr. Ben Perry, at the Georgia Tech Research Institute (GTRI) once told me that all equations that attempt to predict the performance of radar were wild and fanciful musings. As an experimental physicist and radar test engineer, he had come to understand that radar performance was whatever the radar system would allow on any given day, and he knew that performance could change from day to day or even hour to hour. After numerous experiences with radar testing, I grew to appreciate Ben's healthy skepticism about radar performance prediction; however, my education in mathematics and years working in radar have allowed me to develop an appreciation for systems analysis that develops a fundamental understanding of radar performance. After all, the objective of science and engineering is to develop a framework that can predict the outcome of experiments, and radar should be no exception. I have dedicated this book to the memory of Ben Perry in that we should always remember that radar analysis can be wild and fanciful musings unless we understand the assumptions that go into the analysis.

This book represents 26 years of work at GTRI and 12 years at Propagation Research Associates, Inc., (PRA) where I am currently located. The book focuses on my work in angle-of-arrival estimation using mostly interferometric methods but also discusses other methods for angle estimation for comparison. I am most grateful for the support that I received over the several years that this theory was formulated and years prior where I established the fundamental understanding that permitted this undertaking. In particular, I would like to thank Profs. Mike Reed and Dave Schaefer at Duke University for instilling a basic appreciation for mathematical rigor, Dr. Daniel H. Wagner for an appreciation of honesty and integrity in research, and Dr. Ed Reedy who had the confidence to hire me at GTRI and provide opportunities during my tenure. I would like to thank all the people that I have worked with associated with the U.S. Army that provided me the means to continue my pursuit of radar interferometry. Some of these include Col. (retd.) Charles (Chuck) Driessnack, Dr. Jim Baumann, Mr. Jim Mullins, Mr. Heinz Sage, Mr. Chris Hamner, Mr. Ron Smith, and Mr. Mark Shipman. I would also like to thank Mr. Jim Williams, the president of Technovative Applications, Inc., who, as a pioneer in radar interferometry, provided insight and guidance over the years.

I also thank all of my colleagues at PRA who helped me develop the understanding and appreciation for the subject. In particular, I thank Ms. Susan Dugas who provided the motivation and foundations that allowed this book to happen, Dr. Martin Hall who continually challenged my technical arguments allowing me to develop a deeper understanding of the subject matter, and Dr. Bonnie Valant-Spaight who provided review and material for Chapter 9.

If I have left anyone out, it is surely not intentional because this was truly a collaborative effort over several years in the making. I cannot emphasize enough the value in developing technology and building hardware in terms of developing an understanding of how theoretical principles are applied to real-world problems. My work at GTRI allowed me to develop a theoretical framework for radar, but my work at PRA really molded and validated this framework through hardware development. I want to thank Dr. Preston Geren, an Associate Technical Fellow of the Boeing Company, for his thorough review, resulting in valuable corrections and comments and Mr. Mel Belcher of the Georgia Tech Research Institute for his suggestions on the organizational structure. Both reviewers have made the book a more complete and accurate exposition of the topic.

Finally, I thank my wife, Deborah, and my children, Jason, Catherine, and Elizabeth, who put up with all my idiosyncrasies during the process of completing the book.

Chapter 1
Applications of RF Interferometry

The first known use of an interferometer was in the work of Michelson and Morley in 1890 and again in 1920, where they used interference patterns from light emitted by stars to measure the diameter of large stars. The stellar image creates an interference pattern related to the diameter of the star and the size of the optical aperture. Shortly thereafter, in 1946, Ryle and Vonberg transferred the principles of optical interferometry to radio waves for solar observations. These early interferometers combined coherent analog signal to create amplitude interference patterns to achieve enhanced angular resolution for stellar measurements and imaging. With advances in radio frequency transmitter and receiver technology and in analog-to-digital converters, interferometers have entered the digital age. The digital interferometer spatially samples signals at rates greater than or equal to the Nyquist rate for the bandwidth limited signals and uses phase, as opposed to amplitude, information to measure angle-of-arrival with high precision. As a result, these digital radio frequency (RF) interferometers have found application in several areas that include military, commercial, and scientific endeavors. In this chapter, five applications are presented that illustrate the diversity and versatility of interferometry: military, sports, synthetic aperture radar (SAR), radio astronomy, and geostationary satellite tracking. Both SAR and radio astronomy are imaging interferometer techniques that take advantage of large aperture separations, whereas military and sports applications are tracking radars that use interferometry to achieve high angle accuracy as opposed to high angle resolution. In addition, radio astronomy is an example of a passive interferometer whose signal is generated from an external source (stellar objects). The other three interferometer applications are active radars that generate specific waveforms or signals that facilitate interferometer angle estimation. This book focuses on active tracking interferometers where angle accuracy is the driving requirement. The distinction between tracking and imaging interferometers is made in this chapter using the five examples.

1.1 Military Applications

The advantage of an interferometer operating at radar frequency (RF) is its potential for enhanced angle measurement accuracy in an all-weather environment, and for application to military systems that typically need enhanced angle accuracy are fire control radars. A fire control radar is employed to direct fire, which can be missiles, projectiles, laser energy, or high-power RF, on an intended target.

Figure 1.1. Hypervelocity Weapon System Fire Control X-Band Radar Structure Developed by Technovative Applications Using a 10-m Baseline

In each case, an accurate aiming point must be determined to direct the response so as to achieve the desired effect. The RF interferometer provides a good trade between angle accuracy and cost due to its increased aperture size but limited number of total antenna elements.

Figure 1.1 shows an interferometer concept that was designed in the early 1990s for the Hyper-Velocity Weapon System that was developed for the Space Defense Initiative (SDI) by Technovative Applications, Inc. The picture is actually a mock-up of what the system would have looked like if the individual array antennas had been fully populated with elements. The program was cancelled when the SDI was terminated. According to the requirements of the SDI program, the separation of the antennas was several hundreds of the wavelength in order to enhance angle accuracy, but this architecture also created multiple angle ambiguities. The central antenna would have been used to transmit signals, whereas the vertex antennas would have been used to receive the signal. This setup of a single transmit and three receive antennas was developed by Jim Williams, the founder of Technovative Applications, Inc., and has since been replicated for numerous applications.

In addition to directing fire to engage airborne threats, RF interferometers provide excellent radars for indirect fire weapons. The indirect fire mission calls for sufficient accuracy to locate the weapon impact point for fire adjustment, and thus the radar must be able to track the weapon accurately from launch to impact.

Figure 1.2. Projectile Tracking System (Technovative Applications)

Figure 1.2 shows an interferometer system developed by Technovative Applications in Brea, California, to improve the accuracy of artillery. Again, the system is comprised of three receive antennas and one transmit antenna. Another similar design is shown in Figure 1.3, which is designed to track outbound mortar fire. In one of their more recent interferometric radars, Technovative Applications developed the radar shown in Figure 1.4 for counter rocket, artillery, and mortar fire. In all of these designs, the separation of the receive antennas provides enhanced angle accuracy at a reduced cost over a fully populated antenna with the same aperture dimensions.

1.2 Sports Applications

The RF interferometer has made its entry into several sports applications for measuring ball position and other parameters. The first radar applications in sports were simple velocimeter radars that, like police radars, measured only the velocity of a ball. This type of radar was used to measure the velocity of pitches thrown in baseball games, of served tennis balls during matches, and of golf balls after being struck with a driver off the tee. The distance a ball traveled could be determined by integrating velocity over the time of flight. However, recently these systems have all been enhanced with designs that include interferometry for tracking balls in three dimensions (range and two angles) to determine the precise ball trajectory.

Interferometer radars can now track baseballs, tennis balls, soccer balls, cricket balls, hockey pucks, and golf balls to determine their velocity and trajectory. For example, after a home run in a baseball game, the trajectory of the ball, as measured with an interferometer, is superimposed on the screen. However, the sport that has

Figure 1.3. Mortar Tracking System (Technovative Applications)

Figure 1.4. Counter Rocket, Artillery, and Mortar Interferometric Radar (Technovative Applications)

probably made the most use out of an interferometer is golf. In golf, interferometers are used for training aids to measure not only ball parameters but also golf club parameters.

The TrackMan system is an RF interferometer that can measure ball velocity, clubhead velocity, swing plane, ball spin in two axes, and ball trajectory parameters, including maximum ball height and distance to ball impact. TrackMan is a sophisticated radar system that uses highly technical processing to extract information about golf balls in order to help golfers optimize their swings and equipment. In addition to TrackMan, other radar systems, such as the Zelocity and FlightScope, are on the market for golf applications that purportedly measure similar parameters. Figure 1.5 shows the products for the TrackMan, Zelocity, and Flight Scope launch monitor systems, all of which utilize some form of radar interferometry. In addition, other radars can be found in indoor training simulators that are used for club fitting in major golf equipment stores. It is not uncommon to find golf radar systems in pro shops and golf stores. Figure 1.6 shows radar technologies being used by golfers to track the trajectory of the golf ball.

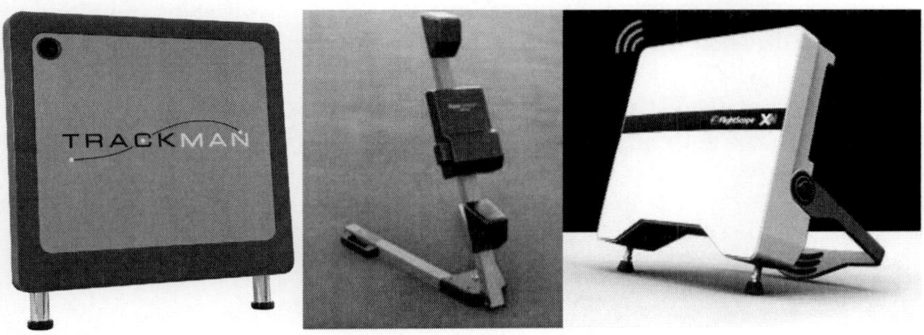

Figure 1.5. Radar Golf Launch Monitors: TrackMan III (left), Zelocity PureLaunch™ (center), and Flight Scope X2 (right)

Figure 1.6. Golf Radar Monitoring Devices in Use

The waveform of choice for sports applications is usually a frequency-modulated continuous-wave (FMCW) waveform for cost-effectiveness. Because ball velocity and ball spin are the parameters of interest, the waveform consists of a number of FMCW pulses to create a coherent dwell of pulses to measure Doppler frequency and Doppler bandwidth. In monitoring golf, the waveform can also be used to measure the maximum clubhead speed because clubhead speed is a factor of about 1.4–1.5 slower than the ball velocity. Angle-of-arrival resolution is determined by the interferometer antenna separation and the operating frequency. Operating frequencies in the X-band and Ku-band are typically chosen to maintain angle resolution for reasonable size and cost. The advantage of these sports radar interferometer systems over conventional radar is that they generally have *a priori* knowledge about the initial location of the ball relative to the radar.

As seen in Figure 1.6, the distance between the initial ball position and the monitoring radar is established during setup. Knowledge of the initial ball position facilitates angle ambiguity resolution for these systems. The antenna is set up so that the ball trajectory is contained within the unambiguous field of view. For a golf radar interferometer, the antenna separation can be approximately 20 cm, and for a 10-GHz operating frequency, the unambiguous field of view is about ±4.5°, which at a 300-yd distance translates into a 90-yd cross-range distance. For most golf shots, the ball will likely be contained in the unambiguous field of view but will certainly be contained within the first angle ambiguity, which is ±9°. As such, these systems are not required to deal with resolving angle ambiguities as with conventional military radar interferometers.

1.3 Synthetic Aperture Radar

Synthetic aperture radar (SAR) is a technique that uses the motion of the radar platform, such as an airborne platform or a satellite, to achieve high angle resolution by creating a large baseline synthetic aperture along a determined path. By also using high range resolution waveforms, SAR is able to develop two-dimensional images with enhanced range and cross-range resolution. The platform motion dictates the nature of the SAR image. Linear motion allows the SAR system to illuminate and image a strip along the ground (strip-map mode SAR), whereas a circular motion illuminates a more localized area on the ground (spot mode SAR). More recently, SAR has used interferometry to create three-dimensional images. A typical application is the use of multiple satellite orbits to create a baseline that provides angle resolution in a third dimension, which, combined with SAR processing, creates a 3-D SAR image in the down-range, cross-range, and up-range dimensions. Figure 1.7 shows an example geometry in which two orbits of a satellite create an interferometer baseline separation for 3-D SAR imaging.

The SAR interferometer uses differential phase measurements from a reference point on the ground and a point of interest located within the same range cell but at an elevation above the reference point. SAR interferometry thus provides relative locations in the vertical dimension because the phase of a signal reflected from the

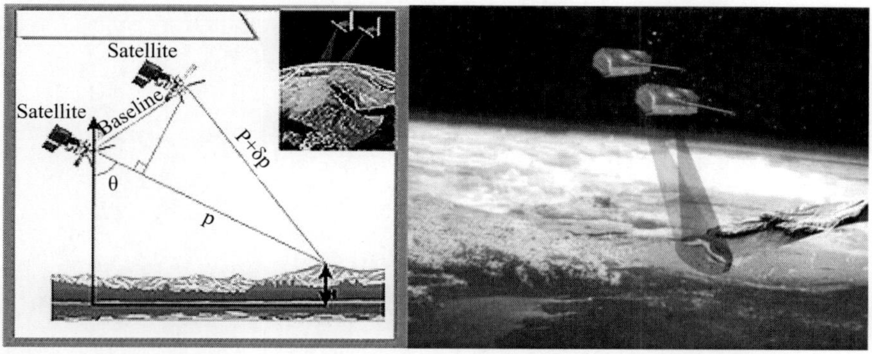

Figure 1.7. SAR Interferometry Imaging in the Vertical Dimension Using Two Orbital Satellite Locations and a Reference Point P on the Ground

reference point at both orbital positions is assumed to be known to a high degree of accuracy. To determine with sufficient accuracy the relative locations of the reference point and a point of interest, the radar must maintain coherency for the two orbital locations. Also, the scene must remain stationary over the time between orbits. In Chapters 6 and 7, a more detailed discussion of SAR interferometry imaging is presented.

Essentially, SAR interferometry was first developed in the late 1960s using an aircraft platform. By the 1980s, the technology had matured to provide topographic imaging with accuracies between 10 and 30 cm. The advent of the U.S. space shuttle created the capability to use multiple satellite passes (repeat pass method) for 3-D topographic imaging. The improved accuracy of the platform locations and phase measurements increased SAR interferometry accuracy by an order of magnitude. As a result, a new application emerged: deformation mapping. Among the deformation mapping applications are earthquake effects such as faults and tectonics, volcano monitoring, land subsidence, glacier and ice motion, and atmospheric refractivity variation. A reasonably complete history of the development of SAR interferometry is contained in [1]. In general, atmospheric refraction variations degrade interferometer accuracy. However, when the imaging surface is well understood, the variations in the differential phase from the expected differential phase provide a means to measure atmospheric variation effects. Figures 1.8, 1.9, and 1.10 show the results of SAR interferometry applications.

1.4 Radio Astronomy

The principles and applications interferometry to radio astronomy have developed and proliferated over the last few decades [2,3]. The need to achieve higher resolutions led to the development of radio interferometry. The original developers of radio interferometry were British radio astronomer Martin Ryle and Australian-born engineer, radio physicist, and radio astronomer Joseph Lade Pawsey, and Ruby Payne-Scott in 1946. The first time that radio astronomy was used for an

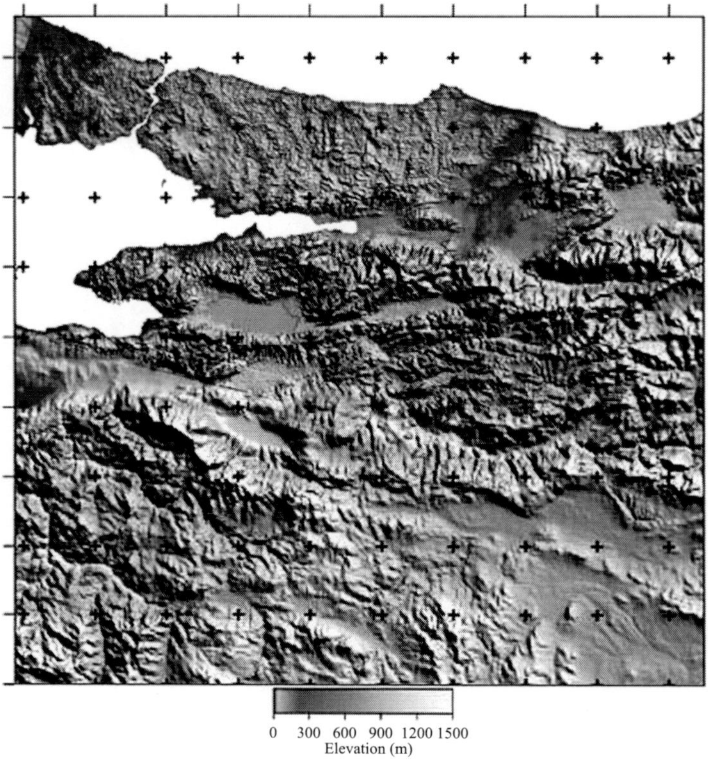

Figure 1.8. Terrain Mapping Using SAR Interferometry

astronomical observation was an experiment carried out by Payne-Scott, Pawsey, and Lindsay McCready on 26 January 1946 using a converted radar dipole antennas near Sydney, Australia. These researchers implemented an array of World War II radars on sea-cliff and observed the sun at sunrise with interference arising from the direct radiation from the sun and the reflected radiation from the sea. The radar frequency was 200 MHz and the baseline was about 200 meters. Payne-Scott, Pawsey, and McCready (PS&M) concluded that the solar radiation during the burst phase was much smaller than the solar disk and emanated from a region associated with a large sunspot group. PS&M developed the principles of aperture synthesis in a seminal paper published in 1947. During World War II, sea-cliff interferometers were developed and demonstrated in Australia, Iran, and the United Kingdom, and observed interference patterns from incoming aircraft created from the direct radar signal and the reflected signal from the sea.

In 1946 Ryle and Vonberg observed the sun at 175 MHz with a Michelson interferometer consisting of two radio antennas separated by a distance of 240 meters [3]. They concluded that the radio radiation source was smaller than 10 arc min in size and that certain burst of radiation consisted of circular polarized signals. Two other groups, David Martyn in Australia and Edward Appleton in the United Kingdom, also detected circular polarization at about the same time.

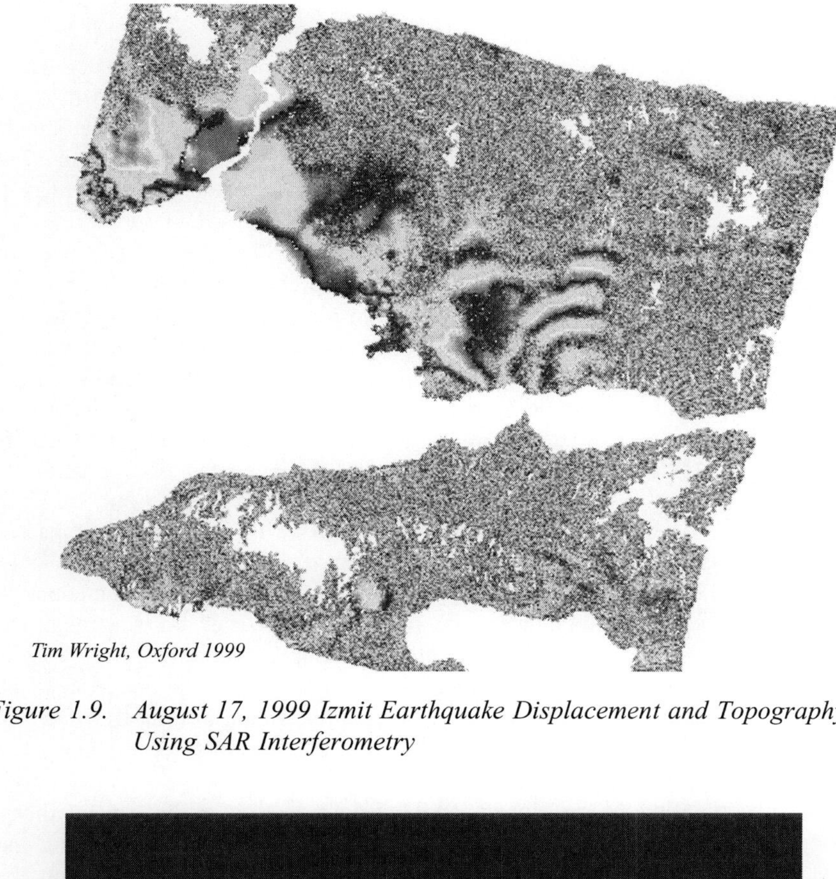

Figure 1.9. August 17, 1999 Izmit Earthquake Displacement and Topography Using SAR Interferometry

Figure 1.10. Topography Imaging Using SAR Interferometry (Eric J. Fielding/NASA/JPL)

1.4.1 Stellar Imaging Using Radio Astronomy

Early applications of radio astronomy used interference patterns created from separated antennas to determine the size of stars. A major development in radio astronomy was using the earth's rotation to synthesize images of stellar objects such as galaxies [2]. Earth rotation synthesis was first introduced by Ryle in 1962 using two antennas separated by a few thousand meters. By placing two antennas in an east–west orientation and making measurements over a 12-hour period the relative angular orientation of the antennas rotates over 180 degrees when projected onto the plane normal to the source. The integrated effect of these observations creates a two-dimensional pattern that can be transformed to create a target image. By increasing the separation of the two antennas the image resolution can be improved with a penalty of increasing the interference pattern frequency. These additional lobes in the pattern are grating lobes due to the violation of the spatial Nyquist spacing. It was realized that if the distance between the antennas is increased that additional antennas can be located between the two antennas to reduce the frequency of the lobing structure. As a result the Very Long Baseline Interferometer (VLBI) arrays came into being around 1970. These arrays consisted of long baselines with multiple antennas to improve images. Figures 1.11–1.13 show examples of existing VLBI arrays consisting of fourteen (14) antennas at the VLBI at Westerbork Netherlands, six (6) antennas at Narrabri Australia, and eight (8) antennas at the VLBI at Cambridge England, respectively.

The projection of the VLBI array onto the plane normal to the direction of the source creates a linear representation of the two-dimensional Fourier transform of the image. The rotation of the earth creates multiple linear representations. The Projection Slice Theorem [2] basically allows total two-dimensional image construction by computing the two-dimensional Fourier transform of these linear slices. The development of the computer and the accessibility of computer processing and memory made possible the precise alignment and timing of receivers and the computation of the number of Fourier transforms. Figure 1.14 shows a point spread function created from a three-antenna system. The image is constructed by taking the Fourier transform of the point spread function. Figure 1.15 shows a VLBI array image of the Whirlpool Galaxy.

The VLBI antennas are not physically connected, but instead, the data received at each VLBI antenna is precisely stamped with timing information, determined by a local atomic clock [3]. The data is then stored for later analysis on magnetic tape or hard disk. Data processing consists correlating the data from all of the antennas to produce high resolution images by applying the Projection Slice Theorem. As a result it is possible to achieve an effective antenna aperture nearly the size of the Earth which enables very high angular resolution stellar imaging that is much higher resolution than optical telescopes. For example, radio telescopes that operate at the highest frequencies can resolve as small as 1 milli-arcsecond.

The primary VLBI arrays that are in operation today are the Very Long Baseline Array (with telescopes located across the North America) and the European VLBI Network (with telescopes in Europe, China, South Africa, and Puerto Rico) [3]. Normally each array usually operates independently, but occasionally the

Figure 1.11. Westerbork Netherlands 14-Antenna Array

arrays operate in combination as the Global VLBI which increases sensitivity to improve accuracy. Until the advent of optical fiber, recording data onto hard media that was physically transported to a site for data processing was the only way to correlate the data. However, since high-bandwidth optical fiber is now available worldwide it is possible to do Global VLBI in real time through optical networks. VLBI telescopes have been responsible for a number of astronomical discoveries such as the detection of the motion of the sun around the Galactic center from the proper motion of Sagittarius A-star, which is believed to be the black hole that is the center of the Milky Way [2].

1.5 Near-Geostationary Interferometric Tracking

Tracking of geostationary satellites has become increasingly important due to the number of these satellites that have been deployed for commercial and military uses [4].

Figure 1.12. Narrabri Australia Six-Antenna Array

Figure 1.13. Cambridge UK Eight-Antenna Array

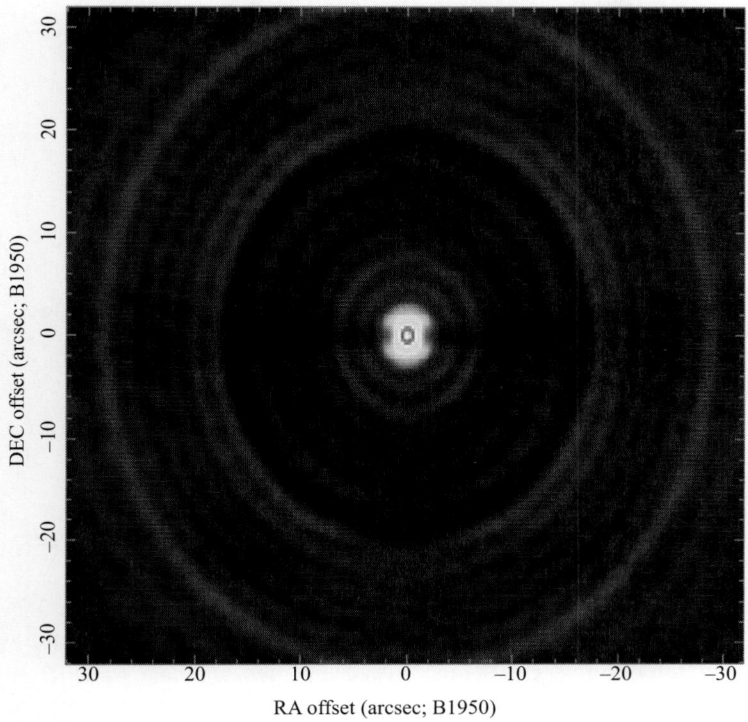

Figure 1.14. Point Spread Function Using Three Antennas and Three Baselines

Figure 1.15. Image of the Whirlpool Galaxy Using a VLBI Array

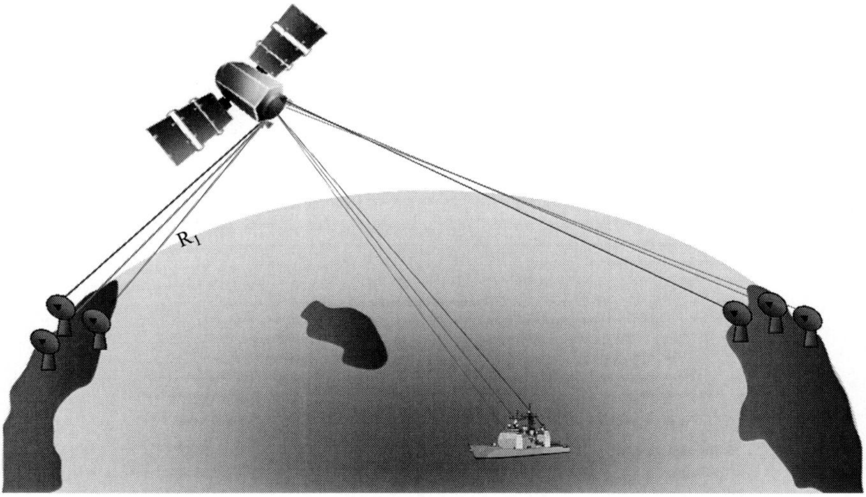

Figure 1.16. Near-Geostationary Satellite Tracking Using Earth-Based Interferometers

Figure 1.16 is a depiction of a geostationary satellite being tracked by multiple earth-based interferometer systems. Because a geostationary satellite must be located directly above the equator, the increased number of these satellites means that their locations are getting closer to one another, thereby requiring improvements in orbit estimation accuracy.

The radar interferometer is the ideal instrument to measure the orbits of near-geostationary satellites due to this type of satellite's restrictions in orbital motion. Even though the intent is to place a satellite in geostationary orbit, forces act on it to push it into a near-geostationary orbit. These near-geostationary orbits remain in nearly fixed positions relative to the rotating earth. The angular movement relative to a fixed point on the earth directly below the satellite position is usually less than 1°, and thus the radar interferometer is not required to scan over a large field of view. For the most part, the interferometer can consist of dish antennas that can mechanically scan to point to the satellite of interest.

The key idea is that an interferometer that measures only azimuth and elevation angle can determine the six parameters that define a near-geostationary orbit [4]. Figure 1.17 illustrates an idealized azimuth-elevation interferometer with baselines AB for azimuth and AC for elevation. The near-geostationary coordinate frame is defined by the longitudinal axis L, the range axis R, and the north axis Z the origin located at the point defined by the intended geostationary satellite location. The one-dimensional azimuth-only interferometer, defined by AB, actually determines four of the orbital parameters that can provide useful longitudinal satellite relative tracking information. The addition of the elevation baseline provides observability to the other two parameters that include the orbit inclination. The addition of range information provides observability to the other two

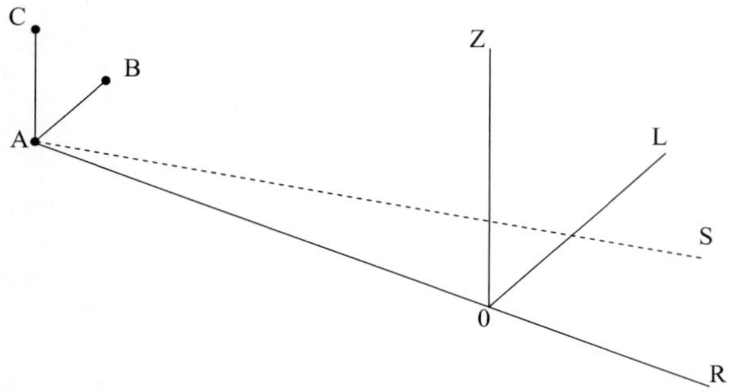

Figure 1.17. Idealistic Interferometer with Baselines AB and AC and the R-L-Z Coordinate Frame for Near-Geostationary Orbit Tracking

parameters; thus, a range-azimuth interferometer also measures all six orbital parameters.

One of the objectives for near-geostationary interferometer tracking is to determine the relative locations of two closely spaced satellites. Differential interferometric tracking provides additional accuracy in estimating relative satellite locations. Differential tracking places the two satellites in the same antenna beam for each of the interferometric antennas. As a result, system errors that are common to differential phase measurements for both azimuth and elevation cancel out in the differential location estimate. The satellites need to be separated in either range or Doppler because the relative angle between them is less than the antenna beamwidth.

References

1. R. F. Hannsen, *Radar Interferometry*, Kluwer Academic Publishers, Dordrecht, Netherlands, 2001.
2. A. R. Thompson, J. M. Moran, and G. W. Swenson Jr., *Interferometry and Synthesis in Radio Astronomy*, John Wiley & Sons, Inc., New York, 2001.
3. http://en.wikipedia.org/wiki/Radio_astronomy
4. S. Kawase, *Radio Interferometry and Satellite Tracking*, Artech House, Norwood, Massachusetts, 2012.

Chapter 2
Probability Theory

The theory of probability is a special case of general measure theory in mathematics and as such has its fundamental definitions and basic results derived from real analysis. In this chapter, we define the fundamentals of probability theory only to the extent necessary for its application to radar interferometry error analysis, while maintaining some fidelity to measure theory. For more detailed expositions of probability theory, the reader can refer to the references or any number of textbooks on the subject [1,2]. The Cramer-Rao lower bound is introduced and will be used in later chapters to provide an estimate for unbiased angle-of-arrival estimates. The Weiss-Weinstein, Ziv-Zakai, and Bhattacharyya lower bounds for biased random processes are defined and applied in Chapter 7 to estimate the performance of angle-of-arrival in the presence of angle ambiguities.

2.1 Random Variable

In essence, a random variable maps events to a number. In this book, the events are estimates of a radar signal parameter, such as phase or angle-of-arrival, and the number is the actual measurement of the radar parameter. Let Ω be the set of possible observable elementary events for a given observation of a radar signal parameter, and let U denote the *sigma algebra* generated by the family of subsets of Ω. The sigma algebra is the set of all possible finite unions and intersections of sets, as well as the countable unions of sets contained in the family of sets belonging to Ω. The elements of U are called measurable sets, and the pair (Ω, U) is called a *measurable space*.

A *random variable* X is a mapping from one measurable space to another that preserves preimages of measurable sets as measureable sets. Let $X{:}\Omega\rightarrow\Omega'$ be a random variable; then

$$X^{-1}(A') \in \mathrm{U} \quad \text{for all } A' \in \mathrm{U}' \tag{2.1}$$

Let (Ω, U) denote a measurable space; then a function μ defined on U is a *probability measure* or *probability* if

$$0 \le \mu(A) \le 1 \quad \text{for all } A \in \mathrm{U} \tag{2.2}$$

$$\mu(\phi) = 0 \quad \text{where } \phi \text{ is the null set, and} \tag{2.3}$$

$$\mu\left(\bigcup_{n=1}^{\infty} A_n\right) = \sum_{n=1}^{\infty} \mu(A_n) \qquad \text{for } A_n \in \mathrm{U} \text{ and } A_n \cap A_m = \phi \tag{2.4}$$

The triple (Ω, U, μ) is called a *probability space.*

For the applications of interest in this book, we are concerned only with the *Lebesque measure* λ [3], defined on the sigma algebra of sets in R^d generated by all d-dimensional intervals of the form $a_k \le x_k \le b_k$ for $k = 1, 2, \ldots, d$. A sigma algebra defined in this way is called the *sigma algebra* B^d *of Borel sets* [4].

2.2 Probability Density

Let (Ω, U, P) be a probability space and (Ω′, U′) be a measurable space, and let X be a random variable with values in Ω′. Now define the function P_X as follows:

$$P_X = (A') = P\big(X^{-1}(A')\big) = P\{\omega\colon X(\omega) \in A'\} \quad \text{for all } A' \in U \tag{2.5}$$

The function P_X is called the *distribution* of P.

For an R^d valued random variable X, the distribution P_X is uniquely defined on B^d by its distribution function, defined as follows:

$$F(x) = F(x_1, x_2, \ldots, x_d) = P\{\omega\colon X_1(\omega) \le x_1, \ldots, X_d(\omega) \le x_d\} = P(X \le x) \tag{2.6}$$

The distribution function F shows how likely it is that X will assume vales to "less than" the point $x \in R^d$. The function F is also called the *joint distribution* of the scalar random variable $X_1, X_2, \ldots, X_d$ that are the components of the d-dimensional random variable X. When $d = 1$, the distribution function F is an increasing and continuous function with the property

$$\begin{aligned} F(-\infty) &= \lim_{x \to -\infty} F(x) = 0 \\ F(\infty) &= \lim_{x \to \infty} F(x) = 1 \end{aligned} \tag{2.7}$$

The *probability density* for a distribution function F is defined as an integrable function f such that

$$F(x) = \int_{-\infty}^{x_1} \int_{-\infty}^{x_2} \cdots \int_{-\infty}^{x_d} f(y_1, y_2, \ldots, y_d) dx_1 dx_2 \ldots dx_d \tag{2.8}$$

As can be seen, F is differentiable and hence continuous, and

$$\frac{\partial^d F}{\partial x_1 \partial x_2 \ldots \partial x_d} = f(x_1, x_2, \ldots, x_d) \tag{2.9}$$

An R^d-valued random variable has *normal density* (or *Gaussian density*) if

$$f(x) = \left[(2\pi)^d \det(C)\right]^{-1/2} \exp\left(-\frac{1}{2}(x - m)C^{-1/2}(x - m)^T\right) \tag{2.10}$$

where $m \in R^d$ and C is a positive definite $d \times d$ matrix. The parameters m and C have significance as first- and second-order statistical moments, which will be defined in the next section.

2.3 Mean and Covariance

For a given probability density function $p(x)$, we would like to characterize the density with only a few parameters. With that in mind, we define the *expected value* or *mean* (m) of an R^d-valued random variable X with density p.

$$m = E(X) = \int_{R^d} XdP = \int_{R^d} xp(x)dx \tag{2.11}$$

The mean is also called the *first moment*, which is designated as m_1. When $d=1$, higher-order moments can be defined by

$$m_k = \int_R x^k p(x)dx \qquad \text{for } k = 1, \ldots, \infty \tag{2.12}$$

It should be noted that for some distributions the integral above does not exist for some higher-order moments.

Another important parameter is the centralized second-order moment, or the covariance for the multivariate distribution. For an R^d-valued random variable X, the *covariance* is a $d \times d$ matrix defined by the following:

$$C = \int_{R^d} (x - m)^T (x - m)p(x)dx \tag{2.13}$$

When $d=1$, the covariance is simply called the variance due to the absence of correlated terms, and (2.13) reduces to

$$C = \text{var}(X) = \int_R (x-m)^2 p(x)dx = \int_R (x^2 - 2xm + m^2)p(x)dx = m_2 - m^2 \tag{2.14}$$

For a Gaussian density, it turns out that all higher-order moments can be expressed as a function of just the mean and variance. Thus, a Gaussian is completely defined by just the two parameters, mean and variance, and the terms that define the preceding Gaussian density are exactly what we have shown. We restate this important expression for the Gaussian density.

$$f(x) = \left[(2\pi)^d \det(C)\right]^{-1/2} \exp\left(-\frac{1}{2}(x - m)C^{-1/2}(x - m)^T\right) \tag{2.15}$$

where m is the mean and C is the covariance of the Gaussian density. For the majority of this book, we will assume that errors have a Gaussian distribution.

2.4 Maximum Likelihood

We now extend the density function so that the density is a function of a parameter θ and we write $f(x|\theta)$. For n samples $x_1, x_2, \ldots, x_n$, we define the *likelihood function* as

$$\mathcal{L}(\theta|x_1, x_2, \ldots, x_n) = \prod_{k=1}^{n} f(x_k|\theta) \tag{2.16}$$

In general, the likelihood function is not a probability density function; however, it is useful in estimating certain parameters that densities depend on. Relating the likelihood function back to a Gaussian density, we see that

$$\begin{aligned}\mathcal{L}(\theta|x_1, x_2, \ldots, x_n) &= \prod_{k=1}^{n} \left[(2\pi)^d \det(C)\right]^{-1/2} \exp\left(-\frac{1}{2}(x_k - m)C^{-1/2}(x_k - m)^T\right) \\ &= \left[(2\pi)^d \det(C)\right]^{-n/2} \prod_{k=1}^{n} \exp\left(-\frac{1}{2}(x_k - m)C^{-1/2}(x_k - m)^T\right)\end{aligned} \tag{2.17}$$

In estimating parameters, such as m or C in (2.17), it is sufficient to find the estimates that maximize the likelihood function. Because the expression contains exponential functions, it is convenient to maximize the log of the likelihood function, called the *log-likelihood function*.

$$\ln \mathcal{L}(\theta|x_1, x_2, \ldots, x_n) = \ln\left[(2\pi)^d \det(C)\right]^{-n/2} - \frac{1}{2}\sum_{k=1}^{n} (x_k - m)C^{-1/2}(x_k - m)^T \tag{2.18}$$

Since the first term in the summand is a constant, maximizing the log-likelihood function is equivalent to minimizing the second term in the summand, which is the scaled log-likelihood function.

$$\ln \mathcal{L}_s(\theta|x_1, x_2, \ldots, x_n) = \sum_{k=1}^{n} (x_k - m)C^{-1/2}(x_k - m)^T \tag{2.19}$$

Note that for a Gaussian distribution, minimizing the scaled log-likelihood function to determine the mean for a known covariance is equivalent to finding a weighted least squares fit of the observed data. In general, the log-likelihood and scaled log-likelihood functions are defined for any distribution function in order to estimate density parameters.

2.5 Cramer-Rao Lower Bound

Suppose we have a density function that is also a function of a parameter $f(x|\theta)$. We would like to estimate the parameter θ, but we would also like to know how good our estimate is. The *Cramer-Rao lower bound (CRLB)* [2,5] provides a lower bound for unbiased (mean zero) estimators where the error density is Gaussian. We state the general result and provide a proof for the one-dimensional case.

$$\mathrm{var}(\theta) \geq \frac{1}{I(\theta)} = \mathrm{CRLB} \tag{2.20}$$

where $I(\theta)$ is the *Fisher information matrix*, defined as

$$I(\theta) = E\left(\left[\frac{\partial}{\partial\theta}\ln(f(x|\theta))\right]^2\right) = -E\left[\frac{\partial^2}{\partial\theta^2}\ln(f(x|\theta))\right] \tag{2.21}$$

We start by defining the following functions for convenience:

$$T = t(x)\frac{\partial}{\partial\theta}\ln(f(x|\theta)) \tag{2.22}$$

and

$$V = \frac{\partial}{\partial\theta}\ln(f(X|\theta)) \tag{2.23}$$

Then we can write

$$\begin{aligned}
\langle T, V\rangle &= E\left(T\frac{\partial}{\partial\theta}\ln(f(X|\theta))\right) \\
&= \int_R t(x)\frac{\partial}{\partial\theta}\ln(f(x|\theta))f(x|\theta)dx \\
&= \int_R t(x)\frac{1}{f(x|\theta)}\left[\frac{\partial}{\partial\theta}f(x|\theta)\right]f(x|\theta)dx \\
&= \int_R t(x)\left[\frac{\partial}{\partial\theta}f(x|\theta)\right]dx \\
&= \frac{\partial}{\partial\theta}\int_R t(x)f(x|\theta)dx \\
&= \frac{\partial}{\partial\theta}E(T)
\end{aligned} \tag{2.24}$$

Applying the Schwartz inequality, we have

$$|\langle T, V\rangle|^2 \leq \mathrm{var}(T)\cdot\mathrm{var}(V) \tag{2.25}$$

And for an unbiased estimator $E(T) = \theta$, thus we have

$$\mathrm{var}(T)\cdot\mathrm{var}(V) \geq |\langle T, V\rangle|^2 = \left|E\left(T\frac{\partial}{\partial\theta}\ln(f(X|\theta))\right)\right|^2 = \left|\frac{\partial}{\partial\theta}E(T)\right|^2 = 1 \tag{2.26}$$

Thus

$$\mathrm{var}(T) \geq \frac{1}{\mathrm{var}(V)} = \frac{1}{E\left(\left(\frac{\partial}{\partial\theta}\ln(f(X|\theta))\right)^2\right)} = \frac{1}{I(\theta)} = \text{CRLB} \tag{2.27}$$

Also,

$$\begin{aligned}
E\left(\left(\frac{\partial}{\partial\theta}\ln(f(X|\theta))\right)^2\right) &= \int_R \left(\frac{\partial}{\partial\theta}\ln(f(x|\theta))\right)^2 f(x|\theta)dx \\
&= \int_R \left(\frac{\partial}{\partial\theta}\ln(f(x|\theta))\right)\left(\frac{1}{f(x|\theta)}\left[\frac{\partial}{\partial\theta}f(x|\theta)\right]\right)f(x|\theta)dx \\
&= \int_R \left(\frac{\partial}{\partial\theta}\ln(f(x|\theta))\right)\left[\frac{\partial}{\partial\theta}f(x|\theta)\right]dx \\
&= -\int_R \frac{\partial}{\partial\theta}\left(\frac{\partial}{\partial\theta}\ln(f(x|\theta))\right)f(x|\theta)dx \\
&= -E\left(\frac{\partial^2}{\partial\theta^2}\ln(f(X|\theta))\right)
\end{aligned} \tag{2.28}$$

Thus,

$$I(\theta) = E\left(\left[\frac{\partial}{\partial\theta}\ln(f(x|\theta))\right]^2\right) = -E\left[\frac{\partial^2}{\partial\theta^2}\ln(f(x|\theta))\right] \tag{2.29}$$

2.6 Lower Bounds for Biased Estimators

The Cramer-Rao lower bound is a tight estimator of the variance when *SNR* is reasonably high, but for low *SNR* other bounds can provide a tighter bound. In the preceding derivation for the CRLB, it was assumed that the estimator was unbiased ($E(T) = \theta$). In this section, we remove the unbiased assumption and derive other lower bounds for estimators that provide tighter bounds when h estimator may be biased, such as when *SNR* is low. First, we revisit the derivation of the CRLB and discuss the impact of bias. For radar interferometry, bias can be introduced as a result of incorrect ambiquity resolution, in which case these lower bounds can provide insight into bounds on angle accuracy. From the derivation we have

$$\mathrm{var}(T)\cdot\mathrm{var}(V) \geq |\langle T, V\rangle|^2 = \left|E\left(T\frac{\partial}{\partial\theta}\ln(f(X|\theta))\right)\right|^2 = \left|\frac{\partial}{\partial\theta}E(T)\right|^2 \tag{2.30}$$

The general result derived for any estimator is

$$\mathrm{var}(T) \geq \frac{\left|\frac{\partial}{\partial\theta} E(T)\right|^2}{I(\theta)} \tag{2.31}$$

For a biased estimator, let $b(\theta) = E(T) - \theta$, and we have

$$\mathrm{var}(T) \geq \frac{\left|\frac{\partial}{\partial\theta} E(T)\right|^2}{I(\theta)} = \frac{(1 + b'(\theta))^2}{I(\theta)} \tag{2.32}$$

Since $b'(\theta)$ could be negative, clearly a biased estimator could have a variance less than the CRLB. In fact, if the estimator is constant but biased, the variance is zero, but the mean square error of a biased estimator is bounded by

$$\begin{aligned} E\left((T-\theta)^2\right) &= E\left((T - m + m - \theta)^2\right) \\ &= E\left((T-m)^2\right) + E\left((\theta - m)^2\right) \geq \frac{(1 + b'(\theta))^2}{I(\theta)} + b(\theta)^2 \end{aligned} \tag{2.33}$$

where $m = E(T)$. In the development of estimators, there can be a trade-off between allowing some small bias while reducing the variance as seen in the right-hand side of (2.33).

Although (2.31) provides a lower bound for estimators, the equation has limitations when applied to errors other than Gaussian. As a result, other lower bounds are required when the errors are non-Gaussian, which is the case for nonlinear estimation. In order to discuss these estimators, we first observe that the expression on the right-hand side of (2.29) can be written in the form

$$\mathrm{var}(T) \geq V(T) \cdot I(\theta)^{-1} \cdot V(T) \tag{2.34}$$

where

$$V(T) = E(T \cdot \psi(\theta)) \tag{2.35}$$

$$I(\theta) = E\left(\psi(\theta)^2\right) \tag{2.36}$$

$$\psi(\theta) = \frac{\partial}{\partial\theta} \ln(f(x|\theta)) \tag{2.37}$$

Note that, for a Gaussian density, the choice of ψ leads to least squares estimation and as such is a natural choice for Gaussian errors. For non-Gaussian errors, other choices of ψ provide better lower bounds for estimators. We will now discuss a few of these choices and the lower bounds that result from applying other choices of ψ.

2.6.1 Bhattacharyya Bound

The general Bhattacharyya lower bound is obtained by choosing ψ in the following manner [10]:

$$\psi(\theta|x) = \sum_{k=1}^{N} a_k \frac{1}{f(x|\theta)} \frac{\partial^k f(x|\theta)}{\partial \theta^k} \tag{2.38}$$

such that

$$\lim_{\theta \to \pm\infty} \theta \frac{\partial^k f(x|\theta)}{\partial \theta^k} = 0, \quad \text{for all } k \tag{2.39}$$

where N is an arbitrary natural number, $a_1 = 1$, a_k is a real number for all k, I is nonsingular, and $\partial^k f(x|\theta)/\partial\theta^k$ is absolutely continuous for all k. Note that when $a_k = 0$ for $k > 1$, then the Bhattacharyya bound is equivalent to the generalized CRLB. The higher-order derivatives of the density function allow for nonzero mean, non-Gaussian density functions.

2.6.2 Bobrovsky-Zakai Bound

To obtain the Bobrovsky-Zakai lower bound, the following is chosen:

$$\psi(\theta) = \frac{f(x+h|\theta) - f(x|\theta)}{\|h\| f(x|\theta)} \tag{2.40}$$

For the multivariate density, the vector h is a test point vector in the subspace of x. The test point vector effectively replaces the need to calculate partial derivatives of the probability density required in the calculation of other bounds by calculating the gradient between locally sampled distribution points (for further details, see [10], for example). In the limit $h \to 0$, the Bobrovsky-Zakai lower bound approaches the Cramer Rao bound when p is differentiable. The bound is applicable provided $p(z; x+h) = 0$ whenever $p(z; x) = 0$.

2.6.3 Weiss-Weinstein Bound

To obtain the Weiss-Weinstein lower bound, the following ψ is chosen:

$$\psi(\theta) = \frac{L^s(\theta|x+h, x) - L^{1-s}(\theta|x+h, x)}{E(L^s(\theta|x+h, x))} \tag{2.41}$$

where

$$L(\theta|x^{(1)}, x^{(2)}) = \frac{f(x^{(1)}|\theta)}{f(x^{(2)}|\theta)} \tag{2.42}$$

The exponent can be chosen to be $s = ½$, as recommended by [10]. Again, in the limit $h \to 0$, the Weiss-Weinstein lower bound becomes the Cramer Rao bound when p is differentiable. The Weiss-Weinstein bound can be extended to include more than one test point (for details, see [9]). This significantly increases the computational complexity of the approach.

2.6.4 Ziv-Zakai Bound

The ZZB is derived for a uniformly distributed random variable starting from the following general identity for mean square error (MSE) estimation [6]:

$$E(\varepsilon^2) = \frac{1}{2}\int_0^{\infty} z \cdot P\left(|\varepsilon| \geq \frac{z}{2}\right) dz \tag{2.43}$$

Then a lower bound on $P\{|\xi| \geq z/2\}$ is found, where $\xi = \hat{\tau} - \tau$ is the estimation error [7,10]. In particular, $P\{|\xi| \geq z/2\}$ is related to the error probability of a classical binary detection scheme with equally probable hypotheses:

$$\begin{aligned} &H_1: r(t) \text{ has distribution } p(r(t)|\tau) \\ &H_2: r(t) \text{ has distribution } p(r(t)|\tau + z) \end{aligned} \tag{2.44}$$

when using a suboptimum decision rule, as described in [8]. It can be shown that a lower bound to (2.43) can be generated using the error probability corresponding to the optimum decision rule based on the likelihood ratio test (LRT):

$$\Lambda(r(t)) = \frac{p(r(t)|\tau)}{p(r(t)|\tau + z)} \tag{2.45}$$

When τ is uniformly distributed in $[0, T_a)$, the ZZB is given by [7,8].

$$ZZB = \frac{1}{T_a}\int_0^{T_a} z(T_a - z)P_{\min}(z)dz \tag{2.46}$$

where $P_{min}(z)$ is the error probability corresponding to the optimum decision rule.

The primary issue with all these bounds is computational complexity. Evaluation of the WWB requires choosing test points and inverting a matrix, whereas evaluating the ZZB requires numerical integration. All of the definitions for these one-dimensional lower bounds can be extended for a multivariate density (see [7,8]).

References

1. L. Arnold, *Stochastic Differential Equations: Theory and Applications*, Krieger Publishing Co., Malabar, FL, 1992.
2. A. Papoulis and S. U. Pillai, *Probability, Random Variables, and Stochastic Processes*, McGraw-Hill, New York, 2002.
3. H. L. Royden, *Real Analysis*, 2nd Edition, The Macmillan Company: Collier-Macmillan Ltd., London, 1968.

4. W. Rudin, *Real and Complex Analysis*, McGraw-Hill, Inc., New York, 1966.
5. H. L. Van Trees, *Detection, Estimation, and Modulation Theory*, John Wiley & Sons, Inc., New York, 1968.
6. S. Bellini and G. Tartara, "Bounds on error in signal parameter estimation," *IEEE Trans. Commun.*, vol. 22, no. 3, pp. 340–342, Mar. 1974.
7. D. Chazan, M. Zakai, and J. Ziv, "Improved lower bounds on signal parameter estimation," *IEEE Trans. Inf. Theory*, vol. 21, no. 1, pp. 90–93, Jan. 1975.
8. B. Z. Bobrovsky, E. Mayer-Wolf, and M. Zakai, "Some classes of global Cramer-Rao bounds," *Ann. Stat.*, vol. 15, pp. 1421–1438, 1987.
9. E. Weinstein and A. J. Weiss, "A general class of lower bounds in parameter estimation," *IEEE Trans. Inf. Theory*, vol. 34, pp. 338–342, Mar. 1988.
10. S. Reece and D. Nicholson, "Tighter alternatives to the Cramer-Rao lower bound for discrete-time filtering," *Proc. IEEE Conf. on Decision and Control*, vol. 1, pp. 101–106, Philadelphia, PA, Jul. 2005.

Chapter 3
Radar Fundamentals

In the mid–nineteenth century, James Clark Maxwell, a British mathematician, developed a unified theory of electricity and magnetism. The equations predicting the behavior of electromagnetic signals, known as Maxwell's equations, led to discoveries that include radar waves. Maxwell's equations can be used to derive representations of radar signals propagating in various media, and for this purpose we can use the equations to derive a free space propagation of radar signals. These signal representations will be useful in understanding the impact of error sources on angle-of-arrival estimation using interferometric radar.

To derive interferometer angle accuracy, we need to define the relationship of radar signal power to thermal noise and interference power. This relationship is expressed in terms of the signal-to-noise ratio (*SNR*) or the signal-to-interference-plus-noise ratio (*SINR*). Both the *SNR* and *SINR* depend on numerous radar design parameters, as well as on the target radar cross section (RCS) and environmental parameters. In this chapter, we develop the basic fundamental radar theory to define *SNR* and *SINR*, which will be used in the next chapter when interferometer angle accuracy is derived. In addition, the fundamentals of radar processing are introduced in order to understand how phase is extracted from radar signals and the importance of waveform selection. For a more detailed discussion of radar principles refer to references [1–9].

3.1 Signal Propagation and Representation

Maxwell visualized electricity and magnetism as fluids flowing through a medium and used the property of incompressibility, similar to fluid flow, for both the electric field E and the magnetic field H. The first two equations combine Gauss's law for electricity with Gauss's law for magnetism stating that the magnetic flux through any Gauss surface is zero. Assuming free space propagation with a zero density charge, the first two Maxwell equations can be expressed as

$$\nabla \cdot E = 0 \qquad \nabla \cdot H = 0 \tag{3.1}$$

In addition to this incompressibility condition, Maxwell coupled the electric and magnetic fields through vectors that spin in a region of the field. The electric field spins around an axis that is perpendicular to the direction of propagation, and the magnetic field spins at an angle that is perpendicular to the spin of the electric field circle. The equations that Maxwell used to describe this interaction of electric and magnetic fields are

$$\nabla \times E = -\frac{1}{c}\frac{\partial H}{\partial t} \qquad \nabla \times H = \frac{1}{c}\frac{\partial E}{\partial t} \tag{3.2}$$

The electric and magnetic fields point in opposite directions as dictated by the opposite signs on the right side of the equations. Notice that the equations involve a time derivative of the electric and magnetic field vectors. This time derivative describes the flow, or time propagation, of each field. If we plug the second equation of (3.2) into the first, we have the following expression for the propagating electric field:

$$\nabla \times (\nabla \times E) = -\nabla \times \left(\frac{1}{c}\frac{\partial H}{\partial t}\right) = -\frac{1}{c}\frac{\partial}{\partial t}(\nabla \times H) = -\frac{1}{c}\frac{\partial}{\partial t}\left(\frac{1}{c}\frac{\partial E}{\partial t}\right) \tag{3.3}$$

$$\nabla \times (\nabla \times E) = \nabla(\nabla \cdot E) - \nabla^2 E = -\nabla^2 E \tag{3.4}$$

This expression can be simplified mathematically to the wave equation as follows:

$$\frac{\partial^2 E}{\partial t^2} - c^2 \nabla^2 E = 0 \tag{3.5}$$

To derive a solution to the wave equation, we separate the space and time variables.

$$E(r, t) = A(r)U(t) \tag{3.6}$$

Substituting into the wave equation and simplifying, we derive

$$\frac{\nabla^2 A}{A} = \frac{1}{c^2 U}\frac{\partial^2 U}{\partial t^2} \tag{3.7}$$

The left-hand side of (3.7) depends only on the variable r, while the right-hand side depends only on the variable t. As such, both sides must equal a constant. Without loss of generality, we choose the constant to be $-k^2$.

$$\frac{\nabla^2 A}{A} = -k^2 \qquad \frac{1}{c^2 U}\frac{\partial^2 U}{\partial t^2} = -k^2 \tag{3.8}$$

Or we have the Helmholtz equation for the spatial variable:

$$\nabla^2 A + k^2 A = 0 \tag{3.9}$$

If we define $\omega = ck$, then the second equation becomes a time version of the Helmholtz equation.

$$\frac{\partial^2 U}{\partial t^2} + \omega^2 U = 0 \tag{3.10}$$

The justification for representing signals as complex exponentials is that these signals are solutions to the Helmholtz equation, which describes electromagnetic propagation in free space. It is fairly straightforward to develop harmonic solutions to the time-varying Helmholtz equation. The solution in time is a linear combination of sine and cosine functions with angular frequency ω; the spatial solution depends on the boundary conditions. For free space propagation, we represent electromagnetic signals as complex exponentials.

$$s(t) = a(t)e^{j\omega t} = a(t)\cos(\omega t) + ja(t)\sin(\omega t) \tag{3.11}$$

where $a(t)$ is the amplitude and $\phi = \omega t$ is the phase.

3.2 Continuous Wave Doppler Waveforms

The simplest waveform is a sinusoidal wave that continues to propagate in time. This type of continuous sinusoidal wave is called a *continuous wave* (CW) waveform. The applications for the CW waveform are usually focused on detecting moving targets with a Doppler signature; as a result, a radar that implements the CW waveform is sometimes referred to as a *CW Doppler* radar. CW Doppler radars have found applications in sports tracking, intrusion detection, and heart monitoring due to its ability to track moving objects even at low velocity with long dwell times. The Doppler resolution is inversely related to the dwell time, which for a CW signal is essentially as long as the time window over which data is sampled. The transmitted CW waveform is modeled as a complex exponential as follows:

$$s_{\text{cw}} = e^{2\pi jft} \tag{3.12}$$

where f is the frequency of the sinusoidal signal and t is time. For a moving target with Doppler frequency the return signal becomes

$$s_{\text{return}} = e^{2\pi j(f+f_{\text{Dopp}})t} \tag{3.13}$$

where f_{Dopp} is the Doppler frequency induced by the target motion. The homodyne receiver mixes the return signal with the reference signal as follows:

$$\begin{aligned} s_{\text{return}} \otimes s_{\text{reference}} &= s_{\text{return}} s_{CW} e^{2\pi j\varphi} e^{2\pi j(f+f_{\text{Dopp}})t} \\ &= e^{2\pi j(f+f_{\text{Dopp}})t} e^{-2\pi jft - 2\pi j\varphi} \\ &= e^{2\pi jf_{\text{Dopp}}t - 2\pi j\varphi} \end{aligned} \tag{3.14}$$

where φ is a random phase. Note that, by sampling the signal to recover the in-phase and quadrature components of the signal, the Discrete Fourier transform (DFT, see Appendix A) of the output of the homodyne process yields the Doppler frequency of the target.

3.3 Pulse Doppler Waveforms

The *pulse Doppler* (P-D) waveform is the fundamental radar waveform that measures the range, angle, and radial velocity of a moving target. Practically every pulsed radar uses this waveform as the operational waveform of choice because it is simple to transmit and process. Also, the P-D waveform has many variations that can be tailored to specific mission applications, and so, for RF interferometry, the

P-D waveform is generally the waveform of choice. In this section, we describe the fundamentals of the P-D waveform.

3.3.1 Basic Pulse-Doppler Parameters

The invention of the magnetron in the 1930s allowed the development of what we now know as radar because it could transmit and receive short time pulses that are capable of measuring the distance (range) to a target. Today we have sophisticated phased array radars with solid-state components that can generate a variety of pulsed waveforms, providing enhanced range and Doppler resolution. The first pulses generated were narrowband with a long pulse width in order to maintain reasonable signal-to-noise ratio (*SNR*) at the observed operating ranges. These long pulse widths did not provide good range accuracy for early radar applications. The P-D waveform allows pulse widths to be much shorter but integrates multiple pulses to achieve the required *SNR* for detection. This coherent pulse integration can be achieved through the fast Fourier transform (FFT) or discrete Fourier transform (DFT), which also determines the Doppler frequency shift (f_{Dopp}) due to target radial velocity:

$$f_{\text{Dopp}} = \frac{2V}{\lambda} \tag{3.15}$$

where V is the target radial velocity, and λ is the wavelength of the narrowband pulse modulation. Modern radars now implement P-D waveforms with medium- and wideband frequency modulation, as well as phase and amplitude modulation. The basic P-D waveform parameters are defined in Figure 3.1.

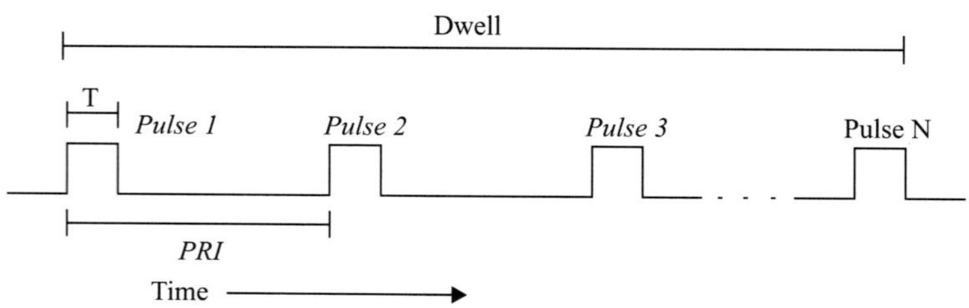

T = Pulse time duration
PRI = Pulse repetition interval
PRF = Pulse repetition frequency = *1/PRI*
Duty = *T/PRI*
N = Total number of pulses
Dwell = *N*PRI*

Pulse modulation methods:
- Narrow band tone
- Linear frequency modulation
- Nonlinear frequency modulation
- Phase code
- Frequency codes

Constant PRF: $PRF_1 = PRF_k$
Staggered PRF: $PRF_1 \neq PRF_k$

Figure 3.1. Basic Pulse-Doppler Waveform Parameters

The *pulse repetition frequency* (PRF) describes how often a pulse is transmitted but is often limited by the specific mission due to range ambiguities reflected by targets beyond the range window of interest. The range ambiguity is defined by

$$R_{\text{amb}} = \frac{c}{2PRF} \tag{3.16}$$

Thus, when a returned pulse is detected, the true range to the target (R_{true}) is given by

$$R_{\text{true}} = \frac{ct}{2} + kR_{\text{amb}} \tag{3.17}$$

where t is the time between the transmission and reception of the pulse, and k is an integer. Although R_{amb} is the maximum unambiguous range for a given PRF, it is possible to distinguish targets at ranges greater than R_{amb} by staggering the PRF or by sequentially transmitting multiple PRFs. Knowledge of the PRF of a transmitted P-D waveform provides information about the unambiguous range interval for a particular radar operating with that P-D waveform.

As discussed, the FFT or DFT integrates the pulses in a P-D waveform. The Doppler resolution (Δf_{Dopp}) of the P-D waveform is inversely related to the *dwell* time T_{dwell}.

$$\Delta f_{\text{Dopp}} = \frac{1}{T_{\text{Dwell}}} \tag{3.18}$$

Just like range ambiguity in the time domain, there is a Doppler ambiguity in the frequency domain. The Doppler ambiguity ($Doppler_{\text{amb}}$) is determined by the PRF:

$$Doppler_{\text{amb}} = PRF \tag{3.19}$$

The impact of the Doppler ambiguity is to create an ambiguity in target radial velocity. Using (3.15), we have

$$V_{\text{amb}} = PRF\frac{\lambda}{2} \tag{3.20}$$

Note that V_{amb} is directly related to PRF and that R_{amb} is inversely related to PRF. This trade-off is typically made between low-, medium-, and high-PRF radar waveforms where low-PRF radars operate range unambiguously, high-PRFs operate Doppler unambiguously, and medium-PRFs are usually ambiguous in both range and Doppler. As with range ambiguity, PRF agility can extend the velocity ambiguity region.

3.3.2 Pulse Modulation and the Time-Bandwidth Product

As discussed, the early radar pulses were narrowband modulation tones. For these pulses, the frequency of the pulse is the Fourier transform of a rectangular time pulse or a sinc function [10] in the frequency domain, and, for a matched filter, the bandwidth (B) is inversely related to the pulse duration (T).

$$B = \frac{1}{T} \tag{3.21}$$

Modern radars can implement modulation schemes that enhance range resolution through pulse compression. The most common frequency-modulated pulses are linear frequency modulation (LFM) and nonlinear frequency modulation (NLFM). LFM pulses are characterized by a quadratic phase; NLFM pulses are characterized by a phase that is described by a higher-order polynomial or sinusoidal basis function. These modulations change the frequency over some bandwidth (B) for the duration of the pulse (T), and thus the time-bandwidth (TB) product is the determining factor for pulse performance.

For the narrowband pulse, $TB = 1$, but for LFM the time-bandwidth product can be 10, 100, or even 1000. The TB product essentially determines the degree of range resolution improvement over a narrowband pulse that can be achieved with LFM and NLFM pulses. For example, for an LFM pulse with $TB = 100$, the range resolution after pulse compression is 100 times smaller than that for a narrowband pulse.

It should be noted that time (range) sidelobes must be dealt with for LFM pulses using an amplitude weighting that reduces SNR and increases range resolution. However, NLFM is designed to have low time sidelobes without impacting SNR.

Phase modulation can also improve range resolution by introducing a chip that is essentially a very short time duration pulse with $TB = 1$. The pulse is formed by stacking several chips together in time. The TB product is then the product of the chip bandwidth and the number of chips.

3.3.3 Pulse-Doppler Waveform Processing and Pulse Compression

By definition, a *matched filter* is a filter that maximizes the SNR [1–9] (see Appendix B). For input x, the output y of the filter h is expressed by the convolution

$$y[n] = \sum_{k=-\infty}^{\infty} h[n-k]x[k] \tag{3.22}$$

For radar, when x is the return signal (s) plus noise (η), the optimal matched filter is

$$x = s + \eta \tag{3.23}$$

$$h = R_\eta^{-1} s^* \tag{3.24}$$

where R_η is the covariance of the noise. Thus for a P-D waveform, each pulse is passed through a matched filter that is formed using a stored template of the transmitted signal pulse. The matched filter process compresses the pulse by a factor of the time-bandwidth product. For a square wave narrowband modulation pulse, the output of the ideal matched filter is a triangle function whose half-power

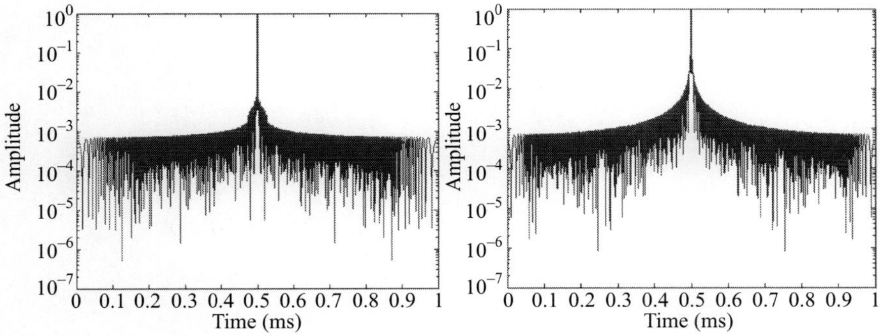

Figure 3.2. Frequency Modulation Pulse Compression with $TB = 5$ for NLFM (left) and LFM (right)

(3 dB) width is equal to the pulse duration. For LFM and NLFM pulses, the half-power matched filter output is equal to the pulse duration T divided by the time-bandwidth product TB. For $TB > 1$, the pulse is compressed by the TB factor at the output of the matched filter, and hence the term "pulse compression" is used to describe the filter output for these cases. Figure 3.2 shows an NLFM and LFM pulse for $TB = 5$. Note that the NLFM is designed to maintain low time sidelobes, whereas the LFM has relatively high sidelobes near the main lobe peak. Adding an amplitude taper weighting to the LFM pulse reduces sidelobes at the cost of decreasing SNR and increasing range resolution.

Applying the Wiener Khinchin theorem [10], the discrete Fourier transform (DFT) (see Appendix B) of the matched filter output in (3.22) becomes

$$Y[n] = H[n]X[n] \tag{3.25}$$

where Y, H, and X represent the DFT of y, h, and x. The transformed output is the product of transforms. Equation (3.25) allows for a convenient and efficient computation of the convolution in (3.22) by computing the inverse discrete Fourier transform (IDFT) of Y. The convolution results from computing three DFTs.

The matched filter output is computed in the digital domain after sampling the return signal and applying the digital filter h. Based on the sampling theorem, conventional sampling requires that the returned signal be sampled twice at the highest signal bandwidth (Nyquist). In principle, sampling at the Nyquist rate for each pulse allows perfect reconstruction of the signal such that no signal components are aliased into the matched filter output. However, Nyquist determines only a sufficient rate at which signals must be sampled, which, as such, represents the maximum rate.

For certain interferometer architectures, multiple simultaneous waveforms are implemented, and the cross-correlation for these waveforms must be minimal when processed with a matched filter. One such class of waveforms results from binary phase codes (binary phase shift key, or BPSK); the Kasami-coded waveforms (to be defined in Chapter 5) are a specific example with low cross-correlation.

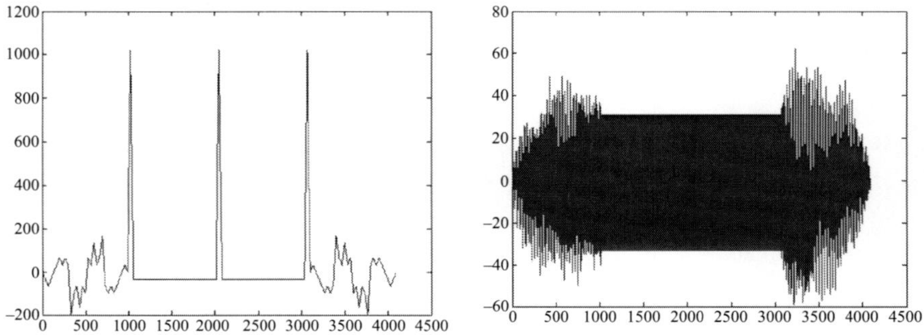

Figure 3.3. Matched Filter for the Autocorrelation and Cross-Correlation for Kasami Codes

The matched filter process compresses the pulse by a factor of the time-bandwidth product, which is equivalent to the code length used.

$$TB = \text{Pulse duration} \times \text{Bandwidth} = \frac{\text{Chip duration} \times \text{Code length}}{\text{Chip duration}}$$
$$= \text{Code length} \tag{3.26}$$

Figure 3.3 shows the auto- and cross-correlation for 1023-length Kasami codes. These low cross-correlation waveforms are important for the orthogonal interferometer architecture defined below in Chapter 5.

3.4 Radar Range Equation

Radar sensitivity is measured in terms of power and expressed in watts. A convenient expression for sensitivity is the *signal-to-noise ratio (SNR)*, which is the ratio of the signal power to the noise power at some point in the receive process. The radar range equation relates *SNR* to relevant radar design parameters, and the equation can be expressed in numerous forms for various applications. Once a radar signal is received, noise is added to it because of the hardware components in the radar receiver. Figure 3.4 shows the fundamental radar processes that happen to a radar signal. At the input of the radar receiver, the signal is typically a small voltage on the order of millivolts. Within the receiver, the thermal heating of electrons adds a random noise voltage to the signal. The noise power is directly related to the operational temperature of the receiver and to the bandwidth of the narrowest bandwidth filter in the receiver chain. For a matched filter, this bandwidth is usually the inverse of the pulse width ($B = 1/T$) or the chip width if the pulse consists of some number of coded chips, as in a binary phase coded waveform.

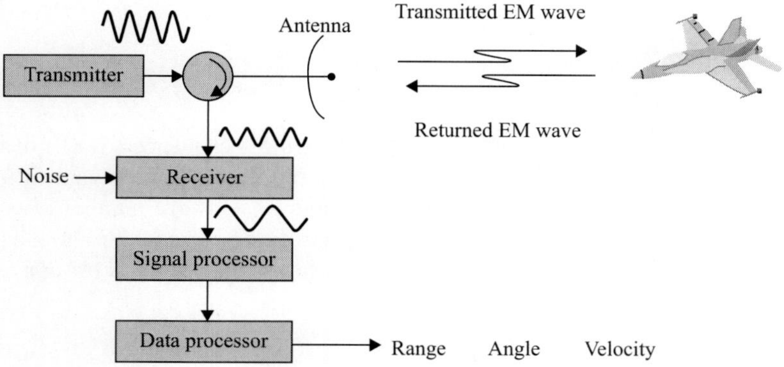

Figure 3.4. Overview of Radar System Processes

For a single pulse or chip, the SNR is expressed as the ratio of the signal power and the power of the thermal noise generated by the receiver.

$$SNR = \frac{S_{\text{power}}}{N_{\text{power}}} \tag{3.27}$$

Numerous radar texts provide a rationale for the expression of signal power. Suffice it to say that signal power is related to transmitted power, antenna gains, radar cross section of the target, wavelength, and propagation losses due to the geometric spreading of the beam. The spreading loss is proportional to the inverse of range (R) squared. The signal power can be expressed as

$$S_{\text{power}} = \frac{P G_T G_R \lambda^2 \sigma_{\text{RCS}}}{(4\pi)^3 R^4 L} \tag{3.28}$$

where

P = peak transmitted power
G_T = gain of the transmit antenna
G_R = gain of the receive antenna
λ = signal wavelength
σ_{RCS} = target radar cross section (RCS)
R = range to the target from the radar
L = system losses

Noise power is solely a function of the system temperature (T_S) and the system bandwidth (B). Let T_0 be the ambient temperature in degrees Kelvin, and let F_N be the receiver noise figure. The receiver noise figure is essentially the ratio of input noise to output noise in the receiver. The system temperature can therefore be expressed as $T_S = T_0 \cdot F_N$, and noise power is expressed as

$$N_{\text{power}} = k_B T_S B = k_B T_0 F_N B \tag{3.29}$$

where $k_B = 1.3806503 \times 10^{-23}$ J K^{-1} (Boltzmann's constant)

Thus, for a single pulse or chip, we have

$$SNR = \frac{P_{\text{ave}} G_T G_R \lambda^2 \sigma_{\text{RCS}}}{(4\pi)^3 R^4 k_B T_S B L} \tag{3.30}$$

Because radar measurements are computed using a sequence (dwell) of radar pulses that are coherently integrated in the signal processor, it is convenient to use an SNR expression that is related to the dwell time T_{dwell} for M number of pulses. The radar duty D is defined as the ratio of the transmitted pulse width T to the pulse repetition interval (PRI). Using radar duty, we can rewrite (3.30) as using average power (P_{ave}) and dwell time T_{dwell}.

$$\begin{aligned} SNR &= \frac{P G_T G_R \lambda^2 \sigma_{\text{RCS}} M}{(4\pi)^3 R^4 k_B T_S B L} \\ &= \frac{P D G_T G_R \lambda^2 \sigma_{\text{RCS}} M}{(4\pi)^3 R^4 k_B T_S \left(\frac{D}{T}\right) L} \\ &= \frac{P_{\text{ave}} G_T G_R \lambda^2 \sigma_{\text{RCS}} M \cdot PRI}{(4\pi)^3 R^4 k_B T_S L} \end{aligned} \tag{3.31}$$

Because the dwell time (T_{dwell}) is the product of the number of pulses integrated and the PRI, we have

$$SNR = \frac{P_{\text{ave}} G_T G_R \lambda^2 \sigma_{\text{RCS}} T_{\text{dwell}}}{(4\pi)^3 R^4 k_B T_S L} \tag{3.32}$$

The specific definition of these parameters can be found in any basic radar text [1–9]. Furthermore, other equivalent formulations for the radar range equation can be found, but the one presented here is tailored for the pulse-Doppler waveform. The *SNR* shows up in expressions that relate angle accuracy to interferometer design parameters through errors in phase due to thermal noise.

For binary phase coded waveforms, the pulse consists of a number of chips N_c, and the system bandwidth is matched to the chip time duration τ. The expression for *SNR* is

$$\begin{aligned} SNR &= \frac{P G_T G_R \lambda^2 \sigma_{\text{RCS}} N_c M}{(4\pi)^3 R^4 k_B T_S B_{Chip} L} \\ &= \frac{P D G_T G_R \lambda^2 \sigma_{\text{RCS}} N_c M}{(4\pi)^3 R^4 k_B T_S \left(\frac{D}{\tau}\right) L} \\ &= \frac{P_{\text{ave}} G_T G_R \lambda^2 \sigma_{\text{RCS}} M \cdot PRI}{(4\pi)^3 R^4 k_B T_S L} \\ &= \frac{P_{\text{ave}} G_T G_R \lambda^2 \sigma_{\text{RCS}} T_{\text{dwell}}}{(4\pi)^3 R^4 k_B T_S L} \end{aligned} \tag{3.33}$$

Note that we have the same result as in (3.32). Assuming that the bandwidth is matched to the chip width, the advantage of using (3.32) for a pulsed-Doppler radar is that the expression for *SNR* does not depend on the system bandwidth, which is sometimes difficult to define. If the radar narrowband filters in the receiver are not matched to the pulse width, then the mismatch loss must be accounted for in the system losses.

3.5 Phase Error

We now consider error sources that affect the phase of the signal as measurement by the radar. These sources include thermal noise resulting form the temperature of the radar receiver, radar energy returns from ground clutter, and unintentional and intentional signal interference.

3.5.1 Thermal Noise

The end goal is to compute the variance of angle-of-arrival for various methods used to estimate angle-of-arrival. Because estimates of angle are flawed due to phase noise, we must first compute the variance of phase error on a signal with additive noise in a way that can be related to radar design parameters. Let the signal plus noise be represented as follows:

$$x = ae^{j\varphi} + a_n e^{j\varphi_n} = s + n \tag{3.34}$$

where

$$n = n_I + jn_Q \tag{3.35}$$

$$n_I \rightarrow N\left(0, \frac{\sigma^2}{2}\right), \qquad n_Q \rightarrow N\left(0, \frac{\sigma^2}{2}\right) \tag{3.36}$$

and

$$n_I = a_n \cos(\varphi_n) \tag{3.37}$$

$$n_Q = a_n \sin(\varphi_n) \tag{3.38}$$

$$\sigma_{n_I}^2 = \sigma_{n_Q}^2 = \frac{\sigma_{a_n}^2}{2} \tag{3.39}$$

Relating these to noise power, we have

$$\sigma_N^2 = \sqrt{\sigma_{N_I}^2 + \sigma_{N_Q}^2} = \sigma_a^2 = N_{\text{power}} \tag{3.40}$$

where N_{power} is the noise power defined by system temperature and system bandwidth. Now, relating phase noise to SNR, we have

$$\tilde{\varphi} = \tan^{-1}\left(\frac{imag(x)}{real(x)}\right) = \tan^{-1}\left(\frac{a\sin(\varphi) + n_Q}{a\cos(\varphi) + n_I}\right) \tag{3.41}$$

Regarding $\tilde{\varphi}$ as a function of the two variables n_I and n_Q and expanding in a first-order Taylor series,

$$\tilde{\varphi} = \varphi + \frac{\sin(\varphi)}{a} n_I + \frac{\cos(\varphi)}{a} n_Q \tag{3.42}$$

Now,

$$E(n_I n_Q) = \sigma_Q^2 \frac{1}{2\pi} \int_0^{2\pi} \sin(\varphi_n)\cos(\varphi_n) d\varphi_n = 0 \tag{3.43}$$

and

$$\begin{aligned} \sigma_Q^2 = E(\tilde{\varphi} - \varphi) &= \frac{\sin^2(\varphi)}{a^2}\sigma_{n_I}^2 + 2\frac{\sin(\varphi)\cos(\varphi)}{a^2} E(n_I n_Q) + \frac{\cos^2(\varphi)}{a^2}\sigma_{n_Q}^2 \\ &= \frac{\sin^2(\varphi)}{a^2}\frac{\sigma_{a_2}^2}{2} + \frac{\cos^2(\varphi)}{a^2}\frac{\sigma_{a_2}^2}{2} \\ &= \frac{1}{2a_n/\sigma_{a_n}^2} = \frac{1}{2S/N} \end{aligned} \tag{3.44}$$

Thus,

$$\sigma_\varphi = \frac{1}{\sqrt{2S/N}} \tag{3.45}$$

The standard deviation of phase error is seen to be inversely proportional to the square root of *SNR* and will be useful when we compute the standard deviation of angle-of-arrival. Thus, phase error can be related to radar design parameters that allow the error to be understood and controlled in the design phase.

3.5.2 Clutter

Radar clutter can have a significant impact on accuracy and resolution. Random clutter is defined as radar returns from stationary or nearly stationary scattering points on the ground that get integrated into the detected signal through integration over the antenna mainlobe and sidelobes. This integrated clutter results in adding to the phase error, and we must understand the phase error floor due to random clutter in order to determine the radar angle accuracy for high *SNR*. It turns out that integrated clutter also degrades the resolution because the integrated clutter returns tend to broaden the antenna beamwidth. Thus, to completely characterize resolution and accuracy, it is necessary to understand the impact of errors other than those due to thermal noise. In the next chapter, we derive expressions for thermal noise–limited angle accuracy using the Cramer-Rao lower bound for various interferometer architectures. In Chapter 9, we investigate the effect of other errors on angle accuracy.

The primary contributors to phase noise for most radar systems are oscillators and power supplies. Figure 3.5 shows a typical phase noise curve for an oscillator.

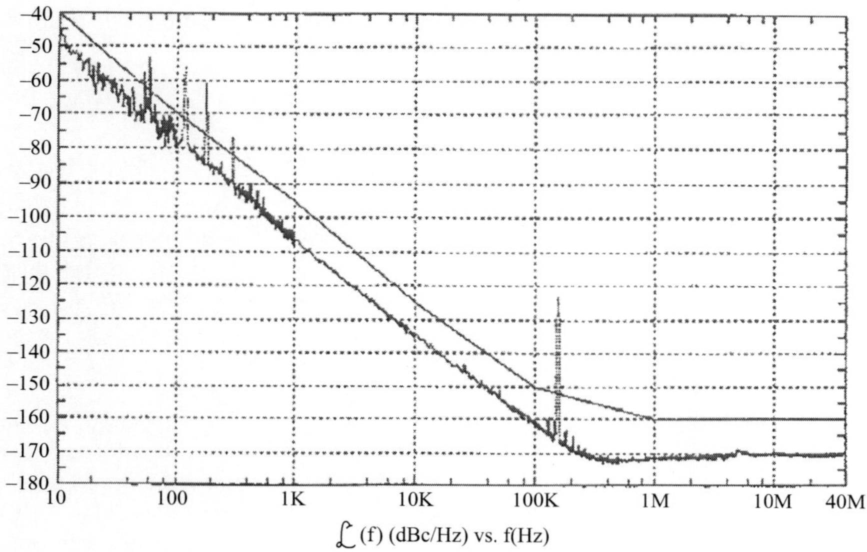

Figure 3.5. Predicted (solid) Versus Measured (jagged) Phase Noise Versus Frequency

The vertical axis represents dBc or dB referenced to the carrier frequency (dBc/Hz). The horizontal axis is referenced to frequency (Hz). To understand the impact of phase noise on clutter power, it is first necessary to understand the zero Doppler contribution of clutter. Numerous references [11,12] have dealt with calculating the total clutter power returned from various types of clutter sources, and it should be noted that this clutter power is a function of several radar parameters such as antenna gain, beamwidth, antenna sidelobes, and pulse bandwidth as well as the angles of illumination and return and the radar cross section of the illuminated clutter. The zero Doppler clutter return computed in Figure 3.5 shows the impact of clutter at frequencies other than zero Doppler [6]. For a target moving at 150 m/s, the Doppler frequency for a 1-GHz carrier is 1 kHz. The curve shows that the phase noise contribution to clutter at 1 kHz is about −105 dB/Hz. Given that zero Doppler clutter power is determined, then the contribution of clutter at 1 kHz can be determined by the following:

$$P_{\text{clutter}} = P_{\text{zero Doppler}} + 10\log(\text{bandwidth}) - 105 \tag{3.46}$$

The impact of clutter on angle-of-arrival is to degrade the overall signal-to-noise ratio with the additional noise due to clutter. Thus the calculation of *SNR* includes the noise contribution due to both thermal noise and clutter.

3.5.3 Multipath and Interference

The radar signal environment consists of other signals that can interfere with the radar signal of interest. This interference can be a structured interference as opposed to a random error and can be coherent with the signal of interest. Some of the primary interference sources are:

- Multipath.
- Structured waveform interference.
- Radio frequency interference (RFI).
- Electronic attack (EA).

The contribution of interference to angle-of-arrival estimation error is due to the increase in overall noise power. The interference acts to increase the noise contribution that can make the signal-to-interference-plus-noise (*SINR*) significantly smaller than the *SNR*.

3.5.3.1 Multipath

Multipath generally occurs due to ground bounce radar signals, but in certain environments, such as urban areas or mountainous terrain, the multipath can occur in three dimensions. Multipath can occur as diffuse or specular scattering. Diffuse multipath creates a random phase at the receive antenna, whereas specular multipath is more structured, and the phase is deterministic. The rougher the scattering surface, the more diffuse is the multipath; conversely, the smoother the scattering surface, the more specular the multipath. Specular ground bounce multipath can create two single-bounce returns and a double bounce return, as shown in Figure 3.6. All of these multipath signals arrive at the radar receiver at slightly later times than the direct path signal, creating a phase interference that can severely degrade the direct path phase integrity, which, in turn, degrades angle accuracy.

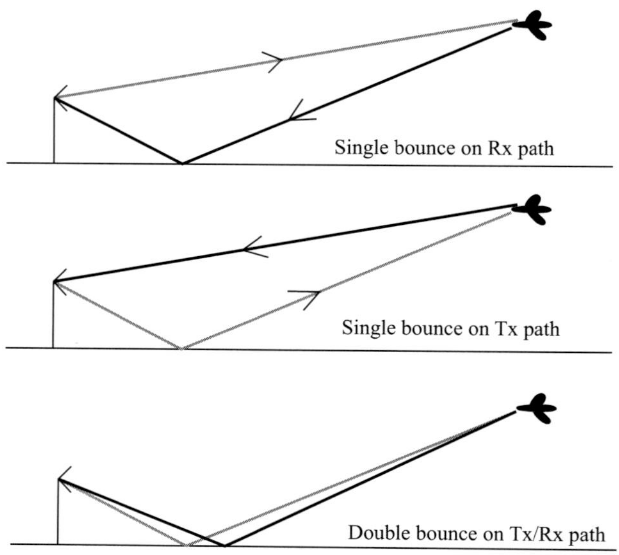

Figure 3.6. Single- and Double-Bounce Multipath Geometry

The relative intensity of the multipath signals compared to the direct path signals is related to the permittivity and permeability of the ground medium. Because multipath results from the coherent addition of deterministic signals, the effect on angle-of-arrival can be to induce an angle bias. However, in a dynamic environment, the multipath geometry can change, and the angle bias decorrelates in time based on the rate of change of the environment. Chapter 8 presents a more detailed discussion of multipath.

3.5.3.2 Structured Waveform Interference

Interference also occurs when multiple simultaneous waveforms are received at nearly the same time at the same antenna. This situation occurs in an orthogonal interferometer when nearly orthogonal waveforms are transmitted simultaneously from multiple antennas and the matched filter output consists of the matched filtered waveform and the sum of cross-correlation with other nearly orthogonal waveforms. Because the waveforms are coherent, the effect on phase degradation is similar to the phase effect due to multipath. For nearly orthogonal waveforms, the cross-correlation power relative to the autocorrelation power is determined by the waveform time-bandwidth product (discussed in detail in Chapter 5). The cross-correlation power relative to the autocorrelation peak is inversely proportional to the time-bandwidth product. The postcompression level of the cross-correlation interference, compared to the thermal noise level, determines the impact of waveform interference on angle accuracy. When the cross-correlation interference is larger than the thermal noise, then the angle accuracy is determined by the waveform interference, and, due to the structured interference angle, averaging is not effective. Otherwise, thermal noise determines angle accuracy, and, due to the random nature of thermal noise, angle averaging can further improve the angle accuracy.

3.5.3.3 Radio Frequency Interference (RFI)

RFI is the interference that occurs from other radiating sources, such as other radars operating at frequencies with spectral components that pass through the receiver to combine with the signal of interest. Again, it is imperative to understand the power level of the RFI relative to the noise level. When the RFI level is larger than the thermal noise level, angle accuracy is determined by the RFI.

3.5.3.4 Electronic Attack (EA)

Electronic attack is the intentional transmission of radiating sources that are designed to degrade radar performance. EA can consist of a random noise jammer or a coherent smart jammer. A random noise jammer essentially radiates noise into the victim radar antenna, thereby increasing the noise level and decreasing the *SNR* of signals of interest. If the jammer has sufficient power, then the SNR can be degraded below detection threshold levels, prohibiting any target detection. For moderate levels of jamming, the degradation in *SNR* degrades angle accuracy. A coherent smart jammer utilizes the victim radar's waveform by receiving and returning a modified waveform designed to confuse the victim radar. The coherent smart jammer can return multiple signals that coherently interfere with waveforms of interest in a manner similar to structured waveform interference. Again, the key performance criterion is the relative power of the coherent EA waveforms relative to the thermal noise level in the receiver.

References

1. D. K. Barton, *Radar Systems Analysis*, Artech House, New York, 1976, second printing 1979.
2. M. I. Skolnik, *Introduction to Radar Systems*, 2nd Edition, McGraw-Hill, New York, 1980.
3. J. L. Eaves and E. K. Reedy, *Principles of Modern Radar*, Van Nostrand Reinhold, New York, 1987.
4. B. K. Barton, *Modern Radar Systems Analysis*, Artech House, Norwood, MA, 1988.
5. M. I. Skolnik, *Radar Handbook*, 2nd Edition, McGraw Hill, New York, 1990.
6. J. A. Scheer and J. L. Kurtz, *Coherent Radar Performance*, Artech House, Norwood, MA, 1993.
7. P. Z. Peebles, *Radar Principles*, John Wiley & Sons, New York, 1998.
8. F. E. Nathanson, *Radar Design Principles*, Scitech Publishing, Mendham, NJ, 1999.
9. M. A. Richards, J. A. Scheer, and W. A. Holm, *Principles of Modern Radar, Basic Principles*, Scitech Publishing, Raleigh, NC, 2010.
10. A. Papoulis, *Signal Analysis*, McGraw-Hill, New York, 1977.
11. J. B. Billingsley, *Low-Angle Radar Land Clutter, Measurements and Empirical Models*, Scitech Publishing and William Andrew Publishing, Norwich, NY, 2002.
12. M. W. Long, *Radar Reflectivity of Land and Sea*, 3rd Edition, Artech House, Boston, MA, 2001.

Chapter 4

Radar Angle-of-Arrival Estimation

The early applications of radar involved detection and ranging, as the acronym (*ra*dio *d*etection *a*nd *r*anging) implies, and the estimation of angle-of-arrival was crude at best, usually being limited to location within a beam. However, World War II brought about significant advances in radar, one of which was determining angle-of-arrival by essentially splitting the radar beam. This measure of angle requires an antenna design or signal processing that can determine a difference between two observations of a signal emanating from a target. This is analogous to how the human eyes work in that each eye has a slightly different view of an object, which the brain processes to determine location. One of the first techniques used in radar to determine angle was termed "monopulse" because angle could be measured using only a single pulse. Monopulse processing is still used today in various forms for radar angle measurement, and so the basics of monopulse are essential in understanding radar angle measurement. Because monopulse angle estimation (as well as all other angle estimation processes) involves a differencing operation, the phase noise is differentiated as opposed to being integrated, as in detection and ranging. Thus, due to this distinction, angle estimation is inherently a less accurate measurement using radar than, say, range estimation. This accuracy distinction is reflected in the fact that target position accuracy usually has a larger cross-range component than range component because cross-range accuracy is directly proportional to angle accuracy.

One of the earliest attempts at characterizing radar measurement errors was made by Barton and Ward [1]. In their classic book, angle-of-arrival errors were characterized for various antenna types including interferometers, and other works [2,3] have built on Barton and Ward. Here we focus on interferometer angle error and on the signal processing required for interferometry, but we also introduce monopulse angle-of-arrival for comparison. In this chapter, we derive the basic equations for monopulse angle estimation and discuss the fundamentals of eigen-based superresolution techniques.

4.1 The Angle-of-Arrival Problem

The determination of angle-of-arrival using radar requires interpreting phase information appropriately. Figure 4.1 shows an illustration of constant phase contours resulting from a radiation source. In theory, the gradient of phase is a vector

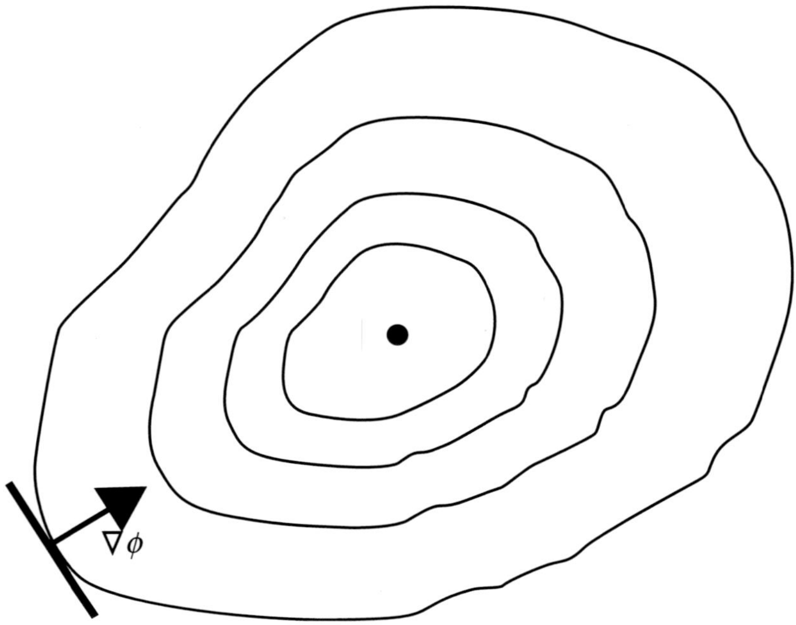

Figure 4.1. Phase Contours and Angle-of-Arrival Estimation Using the Gradient of Phase

that points in the direction of the radiation source. Thus, determining the angle-of-arrival to the radiation source requires differentiating phase, and techniques for estimating angle require multiple antennas making phase difference measurements. This differencing is performed quite simply for an interferometer with the minimal number of antennas; however, the differencing increases sensitivity to noise that affects estimation performance.

4.2 Monopulse Angle Estimation

Monopulse angle estimation has been used in radar for decades [4–6]. The basic monopulse technique for measuring angle-of-arrival divides the antenna into two separate halves, as it were. This division can occur in numerous ways, such as by implementing two or more closely spaced feed horns for a space-fed array, by physically separating an array into two equal subarrays for an analog array, or by simply phasing the elements in a way that achieves a signal difference from one side of an array to the other side.

The simplest weighting for a phased array is to form the difference pattern by dividing the array into two halves where the sum of the elements in one half is subtracted from the sum of the elements in the second half. The sum pattern is the sum of all elements in the array. For an array with N (even) number of elements, the

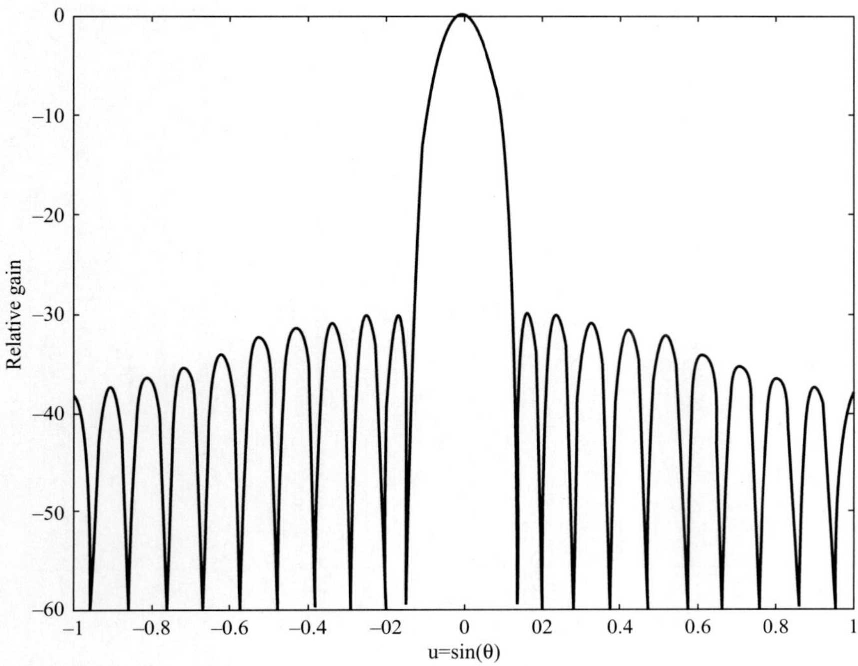

Figure 4.2. Array Sum Pattern with Taylor Weighting (21 elements with $N_{bar} = 4$ and peak sidelobe level = −30 dB)

sum weighting is the vector of all ones of length *N*, and the difference weighting is all ones of the first *N*/2 element and negative ones for the remaining N/2 elements. Although these weightings effectively create sum and difference patterns, the pattern response is a sinc function with relatively high sidelobes. The peak sidelobe is −13.2 dB from the peak of the main beam boresight. As can be expected, more effective weightings have been developed [12].

For phased arrays with amplitude control at each element, one such weighting is a Bayliss weight [15] that can be implemented directly to the output of each element in a phase array to create a differencing. The Bayliss weighting is complemented by a Taylor [15] weighting, which creates the sum pattern for an array. Both the Bayliss and Taylor weights can achieve low antenna sidelobes to reduce unwanted signals from interfering with the mainlobe response. Figures 4.2 and 4.3 show the sum and difference pattern for Taylor [14] and Bayliss [15] weights with $N_{\text{bar}} = 4$ and peak sidelobes = −30 dB down from the maximum sum value. The advantage of these element-weighting functions is the control of antenna sidelobes to mitigate distributed interference or clutter from degrading a signal in the main beam of the antenna pattern. The peak levels of the first N_{bar} number of sidelobes can be made equal at a specified power level below the peak of the mainbeam. The remaining sidelobes asymptotically behave as $1/u$ where $u = \sin(\theta)$ is the sine

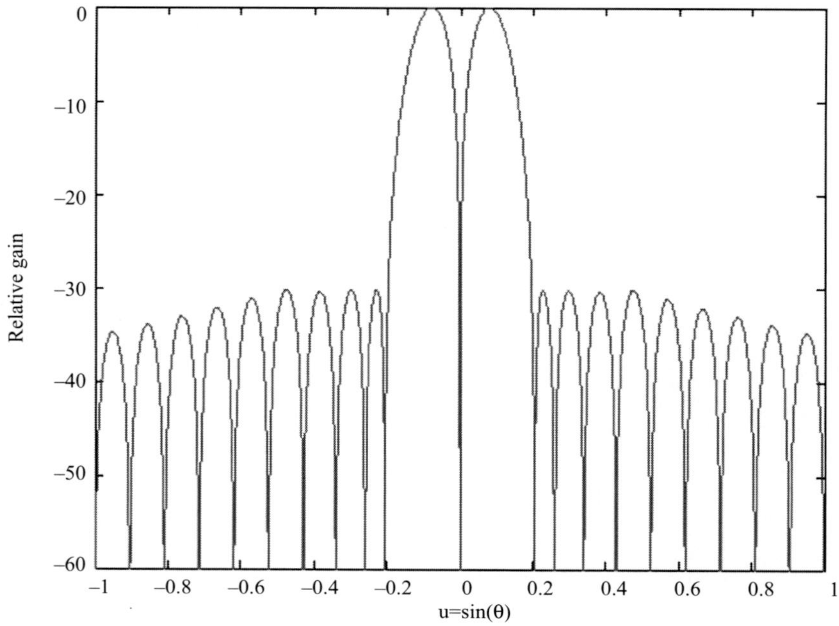

Figure 4.3. Array Difference Pattern with Bayliss Weighting (21 elements with $N_{bar} = 4$ and peak sidelobe level $= -30$ dB)

space coordinate. Again, Figures 4.2 and 4.3 illustrate this ability to achieve control of antenna sidelobes for both the sum and difference patterns in a monopulse phased array application. Hansen [12] defines both the Taylor and the Bayliss weightings as functions of the number of array elements, the parameter N_{bar}, and the desired peak sidelobe level.

Angle-of-arrival is computed using a monopulse function *mp* defined as follows,

$$mp(\widehat{\theta}) = \frac{diff(\widehat{\theta})}{sum(\widehat{\theta})} \tag{4.1}$$

where

$$\widehat{\theta} = \frac{\theta}{\theta_{BW}}$$

θ_{BW} = the 3-dB beamwidth of the array antenna pattern
$diff(\theta)$ = the array difference output
$sum(\theta)$ = the array sum output

Note that *mp* maps angle/angle into volts/volts and is thus unitless. Figure 4.4 shows the monopulse pattern for the Taylor and Bayliss patterns shown in

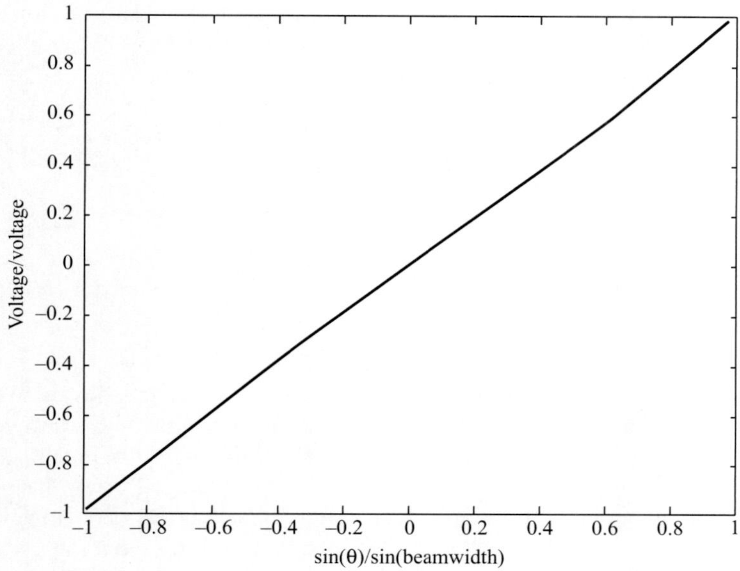

Figure 4.4. Monopulse Response for the Taylor and Bayliss Patterns in Figures 4.2 and 4.3

Figures 4.2 and 4.3. The function *mp* achieves a minimum at the angle occurring at the beam center, which we denote as the zero angle. Thus, without noise, we have

$$mp\colon [-1,1] \rightarrow [-a,a], a = mp(1) \tag{4.2}$$

$$mp\colon (0) = 0 \tag{4.3}$$

To determine angle-of-arrival, we use the inverse of the monopulse function:

$$\theta = \theta_{BW} \cdot mp^{-1}\left(\frac{diff}{sum}\right) \tag{4.4}$$

where *diff* and *sum* are the receive voltages in the difference and sum outputs. For complex arrays, the monopulse function and its inverse are determined empirically by actually scanning a source through the beam to measure the sum and difference responses to create a table of values. For a simple array such as an interferometer, the monopulse function can be determined analytically, as will be shown in the next chapter.

In reality, the monopulse function does not achieve zero at the null but achieves a value limited by noise. Numerous authors have quantified the accuracy of a monopulse antenna [4–6]. The basic result is stated as

$$\sigma_\theta = \frac{\theta_{BW}}{k\sqrt{2SNR}} \tag{4.5}$$

where k is the monopulse slope. The monopulse slope is defined as the slope of the line tangent to the monopulse function at the null of the function. When θ is small and using first-order approximations, we have

$$mp'(0)\,\widehat{\theta} = mp(\widehat{\theta}) = \frac{diff(\theta)}{sum(\widehat{\theta})} = \frac{diff(\varphi(\theta))}{sum(\varphi(\theta))} \approx \varphi \qquad \text{for } \theta \ll 1 \tag{4.6}$$

$$mp'(0)\frac{\sigma_\theta}{\theta_{BW}} = \sigma_\varphi \tag{4.7}$$

$$\sigma_\theta = \frac{\theta_{BW}\sigma_\varphi}{mp'(0)} = \frac{\theta_{BW}}{k\sqrt{2SNR}} \tag{4.8}$$

which shows the preceding result.

Any continuous increasing (or decreasing) function that satisfies these monopulse mapping conditions can be used as a monopulse function. However, the expression for angle accuracy shows that the functions with the highest slope k achieve more improved angle accuracy. For a typical phased array that applies phase weighting, a reasonable slope factor is approximately 1.4–1.6. However, for an interferometer, this slope factor is even greater, as will be shown in Chapter 5. This greater slope factor, along with smaller beamwidth, accounts for the increased angle accuracy performance for an interferometer.

4.3 Phased Array Beam Pointing Error

A monopulse system determines angle-of-arrival using a ratio of difference and sum voltages versus normalized angle. The beam center is assumed to be located at the angle where the voltage ratio is zero, or the null of the difference pattern. Any error in steering the beam affects angle accuracy because the estimate of angle using monopulse is relative to the assumed beam position. For a linear array with uniform excitation, the array beam pointing error variance is [11,12]

$$\sigma_\theta^2 = \frac{3\sigma^2\lambda^2}{\pi^2 N^3 d^2} \tag{4.9}$$

where N is the number of elements in the array, d is the spacing between individual elements, and σ^2 is the variance of the phase or amplitude error affecting each array element. For a linear array, the beamwidth θ_3 is to the array length L by

$$\theta_3 = \frac{0.866\lambda}{L} = \frac{0.866\lambda}{Nd} \tag{4.10}$$

Equation (4.9) can be expressed in terms of the beamwidth.

$$\sigma_\theta^2 = \frac{3\sigma^2\theta_3^2}{0.866^2\pi^2 N} \tag{4.11}$$

Or the angle error standard deviation for a linear array is

$$\sigma_\theta = \frac{\sqrt{3}\sigma\theta_3}{0.866\pi\sqrt{N}} \tag{4.12}$$

For a two-dimensional array with $N \times N$ elements, the beam pointing error is [11,12]:

$$\sigma_\theta = \frac{\sqrt{3}\sigma\theta_3}{0.866\pi N} \tag{4.13}$$

As an example, consider the effect of a phase shifter quantization error on beam pointing. Quantization error is a uniformly distributed error over the least interval of phase quantization Δ.

For B number of bits in a phase shifter,

$$\Delta = \frac{2\pi}{2^B} \tag{4.14}$$

and

$$\sigma = \frac{\Delta}{\sqrt{12}} = \frac{2\pi}{2^B\sqrt{12}} \tag{4.15}$$

As a result, the beam pointing error due to phase shifter quantization for a two-dimensional array is

$$\sigma_\theta = \frac{\theta_3}{0.866 \cdot 2^B N} \tag{4.16}$$

Other errors that affect beam pointing are element-level amplitude and phase errors and antenna errors due to thermal loading effects. Errors due to thermal loading result when the array is heated nonuniformly to cause array elements to respond nonuniformly when excited.

4.3.1 *Effect of Correlated Phase Errors on Phased Array Beam Pointing*

Because the interferometer consists of multiple distributed phase arrays, one effective way to resolve angle ambiguities is to use monopulse estimation in one or more distributed antenna arrays. However, correlated errors across the elements of the phase arrays distort the phase measurements at each array. In [14], Nester derived beam pointing error for two cases: (1) when the phase errors are independent over the elements in the array and (2) when the phase errors are correlated at a distance L. The variance for the beam pointing error for both cases is shown to be

$$\sigma_\theta^2 = \left(\frac{\lambda}{D\cos(\theta)}\right)^2 \frac{4}{N\pi^2}\sigma_\varphi^2 \quad \text{(Case 1)} \tag{4.17}$$

$$\sigma_\theta^2 = \left(\frac{\lambda}{D\cos(\theta)}\right)^2 \frac{4}{\left(\frac{D}{L}\right)\pi^2}\sigma_\varphi^2 \quad \text{(Case 2)} \tag{4.18}$$

where N is the total number of elements in a linear array of length D. In [13], the effect of correlated phase errors on phased arrays was expanded to several other cases. Many phased arrays use a corporate network feed that combines multiple element modules (MEMS) rather than single elements. In particular, it was shown that, for a combiner network with M levels where the total number of elements is $N = 2M$, the beam pointing error is

$$\sigma_\theta^2 = \left(\frac{\lambda}{D\cos(\theta)}\right)^2 \frac{4}{N\pi^2}\left(2^M - 1\right)\sigma_\varphi^2 \tag{4.19}$$

For exponential correlation across array elements, it was shown that

$$\sigma_\theta^2 = \left(\frac{\lambda}{D\cos(\theta)}\right)^2 \frac{4}{\left(\frac{\rho D}{2}\right)\pi^2}\left(1 - \frac{1}{\rho D}\left(1 - e^{-\rho D/2}\right)\left(3 - e^{-\rho D/2}\right)\right)\sigma_\varphi^2 \tag{4.20}$$

where the phase error is exponentially correlated across the array element located at x and x', as follows:

$$R(x, x') = \sigma_\xi^2 e^{-\rho(x - x')} \tag{4.21}$$

and σ_ξ^2 is the variance of the phase error at any one element location.

4.3.2 *Interferometer Accuracy and Beam Pointing Error*

As can be seen, phased array beam point error can be affected by various errors, and correlation on these errors can further degrade beam pointing accuracy. Furthermore, these beam pointing errors degrade monopulse accuracy because monopulse measures angle-of-arrival relative to where the beam is pointed. For an interferometer, the array beam pointing errors do not affect angle accuracy because the interferometer is using phase difference between distributed arrays. The arrays are steered in the direction of the target to enhance the signal gain, and the fact that the beams are not pointed directly at the target is of no consequence in computing angle-of-arrival. If the interferometer is measuring angle in an absolute sense, then the interferometer arrays must be calibrated and aligned with some absolute angle reference, such as true north. For some fire control applications, only the relative angle between an interceptor and the target is of interest, in which case array alignment and beam pointing errors do not affect relative angle measurement performance. However, if the interferometer uses array monopulse estimates to resolve interferometer angle ambiguities, then beam pointing errors can have an effect on interferometer performance if the beam pointing errors are sufficiently large to cause errors in determining the correct integral number of 2π phase increments. It may therefore become necessary to calibrate or eliminate correlation effects in each of the distributed arrays.

4.4 Resolution Versus Accuracy

There is a distinction between angle resolution and angle accuracy. A radar antenna is able to resolve only two targets that are separated by more than a beamwidth. If two targets are separated in angle by less than a beamwidth, then the returns combine to appear as one target with a slightly broadened mainlobe. Thus the radar cannot resolve targets whose angular difference is less than a beamwidth. For an RF antenna, the resolution is equal to the antenna beamwidth. Figure 4.5 shows two resolved targets and two unresolved targets for a 32-element array with a 3-dB beamwidth of 1.6°. The unresolved targets are located at 0° and 1.5°, and the resolved targets are located at 0° and 3°. Note that the separation of the resolved targets is less than a beamwidth and thus can be resolved with sufficient SNR. When target separation is less than a beamwidth, conventional beamforming cannot resolve the two targets. However, unconventional superresolution techniques can resolve targets separated by less than a beamwidth under certain conditions. Section 4.5 will discuss the ability to resolve targets using eigen analysis.

Given that a target is detected in a radar beam, the angular accuracy is defined as the standard deviation of the measured angle σ_θ. The angle accuracy is inversely proportional to root *SNR* due to the phase error and typically can be much less than the beamwidth. Equation (4.5) shows that, for monopulse accuracy, the beamwidth is split by a factor that is the product of the monopulse slope and root *SNR*.

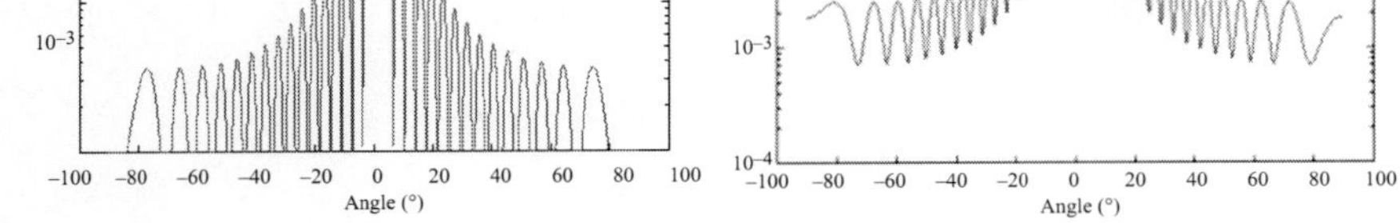

Figure 4.5. Left: Two Resolved Targets Located at 0° and 3° for a 32-Element Array; Right: Two Unresolved Targets Located at 0° and 1.5° for a 32-Element Array (3-dB beamwidth = 1.6°)

This equation represents the noise-limited performance of a monopulse radar, and, in theory, the accuracy can be made small by making *SNR* high. This statement is true only to the degree that the phase error remains noise limited. Because phase error is affected by processes other than phase noise, the phase noise floor must be established to determine the level at which a beam can be split to determine accuracy. For conventional radar systems, a splitting ratio of 20:1 is common, but ratios that approach 100:1 or higher can be achieved only by interferometers.

Radar clutter has the greatest impact on accuracy and resolution. Random clutter is defined as radar returns from stationary or nearly stationary scattering points on the ground, and it gets integrated into the detected signal through integration over the antenna sidelobes. This integrated clutter results in adding to the phase error, and it is necessary to understand the phase error floor due to random clutter in order to determine the radar angle accuracy for high *SNR*. It turns out that integrated clutter also degrades the resolution because the integrated clutter returns tend to broaden the antenna beamwidth. Thus, to completely characterize resolution and accuracy, it is necessary to understand the impact of errors other than those due to thermal noise. In Chapter 5, we derive expressions for thermal noise-limited angle accuracy using the Cramer-Rao lower bound for various interferometer architectures. In Chapter 10, we investigate the effects of other errors on angle accuracy.

4.5 Enhanced Angle Estimation Using the Array Covariance

The signal covariance matrix R is used in numerous methods that estimate certain signal parameters. Many of these methods depend on a good understanding of the eigen-structure for the covariance matrix [2,3]. For example, the MUSIC algorithm [7,8] uses the noise space spanned by the eigenvectors associated with the noise eigenvalues to determine signal characteristics, such as Doppler or angle-of-arrival, but must first define the noise space by separating the eigenvectors associated with noise from those associated with signals. In the absence of noise, the rank of R is equal to the number of signal sources, and the eigenvalues of R can determine the number of signals present in a signal environment by comparing the magnitudes of the eigenvalues to determine which of these eigenvalues is equivalent or not equivalent in some sense to the noise variance. Wax and Kaileth [9] have used information theoretical criteria to separate noise and signal eigenvalues providing rigor to the phrase "in some sense."

4.5.1 Angle-of-Arrival Resolution Performance

We will now show how the structure of the array covariance matrix R is related to the signals and noise. If we assume that an array consists of N number of elements, then the signal and noise structure of the $N \times N$ covariance matrix R can provide insight into the requirements for adequate signal resolution.

4.5.1.1 Signal Model and Eigen-Analysis

Assume that a two-signal environment is sampled with an array of N elements. The voltages x_i for each of these spatial samples are represented as follows:

$$\begin{aligned} x_1 &= a_1 + a_2 e^{j\varphi} + \varepsilon_1 \\ x_2 &= a_1 e^{jd_1\widehat{\theta}_1} + a_2 e^{jd_1\widehat{\theta}_2 + j\varphi} + \varepsilon_2 \\ &\vdots \\ x_N &= a_1 e^{jd_{N-1}\widehat{\theta}_1} + a_2 e^{jd_{N-1}\widehat{\theta}_2 + j\varphi} + \varepsilon_N \end{aligned} \tag{4.22}$$

where ε is normally distributed with mean equal to zero, and variances σ^2 and φ are uniformly distributed over $[0\ 2\pi]$. The array covariance matrix R is defined as follows:

$$\begin{aligned} R &= E\left(x_i x_j^*\right) \\ &= \begin{bmatrix} a_1^2 + a_2^2 + \sigma^2 & a_1^2 e^{jd_1\widehat{\theta}_1} + a_2^2 e^{jd_1\widehat{\theta}_2} & \cdots & a_1^2 e^{j(d_1 - d_{N-1})\widehat{\theta}_1} + a_2^2 e^{j(d_1 - d_{N-1})\widehat{\theta}_2} \\ a_1^2 e^{-jd_1\widehat{\theta}_1} + a_2^2 e^{-jd_1\widehat{\theta}_2} & a_1^2 + a_2^2 + \sigma^2 & & \vdots \\ \vdots & & \ddots & a_1^2 e^{jd_1\widehat{\theta}_1} + a_2^2 e^{jd_1\widehat{\theta}_2} \\ a_1^2 e^{-j(d_1 - d_{N-1})\widehat{\theta}_1} + a_2^2 e^{-j(d_1 - d_{N-1})\widehat{\theta}_2} & \cdots & a_1^2 e^{-jd_1\widehat{\theta}_1} + a_2^2 e^{-jd_1\widehat{\theta}_2} & a_1^2 + a_2^2 + \sigma^2 \end{bmatrix} \\ &= \sigma^2 I + a_1^2 \begin{bmatrix} 1 \\ e^{jd_1\widehat{\theta}_1} \\ \vdots \\ e^{jd_{N-1}\widehat{\theta}_1} \end{bmatrix} \begin{bmatrix} 1 & e^{jd_1\widehat{\theta}_1} & \cdots & e^{jd_{N-1}\widehat{\theta}_1} \end{bmatrix} + a_2^2 \begin{bmatrix} 1 \\ e^{jd_1\widehat{\theta}_2} \\ \vdots \\ e^{jd_{N-1}\widehat{\theta}_2} \end{bmatrix} \begin{bmatrix} 1 & e^{jd_1\widehat{\theta}_2} & \cdots & e^{jd_{N-1}\widehat{\theta}_2} \end{bmatrix} \\ &= \sigma^2 I + R' \end{aligned} \tag{4.23}$$

If we define the signal steering vector S_θ as

$$S_\theta = \begin{bmatrix} 1 & e^{jd_1\widehat{\theta}} & \cdots & e^{jd_{N-1}\widehat{\theta}} \end{bmatrix} \tag{4.24}$$

then (4.24) can be expressed as the defining eigen-structure for a matrix with dimensions equal to number of signals. Let Z represent any eigen-value of R. Then

$$RZ^T = \eta Z^T \tag{4.25}$$

And using the definition of R, we can write

$$S_\theta^* R' Z^T = (\eta - \sigma^2) S_\theta^* Z^T \tag{4.26}$$

$$
\begin{aligned}
(\eta-\sigma^2)S_\theta^* Z^T &= S_\theta^* R' Z^T \\
&= S_\theta^* \left(a_1^2 \begin{bmatrix} 1 \\ e^{jd_1\hat{\theta}_1} \\ \vdots \\ e^{jd_{N-1}\hat{\theta}_1} \end{bmatrix} \begin{bmatrix} 1 & e^{jd_1\hat{\theta}_1} & \dots & e^{jd_{N-1}\hat{\theta}_1} \end{bmatrix} + a_2^2 \begin{bmatrix} 1 \\ e^{jd_1\hat{\theta}_2} \\ \vdots \\ e^{jd_{N-1}\hat{\theta}_2} \end{bmatrix} \begin{bmatrix} 1 & e^{jd_1\hat{\theta}_2} & \dots & e^{jd_{N-1}\hat{\theta}_2} \end{bmatrix} \right) Z^T \\
&= a_1^2 S_\theta^* \begin{bmatrix} 1 \\ e^{jd_1\hat{\theta}_1} \\ \vdots \\ e^{jd_{N-1}\hat{\theta}_1} \end{bmatrix} \begin{bmatrix} 1 & e^{jd_1\hat{\theta}_1} & \dots & e^{jd_{N-1}\hat{\theta}_1} \end{bmatrix} Z^T + a_2^2 S_\theta^* \begin{bmatrix} 1 \\ e^{jd_1\hat{\theta}_2} \\ \vdots \\ e^{jd_{N-1}\hat{\theta}_2} \end{bmatrix} \begin{bmatrix} 1 & e^{jd_1\hat{\theta}_2} & \dots & e^{jd_{N-1}\hat{\theta}_2} \end{bmatrix} Z^T \\
&= a_1^2 S_\theta^* S_{\theta_1}^T S_{\theta_1}^* Z^T + a_2^2 S_\theta^* S_{\theta_2}^T S_{\theta_2}^* Z^T
\end{aligned}
\tag{4.27}
$$

We can express (4.27) in a matrix eigen-structure relationship that involves only a 2×2 dimensional matrix:

$$
\begin{bmatrix} a_1^2 S_{\theta_1}^* S_{\theta_1}^T & a_2^2 S_{\theta_1}^* S_{\theta_2}^T \\ a_1^2 S_{\theta_2}^* S_{\theta_1}^T & a_2^2 S_{\theta_2}^* S_{\theta_2}^T \end{bmatrix} \begin{bmatrix} S_{\theta_1}^* Z^T \\ S_{\theta_2}^* Z^T \end{bmatrix} = (\eta-\sigma^2) \begin{bmatrix} S_{\theta_1}^* Z^T \\ S_{\theta_2}^* Z^T \end{bmatrix} = \gamma \begin{bmatrix} S_{\theta_1}^* Z^T \\ S_{\theta_2}^* Z^T \end{bmatrix} \tag{4.28}
$$

where $\gamma = \eta - \sigma^2$, and

$$
S_{\theta_i}^* S_{\theta_i}^T = N \tag{4.29}
$$

$$
S_{\theta_1}^* S_{\theta_2}^T = \sum_{k=0}^{N-1} e^{jd_k(\hat{\theta}_2-\hat{\theta}_1)} \tag{4.30}
$$

For the general case of K signals, we have the following relationship:

$$
\begin{aligned}
&\begin{bmatrix} a_1^2 N & a_2^2 S_{\theta_1}^* S_{\theta_2}^T & \cdots & a_K^2 S_{\theta_1}^* S_{\theta_K}^T \\ a_1^2 S_{\theta_2}^* S_{\theta_1}^T & a_2^2 N & \ddots & \vdots \\ \vdots & \ddots & \ddots & a_K^2 S_{\theta_1}^* S_{\theta_2}^T \\ a_1^2 S_{\theta_K}^* S_{\theta_1}^T & \cdots & a_{K-1}^2 S_{\theta_2}^* S_{\theta_1}^T & a_K^2 N \end{bmatrix} \begin{bmatrix} S_{\theta_1}^* Z^T \\ S_{\theta_2}^* Z^T \\ \vdots \\ S_{\theta_K}^* Z^T \end{bmatrix} \\
&= (\eta-\sigma^2) \begin{bmatrix} S_{\theta_1}^* Z^T \\ S_{\theta_2}^* Z^T \\ \vdots \\ S_{\theta_K}^* Z^T \end{bmatrix}
\end{aligned}
\tag{4.31}
$$

$$
\tilde{R}_{KxK} Y_{1xK}^T = \gamma Y_{1xK}^T \tag{4.32}
$$

Because σ^2 is an eigenvalue of R, $\gamma = 0$ and $S_{\theta_i}^* Z^T$ must be identically equal to the zero vector since M has a nonzero determinant and is not equal to a scaled version of the identity matrix. Thus, verifying that the eigenvector Z^T associated with the noise eigenvalue σ^2 is orthogonal to all of the signal vectors $S^*_{\theta_i}$, and determining the eigenvalues of R is equivalent to determining the eigenvalues of $\tilde{R}$. A closed-form expression for the eigenvalues of (4.31) and (4.32) exist only for $K < 5$, and, although an expression exists for $K=3$ and 4, the results are somewhat complicated and less intuitive as the $K=2$ case, and therefore only the $K=2$ case is discussed here.

When we consider the eigenvector Z^T associated with the noise eigenvalue σ^2, we see that $\gamma=0$, and the only solution to (4.31) and (4.32) is $Y=0$. Thus, $S^*_{\theta_i}Z^T = 0$ for all signals vectors $S^*_{\theta_i}$, $i=1,\dots,K$, and the noise eigenvector Z^T is orthogonal to all of the signal vectors $S^*_{\theta_i}$. This orthogonality between the noise eigenvectors and the signal eigenvectors is the basic relationship that allows signal extraction for numerous algorithms such as MUSIC [7,8].

4.5.1.2 Two-Signal Case ($K=2$) with Uniform Sampling

First consider the case when $K=2$ and the sampling interval is nonuniform:

$$\tilde{R} = \begin{bmatrix} a_1^2 N & a_2^2 b \\ a_1^2 b^* & a_2^2 N \end{bmatrix} \tag{4.33}$$

$$\begin{aligned} 0 &= \det(\tilde{R} - \gamma I) = (a_1^2 N - \gamma)(a_2^2 N - \gamma) - a_1^2 a_2^2 |b|^2 \\ &= \gamma^2 - N(a_1^2 + a_2^2)\gamma + N^2 a_1^2 a_2^2 - a_1^2 a_2^2 |b|^2 \end{aligned} \tag{4.34}$$

$$\gamma_\pm = \frac{N(a_1^2 + a_2^2) \pm \sqrt{N^2(a_1^2 - a_2^2)^2 + 4a_1^2 a_2^2 |b|^2}}{2} \tag{4.35}$$

$$\eta_\pm = \sigma^2 + \frac{N(a_1^2 + a_2^2) \pm \sqrt{N^2(a_1^2 - a_2^2)^2 + 4a_1^2 a_2^2 |b|^2}}{2} \tag{4.36}$$

where

$$|b|^2 = \left(\sum_{k=0}^{N-1} e^{jd_k(\widehat{\theta}_2 - \widehat{\theta}_1)} \right) \left(\sum_{k=0}^{N-1} e^{-jd_k(\widehat{\theta}_2 - \widehat{\theta}_1)} \right) \tag{4.37}$$

Equations (4.35)–(4.37) define a general result that applies for any two signals where the sample spacing can be nonuniform. To derive a more intuitive relationship between eigenvalue classification and signal resolution, we will consider uniform sample spacing, assume that the signals are closely spaced, and make the following approximations:

$$\begin{aligned} |b|^2 &= \left| e^{j\frac{N-1}{2}d(\theta_2 - \theta_1)} \frac{\sin\left(\frac{Nd}{2}(\widehat{\theta}_2 - \widehat{\theta}_1)\right)}{\sin\left(\frac{1}{2}(\widehat{\theta}_2 - \widehat{\theta}_1)\right)} \right|^2 \underset{\theta_2 \to \theta_1}{\approx} \frac{\sin^2\left(\frac{Nd}{2}(\widehat{\theta}_2 - \widehat{\theta}_1)\right)}{\left(\frac{d}{2}(\widehat{\theta}_2 - \widehat{\theta}_1)\right)^2} \\ &= N^2 \mathrm{sinc}^2\left(\frac{Nd}{2}(\widehat{\theta}_2 - \widehat{\theta}_1)\right) \end{aligned} \tag{4.38}$$

$$\eta_{\pm} = \sigma^2 + \frac{N\left(a_1^2 + a_2^2\right) \pm N\sqrt{\left(a_1^2 - a_2^2\right)^2 + 4a_1^2a_2^2\mathrm{sinc}^2\left(\frac{Nd}{2}\left(\widehat{\theta}_2 - \widehat{\theta}_1\right)\right)}}{2} \tag{4.39}$$

$$\frac{\eta_{\pm}}{\sigma^2} = 1 + \frac{N\left(\frac{a_1^2}{\sigma^2} + \frac{a_2^2}{\sigma^2}\right) \pm N\sqrt{\left(\frac{a_1^2}{\sigma^2} - \frac{a_2^2}{\sigma^2}\right)^2 + 4\frac{a_1^2}{\sigma^2}\frac{a_2^2}{\sigma^2}\mathrm{sinc}^2\left(\frac{Nd}{2}\left(\widehat{\theta}_2 - \widehat{\theta}_1\right)\right)}}{2} \tag{4.40}$$

$$\frac{\eta_{\pm}}{\sigma^2} = 1 + N\left(\mathrm{SNR}_1^S + \mathrm{SNR}_2^S\right) \pm N\sqrt{\left(\mathrm{SNR}_1^S - \mathrm{SNR}_2^S\right)^2 + 4\cdot\mathrm{SNR}_1^S\cdot\mathrm{SNR}_2^S\cdot\mathrm{sinc}^2\left(\frac{Nd}{2}\left(\widehat{\theta}_2 - \widehat{\theta}_1\right)\right)} \tag{4.41}$$

Using a Taylor series expansion in (4.41) and discarding terms with an order higher than 2, the second-order approximation to (4.41) is

$$\frac{\eta_{\pm}}{\sigma^2} \approx 1 + \frac{N\cdot\mathrm{SNR}_1^S\cdot\mathrm{SNR}_2^S}{\mathrm{SNR}_1^S + \mathrm{SNR}_1^S}\left(\frac{\left(\widehat{\theta}_2 - \widehat{\theta}_1\right)}{\widehat{\theta}_0}\right)^2$$

$$= 1 + \frac{N\cdot\mathrm{SNR}_1^S\cdot N\cdot\mathrm{SNR}_2^S}{N\cdot\mathrm{SNR}_1^S + N\cdot\mathrm{SNR}_1^S}\left(\frac{\left(\widehat{\theta}_2 - \widehat{\theta}_1\right)}{\widehat{\theta}_0}\right)^2 \tag{4.42}$$

$$= 1 + \frac{SNR_1\cdot SNR_2}{SNR_1 + SNR_1}\left(\frac{\left(\widehat{\theta}_2 - \widehat{\theta}_1\right)}{\widehat{\theta}_0}\right)^2\left(2^{\mathrm{nd}}\ \mathrm{order}\right) \tag{4.43}$$

$$\mathrm{ENR} = 1 + \frac{\mathrm{SNR}_1\cdot\mathrm{SNR}_2}{\mathrm{SNR}_1 + \mathrm{SNR}_2}\cdot\mathrm{RES}^2 \tag{4.44}$$

where ENR is defined as the eigenvalue-to-noise ratio, $\mathrm{SNR}_i = N\cdot\mathrm{SNR}_i^S$, SNR_i is the coherent integrated signal-to-noise ratio, and SNR_i^S is the single sample signal-to-noise ratio,

$$\mathrm{SNR}_i^S = \frac{a_i^2}{\sigma^2} \tag{4.45}$$

and where $\mathrm{RES} = \frac{\left(\widehat{\theta}_2 - \widehat{\theta}_1\right)}{\widehat{\theta}_0}$ and $\widehat{\theta}_0 = \frac{2}{Nd\sqrt{3}}$.

Equation (4.44) is a second-order approximation to (4.41) when $\mathrm{RES} \ll 1$ and tends to be an overestimate of ENR for $0 < \mathrm{RES} < 1$. Equation (4.44) is therefore modified to provide a reasonable average approximation over the interval $0 < \mathrm{RES} < 1$.

$$\mathrm{ENR} = 1 + \frac{1}{\alpha}\frac{\mathrm{SNR}_1 \cdot \mathrm{SNR}_2}{\mathrm{SNR}_1 + \mathrm{SNR}_2} \cdot \mathrm{RES}^2 \tag{4.46}$$

where α is an approximation parameter used to provide a better average fit of the second-order approximation over $0 < \mathrm{RES} < 1$. For example, $\alpha = 1.25$ provides a reasonable fit over the interval; however, when $\alpha = 1.25$, (4.44) is an underestimate for ENR when $\mathrm{RES} \ll 1$. However, for establishing a bound on resolution performance, the small underestimate provides margin to the bound. When $\mathrm{RES} \ll 1$, it is sufficient to chose $\alpha = 1$.

For general application, the covariance matrix R is estimated averaging M number of samples of the estimated covariance. Nadakuditi and Edelman [10] develop a relation between ENR and the number of samples M for $N, M \gg 1$.

$$\mathrm{ENR} = \frac{\eta_-}{\sigma^2} > 1 + \sqrt{\frac{N}{M}} \tag{4.47}$$

Substituting (4.46) we have

$$\frac{1}{\alpha}\frac{\mathrm{SNR}_1 \cdot \mathrm{SNR}_2}{\mathrm{SNR}_1 + \mathrm{SNR}_2} \cdot \mathrm{RES}^2 > \sqrt{\frac{N}{M}} \tag{4.48}$$

$$\mathrm{RES} > \sqrt{\alpha \frac{\mathrm{SNR}_1 + \mathrm{SNR}_2}{\mathrm{SNR}_1 \cdot \mathrm{SNR}_2}\sqrt{\frac{N}{M}}} \tag{4.49}$$

Equation (4.49) provides a rule of thumb to assess the resolution of the smallest signal eigenvalue relative to the noise eigenvalue in a sample covariance matrix. For the following applications, we chose $1.25 \le \alpha \le 1.5$; however for $\mathrm{RES} \ll 1$, a value of $\alpha = 1$ is sufficient.

4.5.1.3 Application to Angle-of-Arrival Resolution

In angle-of-arrival applications, the sample interval d is the element spacing in an array of antenna elements, and the definition of RES is related to the beamwidth as follows:

$$\widehat{\theta} = \frac{2\pi}{\lambda}\sin(\theta) \tag{4.50}$$

$$\mathrm{RES} = \frac{|\widehat{\theta}_2 - \widehat{\theta}_1|}{\widehat{\theta}_0} = \frac{\pi N d}{\lambda\sqrt{3}}|\theta_2 - \theta_1| = \frac{|\theta_2 - \theta_1|}{\theta_{BW}} \tag{4.51}$$

where

$$\theta_{BW} = \frac{\lambda\sqrt{3}}{\pi N d} \approx 0.55\frac{\lambda}{Nd} \tag{4.52}$$

Substituting into (4.49), we have

$$\frac{|\theta_2 - \theta_1|}{\theta_{BW}} > \sqrt{\frac{\mathrm{SNR}_1 + \mathrm{SNR}_2}{\mathrm{SNR}_1 \cdot \mathrm{SNR}_2}\sqrt{\frac{N}{M}}} \tag{4.53}$$

Equation (4.53) determines the potential resolution performance as a fraction of a beamwidth, which defines the Fourier resolution of an array. Consider the case where $SNR_1 = SNR_2 = SNR$. Equation then becomes

$$\frac{|\theta_2 - \theta_1|}{\theta_{BW}} > \sqrt{\frac{2}{\mathrm{SNR}}\sqrt{\frac{N}{M}}} \tag{4.54}$$

Equation (4.54) shows that, for a large linear array with N elements, the resolution performance, or beamsplit ratio, is inversely proportional to the square root of the SNR and the fourth root of the number of time samples M. Since SNR is proportional to N^2 due to the increase in array gain the resolution increases on the order of $N^{3/4}$ for enhanced resolution algorithms that use the array covariance such as MUSIC.

4.5.1.4 Application to Resolution Performance

Equation (4.49) can be used to establish design requirements to achieve a desired resolution. For example, to establish requirements for the number of samples M used to determine the sample covariance, (4.49) can be rearranged as follows:

$$M > \left(\frac{1}{\mathrm{RES}}\sqrt{\alpha\frac{\mathrm{SNR}_1 + \mathrm{SNR}_2}{\mathrm{SNR}_1 \cdot \mathrm{SNR}_2}\sqrt{N}}\right)^4 \tag{4.55}$$

Note that M scales as the inverse of the fourth power of the resolution RES. Because (4.47) is valid for large values of N and M, define M_0 as the minimum value for M. In addition to (4.55), M must also satisfy $M > M_0$.

Figure 4.6 shows the required performance for M versus SNR_2 for values of $\alpha = 1$ (dark), 1.25 (medium), and 1.5 (light) when $N = 10$, $SNR_1 = 20$ dB, and the desired RES = 0.2 with $M_0 = 50$.

4.5.2 *Signal Versus Noise Eigenvalue Classification*

The objective is to develop a greater understanding of eigenvalue classification by determining criteria for the identification of the smallest signal eigenvalue when compared to noise, which will establish basic limitations on resolving closely spaced signals. For an environment consisting of two closely separated signals, the essential result from (4.44) can be restated.

$$\mathrm{ENR} = 1 + \frac{\mathrm{SNR}_1 \cdot \mathrm{SNR}_2}{\mathrm{SNR}_1 + \mathrm{SNR}_2} \cdot \mathrm{RES}^2 \qquad \mathrm{RES} \ll 1 \tag{4.56}$$

where ENR is the ratio of the magnitude of the smallest signal eigenvalue to the noise variance, SNR_i is the array signal-to-noise ratio for each of the two signals, and RES is the normalized separation or resolution between the two signals. Equation (4.56) provides a fundamental understanding of the sensitivity of the smallest eigenvalue magnitude relative to the noise variance versus signal SNR and normalized resolution. As either SNR or resolution approaches zero, the smallest

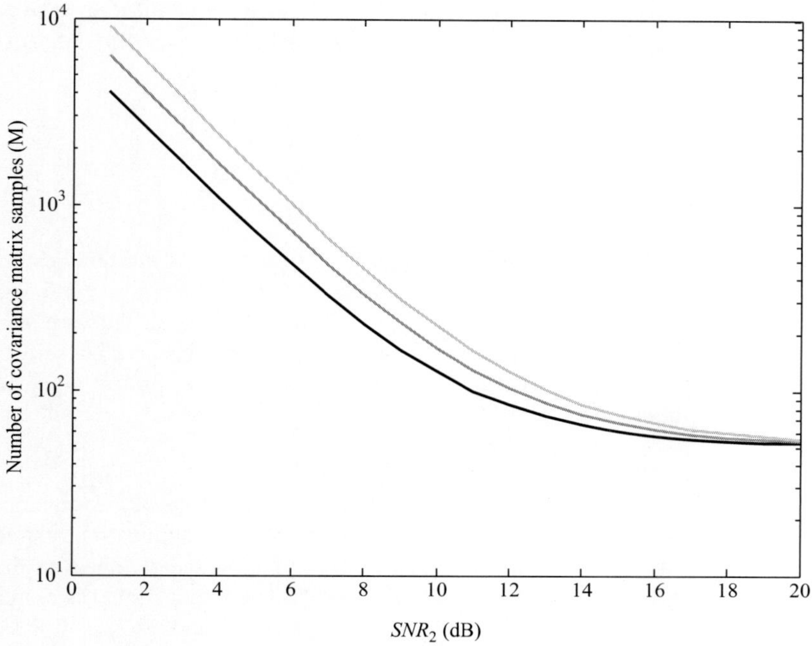

Figure 4.6. Required Number of Covariance Samples M Versus SNR_2 for $\alpha = 1$ (dark), 1.25 (medium), and 1.5 (light) when $N = 10$, $SNR_1 = 20$ dB, and the Desired RES = 0.2

signal eigenvalue approaches the noise floor at a rate determined by (4.56). For angle-of-arrival, we have shown

$$\text{RES} = \frac{|\theta_2 - \theta_1|}{\theta_{BW}} \tag{4.57}$$

As a result, (4.56) becomes

$$\text{ENR} = 1 + \frac{\text{SNR}_1 \cdot \text{SNR}_2}{\text{SNR}_1 + \text{SNR}_2} \cdot \left(\frac{|\theta_2 - \theta_1|}{\theta_{BW}} \right)^2 \tag{4.58}$$

Equation (4.58) shows the relationship between ENR to *SNR* and angle resolution. The probability of identifying a signal eigenvalue correctly can be estimated using a Rayleigh probability distribution as the distribution for noise. We choose the Rayleigh distribution whose mean is equal to the noise variance σ^2.

$$p_{noise}(z) = \frac{\pi z}{2\sigma^4} e^{-\frac{\pi z^2}{4\sigma^4}} \tag{4.59}$$

$$E(Z) = \sigma^2 \tag{4.60}$$

To classify the signal eigenvalue correctly, the noise eigenvalue must be smaller in value. We compute the probability that the noise eigenvalue is less than the signal eigenvalue.

$$\text{Prob}(\eta > \sigma^2) = \int_0^{ENR\sigma^2} \frac{\pi z}{2\sigma^4} e^{-\frac{\pi z^2}{4\sigma^4}} dz = \int_0^{\frac{\pi ENR^2 \sigma^4}{4\sigma^4}} e^{-u} du = 1 - e^{-\frac{\pi ENR^2}{4}} \tag{4.61}$$

In reality, η is a random variable, but for high *SNR*, (4.61) provides a good estimate for the classification of signals in noise. When $ENR = 1$, the noise and signal eigenvalues are equal, and the probability of correctly classifying the signal eigenvalue is 0.5441. To achieve a probability of classification greater than 0.95, the *ENR* must exceed 2, or essentially the magnitude of the smallest eigenvalue must be greater than twice the noise power.

4.5.2.1 Simulation of Three-Element Array

Equation (4.46) extends the approximation to ENR over the entire beamwidth by introducing the parameter α. We simulate the angle-of-arrival application case for a three-element array ($N = 3$) and two signals ($K = 2$) to show the goodness of fit for the ENR approximations over the entire beamwidth. The strong signal is assumed to have $SNR = 20$ dB, and the weak signal is assumed to have $SNR = 10$ dB and 20 dB. The antennas are spaced one wavelength apart ($d = \lambda$). The covariance matrix R is formed using $M = 100$ samples. Figure 4.7 shows the results for the simulated and estimated ENR versus RES for $\alpha = 1.25$. Note the agreement with the simulated results (medium) and the estimated results using (4.35)–(4.37) (dark). Also the second-order approximation (light) using Equation (4.46) is generally valid for RES < 1 and is an underestimate of ENR for RES ≪ 1. Figure 4.8 shows the second order-approximation for ENR when $\alpha = 1.5$ is an underestimate for 0 < RES < 1, which allows for additional margin in (4.49).

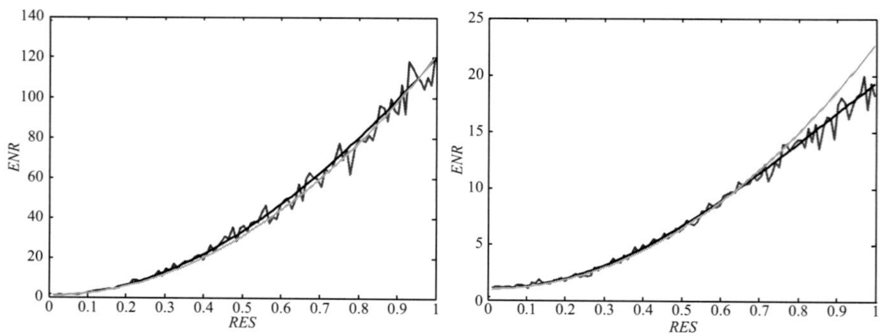

Figure 4.7. Simulated and Estimated Results ($\alpha = 1.25$) for ENR Versus RES for $SNR_1 = 20$ dB and $SNR_2 = 20$ dB (left) and $SNR_2 = 10$ dB (right); Three-Element Simulation (medium), ENR Derived from (4.35)–(4.37) (dark) and Second-Order (light) Approximations

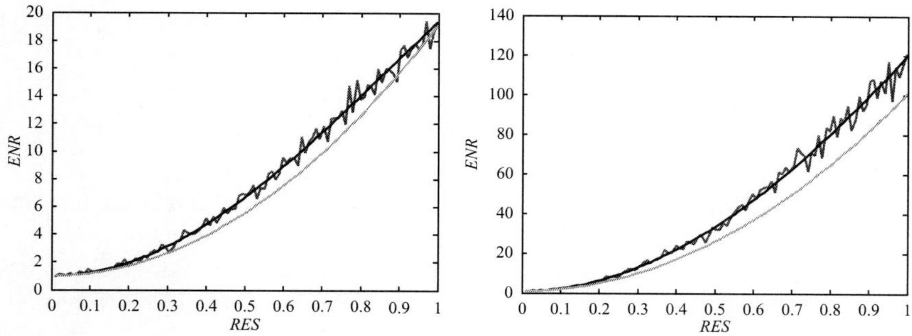

Figure 4.8. *Simulated and Estimated Results (α = 1.5) for ENR Versus RES for $SNR_1 = 20$ dB and $SNR_2 = 20$ dB (left) and $SNR_2 = 10$ dB (right); Three-Element Simulation (medium), ENR Derived from (4.35)–(4.37) (dark) and Second-Order (light) Approximations*

4.6 Enhanced Angle Resolution Algorithms

Starting in the 1970s, algorithms were developed that achieved higher resolution than the Nyquist limits for resolution would predict. These algorithms fall under the name of superresolution algorithms and achieve enhanced angle resolution by making assumptions about the nature of the signal and by implementing nonlinear techniques that estimate angle-of-arrival. Techniques that use the array covariance, discussed in the previous section, assume that the signal space can be separated from the noise space using the eigenvectors of the array covariance matrix. Clearly, there is a trade-off in the array size and the number of samples used to determine the array covariance, which is sometimes referred to as the spatial time-bandwidth product. Other techniques estimate the signal parameters directly, but in all cases the spatial time-bandwidth product determines the quality of the estimate. If M is the number of time samples determining that array covariance and if N is the number of array elements, then, for large N, the asymptotic variance for angle estimation using MUSIC [8] for a uniform linear array with element spacing D can be shown to be

$$\sigma_\theta^2 \geq \frac{6\lambda^2}{4\pi^2 \cdot D^2 \cdot \cos^2(\theta) \cdot M \cdot N \cdot (N^2 - 1) \cdot SNR} \tag{4.62}$$

An interferometer measures the angle after each pulse or dwell that is processed, which essentially trades increased aperture for minimum time to determine angle-of-arrival. However, the digital interferometer achieves only the resolution determined by each of the receive antennas, whereas the analog interferometer achieves the enhanced resolution of the coherently combined antennas. It is more appropriate to define the digital RF interferometer as a superaccuracy technique rather than a superresolution technique, and the fact that measurements are made using a single look at a target have a significant advantage in a highly dynamic environment.

In (4.62), for an interferometer measuring angle-of-arrival with a single look, we have $M = 1$ and $N = 2$, and reduces to

$$\sigma_\theta^2 \geq \frac{\lambda^2}{4\pi^2 \cdot D^2 \cdot \cos^2(\theta) \cdot SNR} \tag{4.63}$$

In the next chapter, we will derive the Cramer-Rao lower bound for the interferometer and show that it is identical to (4.63). The essential benefit that superresolution algorithms bring to angle estimation is the increased accuracy due to time averaging of the array covariance. However, it should be noted that, if the single-look angle measurements are fed into a smoothing filter, then, for M number of independent measurements, the overall angle accuracy is equivalent to $N = 2$, or

$$\sigma_\theta^2 \geq \frac{\lambda^2}{4\pi^2 \cdot D^2 \cdot \cos^2(\theta) \cdot M \cdot SNR} \tag{4.64}$$

In the sense of (4.64), interferometric angle estimation performance is equivalent to superresolution angle estimation for two antennas ($N = 2$). Whereas superresolution uses time averaging to estimate the array covariance, the interferometer can average each single-angle measurement over the same time period to achieve equivalent performance.

References

1. D. K. Barton and H. R. Ward, *Handbook of Radar Measurements*, Artech House, Dedham, MA, USA, 1984.
2. S. Chandran, *Advances in Direction-of-Arrival Estimation*, Artech House, Norwood, MA, USA, 2005.
3. S. Haykin, *Array Signal Processing*, Prentice-Hall, Englewood Cliffs, NJ, USA, 1985.
4. S. M. Sherman, *Monopulse Principles and Techniques*, Artech House, Norwood, MA, USA, 1984.
5. R. C. Johnson and H. Jasik, *Antenna Engineering Handbook*, McGraw Hill, New York, 1984.
6. P. Z. Peebles, *Radar Principles*, John Wiley & Sons, New York, 1998.
7. R. O. Schmidt, "Multiple emitter location and signal parameter estimation," *IEEE Trans. Antennas Propag.*, vol. AP-34, pp. 276–280, Mar. 1986.
8. P. Stoica and A. Nehoral, "MUSIC, maximum likelihood, and Cramer Rao bound," *IEEE Trans. ASSP*, vol. 37, no. 6, pp. 720–741, May 1989.
9. M. Wax and T. Kaileth, "Detection of signals by information theoretic criteria," *IEEE Trans. Acoust. Speech Signal Process*, vol. ASSP-33, pp. 387–392, Apr. 1985.
10. R. R. Nadakuditi and Alan Edelman, "Sample eigenvalue based detection of high-dimensional signals in white noise using relatively few samples," http://arxiv.org/PS_cache/arxiv/pdf/0705/0705.2605v1.pdf, May 2007.

11. K. R. Carver, W. K. Cooper, and W. L. Stutzman, "Beam-pointing errors of planar phased arrays," *Trans. IEEE*, vol. AP-21, pp. 199–202, Mar. 1973.
12. R. C. Hansen, *Phased Array Antennas*, John Wiley & Sons, New York, 1998.
13. J. T. Nessmith, E. J. Holder, L. E. Corey, and R. L. Howard, "The pointing accuracy of phased array radars with correlated phase errors," *Proc. IEE International Radar Conference Radar-87*, pp. 46–49, Oct. 1987.
14. W. H. Nester, "A study of tracking accuracy in monopulse phased arrays," *IRE Trans. AP*, pp. 237–246, May 1962.
15. T. T. Taylor, "Design of line-source antennas for narrow beamwidth and low sidelobes," *IRE Transactions-Antennas and Propagation*, AP-3, pp. 16–28, 1955.
16. E. T. Bayliss, "Design of monopulse antenna difference patterns with low sidelobes," *Bell System Technical Journal*, pp. 623–650, May–June 1968.

Chapter 5
Radar Waveforms

Waveform selection is a critical component of an interferometer architecture design. Achieving angle accuracy in complex target and interference environments can be accomplished only with the proper choice of waveforms. Medium- to high-resolution waveforms can mitigate the effects of clutter and multipath that can degrade phase integrity and angle accuracy. When the clutter and/or multipath are spatially distributed, the range sidelobe performance resulting from the matched filter is critical in reducing the effects of distributed clutter and multipath. Because an interferometer may use small distributed arrays with relatively large beamwidths, the selection of waveforms to mitigate the scattering effect or to resolve target features becomes an important part of an interferometer architecture. In this chapter, several types of frequency- and phase-modulated waveforms are defined, and their impact on post-compression angle accuracy is discussed.

The fundamental concept behind the orthogonal interferometer is the use of an antenna array composed of multiple coherent transmit/receive radar antennas, including separated arrays or contiguous subarrays, that transmit nearly orthogonal coded waveforms. Each receive antenna is also capable of receiving all of the waveforms. The separation of the signals is achieved through the cross-correlation suppression of the waveforms. Near orthogonality in waveforms can be achieved in several ways: phase coding (CDMA), frequency coding (FDMA), time separation (TDMA), and amplitude modulation (AM). However, some methods have less promise than others. For example, for distributed receivers and unknown target geometries, TDMA waveforms may be difficult to implement in radars that use long integration times to track highly dynamic targets.

More recently, research has been conducted in amplitude modulation, such as wavelets and chaotic waveforms, for radar applications. These AM codes can achieve almost zero cross correlations at the expense of reducing SNR due to the amplitude modulation and quadrature filtering schemes. Daubechie wavelets are particularly attractive for radar application due to their compact support structure and orthogonality in scale and time. However, the amplitude modulation reduces the transmitted power affecting SNR, and thus these codes sometimes require significant power to achieve desired performance for radar operations.

Other approaches are randomized in frequency (frequency hopping) or phase and have produced low cross correlation from pulse to pulse so that a delayed pulse has minimal interference with the desired signal pulse. These frequency or phase codes have low inter-code cross-correlation properties, which allow these orthogonal codes to achieve lower cross correlation than can be achieved through conventional matched filters.

5.1 Frequency Coding

Frequency modulation is one of the oldest methods used to generate nearly orthogonal codes. Costas codes afford the possibility of low cross-correlation sidelobes between two interfering pulses, allowing the radar to separate various target signals. LFM and NLFM waveforms that vary the rate of frequency change can be designed for low cross correlation and can be processed efficiently in the signal processor.

Consider the following representation for a radar signal [1]:

$$u(t) = g(t)\exp(j\varphi(t)) \tag{5.1}$$

Its Fourier transform is represented by

$$U(f) = |U(f)|\exp(j\Phi(t)) = \int_{-\infty}^{\infty} g(t)\exp[j(-2\pi ft + \varphi(t))]dt \tag{5.2}$$

The instantaneous frequency is the time derivative of the phase at time t_k.

$$f_k = \frac{1}{2\pi}\varphi'(t_k) \tag{5.3}$$

The stationary phase principle states that the energy spectral density is inversely proportional to the rate of change of frequency. This principle can be intuitively represented by the following expression

$$|U(f_k)|^2 \approx 2\pi\frac{g^2(t_k)}{|\varphi''(t_k)|} \tag{5.4}$$

For LFM signals, the spectrum is shaped through $g(t_k)$ because $|\varphi''(t_k)|$ is constant. For NLFM, the spectrum is shaped through $|\varphi''(t_k)|$ because $g(t_k)$ is kept constant. The stationary phase principle can be inverted to yield

$$\Phi''(f) = 2\pi\frac{V^2(f)}{g^2(t)} \tag{5.5}$$

where $V(f)$ is a function that approximates $U(f)$.

The expressions defined by (5.4) and (5.5) establish the relationships among phase change, amplitude, and energy density for frequency-coded waveforms. The Georgia Tech Research Institute (GTRI) has developed NLFM waveforms using a stationary phase principle methodology that specifies a desired density $V(f)$ and constructs an NLFM that approximates the density [2,3]. Varshney and Thomas [4] show that NLFM waveforms provide enhanced detection and sidelobe performance over LFM and apodization methods.

5.1.1 Costas Codes

Costas waveforms are coherently processed, frequency-hopped uniform pulse trains [1]. Frequency-hopped signals consist of a waveform of period T, subdivided into N subpulses of period $t_1, t_2, \ldots, t_N$, during which a frequency chosen from a set of N available frequencies, $f_1, f_2, \ldots, f_N$, is transmitted, each frequency being unique to a subpulse [1].

In this work, the Costas sequences responsible for the frequency-hopping pattern of the Costas waveforms were generated by the Welsh-Costas Method [7,8]. Using a prime of $p = 11$, Table 5.1 lists the four Costas sequences, or Costas Codes, thus generated. Using a center frequency of f_0 Hz and a bandwidth of B Hz, each element, θ_m, from a Costas code in Table 5.1 can be used to generate a subpulse frequency by using

$$f_n = \left(f_0 - \frac{B}{2}\right) + (\theta_n - 1)\frac{B}{N-1} \tag{5.6}$$

where fn is the calculated frequency of the subpulse.

Table 5.2 shows the total number of Costas codes available for code orders up to 26. Note that the number of codes increases up to order 16, where there are a total of 21,104 Costas codes. In Figure 5.1, two Costas codes of order 29 are used to emulate a target pulse and a jammer pulse. The cross correlation between the two codes allows the matched filter to separate the target pulse from the jammer pulse when the amplitudes are equal. It is not known how many codes of order 29 exist. It should be noted that one issue with Costas codes is that, if the jammer amplifies the signal, the returned jammer signal may overwhelm the matched filter due to the fact that the cross correlation is not low enough.

Table 5.1. Example of Costas Codes (Length = 10)

2	4	8	5	10	9	7	3	6	1
6	3	7	9	10	5	8	4	2	1
7	5	2	3	10	4	6	9	8	1
8	9	6	4	10	3	2	5	7	1

Table 5.2. Number of Costas Codes per Order

Order	**Number**	**Order**	**Number**	**Order**	**Number**
1	1	10	2,160	19	10,240
2	2	11	4,368	20	6,464
3	4	12	7,852	21	3,536
4	12	13	12,828	22	2,052
5	40	14	12,752	23	872
6	116	15	19,612	24	200
7	200	16	21,104	25	88
8	444	17	18,276	26	56
9	760	18	15,096	27	??

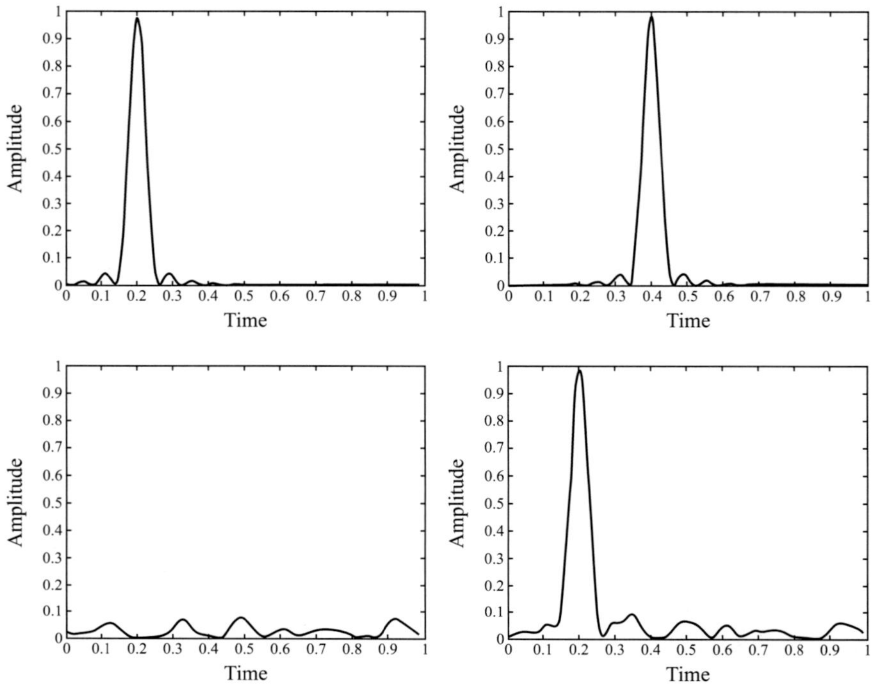

Figure 5.1. Two Costas-Coded Waveforms with N = 29: Upper Left: 0.2-s Delayed Pulse Emulating a Signal; Upper Right: 0.4-s Delayed Jammer Pulse; Lower Left: Cross Correlation Between the Two Waveforms; Lower Right: Effect of the Low Cross Correlation to Suppress the Jammer Pulse While Compressing the Signal Pulse

5.1.2 Linear Frequency Modulation

A linear frequency modulation (LFM) waveform is a waveform that uses frequency modulation where the frequency changes linearly across the transmit pulse [9]. LFM waveforms are also referred to as "chirp" waveforms due to the frequency-chirping modulation. The transmit waveform is generated with a frequency versus time characteristic, such as shown in Figure 5.2.

Because frequency is the time derivative of phase, the time waveform for LFM is a quadratic phase and can be expressed as,

$$x(t) = rect(t/T)e^{j2\pi(f_c t + \alpha t^2)} \tag{5.7}$$

where f_c is the carrier frequency, and

$$rect(t/T) = \begin{bmatrix} 1, & if\ |t/T| < 1/2 \\ 0, & if\ |t/T| > 1/2 \end{bmatrix} \tag{5.8}$$

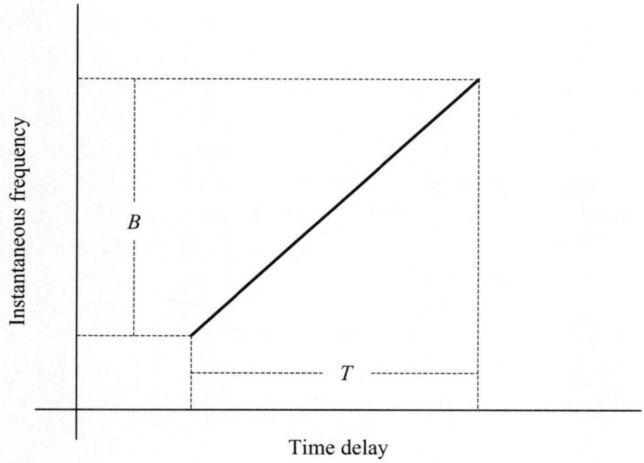

Figure 5.2. Linear Frequency Modulation Frequency Versus Time Delay

and α is a quadratic phase constant. The time bandwidth product for the LFM waveform is the product of T and B. For an LFM, an approximately rectangular spectrum results for time bandwidth products greater than 10 [1].

5.1.2.1 Matched Filter Response

A receiver that maximizes the output peak signal-to-mean noise ratio is called the matched filter [6]. The matched filter for the LFM waveform described by Figure 5.2 is shown in Figure 5.3.

The result of processing the transmit LFM waveform with the matched filter response is to collapse it into a narrow time pulse. The pulse width of the

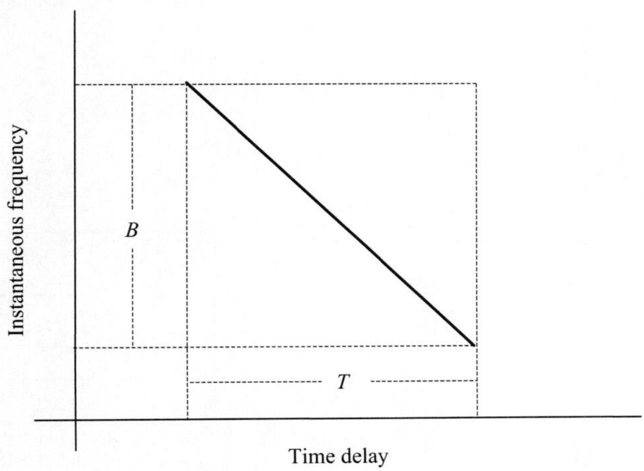

Figure 5.3. LFM Matched Filter Frequency Response

compressed pulse is inversely proportional to the chirp bandwidth. The compressed pulse has a characteristic of $\sin(\pi Bt)/(\pi Bt)$ form with a 13.2-dB first-time sidelobe level below the mainlobe. Weighting function such as Hamming or Hanning weights [14] can be applied to reduce the sidelobes levels well under 13.2 dB below the mainlobe. However, the effect of introducing weighting functions is to cause a mismatch in the receiver response that results in a 1- to 2-dB mismatch loss. For example, a Hamming weighting function produces a −42.8-dB peak sidelobe with a 1.34-dB mismatch loss. A Hanning weighting will produce a −32.3-dB peak sidelobe with a 1.76-mismatch loss [6].

5.1.2.2 Matched Filter Implementation

The output of the matched filter is proportional to the cross correlation of the input signal and a delayed replica of the input signal.

$$R(t) = \int_{-\infty}^{\infty} x(\lambda)y(\lambda - t)d\lambda \tag{5.9}$$

where $x(t)$ is the input signal, and $y(t)$ is the receiver filter time function. For an LFM waveform, $y(t)$ is the time response output of the matched filter. Notice that (5.9) is a convolution operation. Convolution is more efficiently implemented in the frequency domain using the principle that the Fourier transform of a convolution of two functions is equal to the product of the Fourier transform of the two functions. Figure 5.4 shows the implementation of the matched filter using fast Fourier transforms (FFTs). Matched filter processing is sometimes referred to as two-pass processing due to the two-stage implementation of FFTs. Thus the total numerical operations require two FFT's.

The total number of digital samples that are processed in the matched filter implementation is a function of the desired range window. The time samples occur

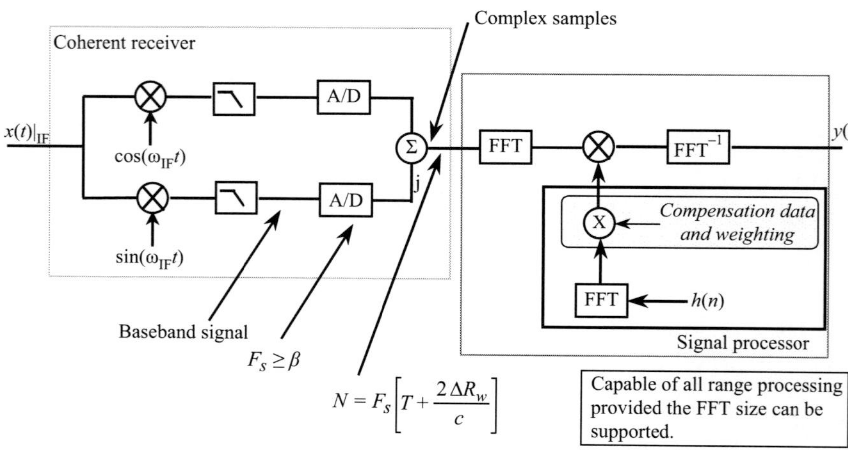

Figure 5.4. Matched Filter Implementation for LFM Processing

over a duration equal to the pulse width plus the receive window so that a loss in signal energy and range resolution does not result. Thus the number of samples is given by

$$N = F_s\left[T + \frac{2\,\Delta R}{c}\right] \tag{5.10}$$

where F_s is the sampling frequency. F_s must satisfy the Nyquist sampling rate of the intermediate frequency (IF) signal that is determined by the desired range resolution. The IF signal bandwidth must be at least equal to the total bandwidth swept by the LFM waveform.

One waveform that has been used in pulse-Doppler applications is the linear frequency modulation (LFM) waveform. An advantage of the LFM waveform is the fact that it is Doppler tolerant. In other words, changes in the target Doppler induce small changes in sensitivity performance. Because frequency is the derivative of phase and the frequency is linear, phase is a quadratic function of time.

$$f(t) = \frac{d}{dt}\frac{\beta}{2\tau}t^2 = \frac{\beta}{\tau}t, \qquad -\frac{\tau}{2} < t < \frac{\tau}{2} \tag{5.11}$$

where τ is the length of the LFM waveform in time, and β is total swept bandwidth. This sweeping of frequency in time produces a range Doppler coupling that creates an ambiguity in the determination of range and Doppler. Certain techniques can be implemented to mitigate range-Doppler coupling, such as using a sequence of up-chirp and down-chirp LFM pulses.

The LFM matched filter output is given by the following:

$$x(t - t_d) = \exp\left(j\pi\frac{\beta}{\tau}(t - t_d)^2\right) \qquad \text{received waveform delayed by } t_d \tag{5.12}$$

$$x(t) = \exp\left(j\pi\frac{\beta}{\tau}t^2\right) \qquad \text{reference waveform} \tag{5.13}$$

$$\begin{aligned} y(t) &= x(t)x(t - t_d) \\ &= \exp\left(j2\pi\frac{\beta}{\tau}t_d t\right)\exp\left(-j\pi\frac{\beta}{\tau}t_d^2\right) \qquad \text{matched filter output} \end{aligned} \tag{5.14}$$

These waveform represents an up-chirp LFM waveform, so-called due to the positive slope of the frequency function. If the slope of the frequency function is negative, then the LFM waveform is called a down-chirp. The cross correlation of the up-chirp and down-chirp LFM waveforms is small in magnitude after bandpass filtering to eliminate the unwanted high-frequency contributions. Thus, up- and down-chirp waveforms are nearly orthogonal. LFM can achieve low cross correlation by varying the rate of change of the linear frequency slope from pulse to pulse. For LFM waveforms, $|\varphi''(t_k)|$ is a constant in (5.10), and thus the spectral density is determined by the amplitude shaping $g(t)$. Typically, g is defined by a weighting function that reduces the spectral density sidelobes.

One possible variation on LFM waveforms is to represent $|\varphi''(t_k)|$ by constant step functions [10]. This has the effect of changing the rate of change of phase over time. According to the stationary phase principle stated by (5.10), changing the rate of phase inversely affects the spectral density. By carefully defining the step function $|\varphi''(t_k)|$ for LFM waveforms, the cross correlation among these waveforms can be made small. For example, the waveform defined when $\varphi''(t_k) = +c$ (up-chirp) and the waveform defined by $\varphi''(t_k) = -c$ (down-chirp) have low cross correlation and are essentially orthogonal. Other phase derivative rate of change step functions can produce low correlation waveforms. Figure 5.5 shows the autocorrelation between two variable-frequency-rate LFM waveforms that are defined by stepped phase second-derivative functions whose steps are determined by two Costas codes in Table 5.1. Note the relatively low cross correlation that allows the matched filter (light gray) response to be detected.

One other application for implementing up-chirp and down-chirp LFM waveforms is to resolve the range-Doppler coupling. Because the frequency is linearly swept for an LFM waveform, the range-Doppler ambiguity region is a ridge with slope depending on the rate of sweep. Furthermore, the time of arrival from a target correlates differently for an up-chirp LFM compared to a down-chirp LFW waveform, and the intersection of the positive and negative sloped ambiguity ridges

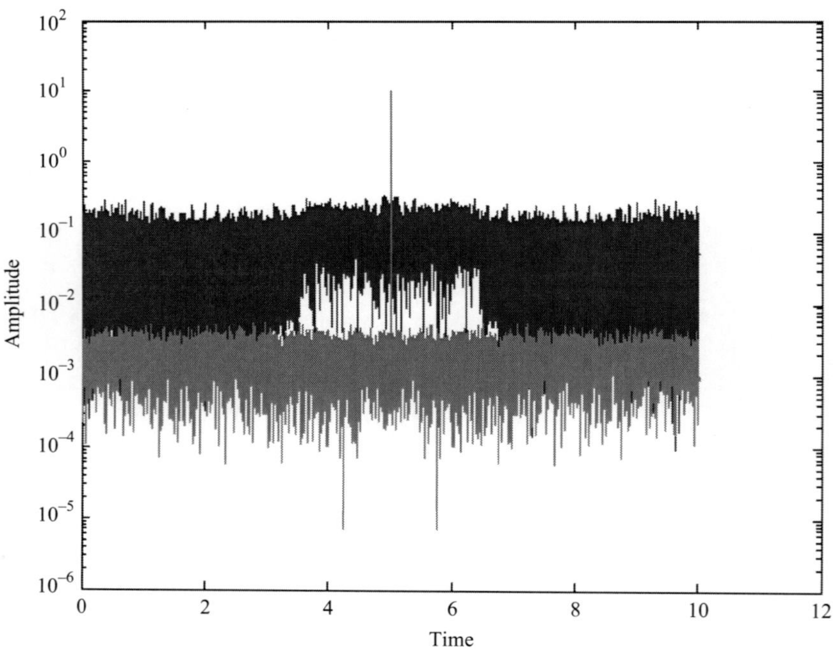

Figure 5.5. Autocorrelation (light gray) and Cross Correlation (dark gray) of Two LFM Waveforms with Stepped Phase Rate of Change Functions Defined by Costas Codes

produces a thumbtack ambiguity region that significantly reduces range-Doppler coupling.

For angle-of-arrival estimation using LFM waveforms, the phase characteristics of the waveform are the dominant contributor. However, in modeling an LFM waveform, it is necessary to capture not only the phase structure but also the amplitude of the LFM waveform. Using a signal definition for an LFM waveform similar to Levanon [1],

$$s(t) = \frac{A}{\sqrt{T}} rect\left(\frac{1}{2} - \frac{t}{T}\right) e^{2\pi ja\left(\frac{T}{2}-t\right)^2} e^{2\pi jf_c t} e^{j\phi} \tag{5.15}$$

where *rect*(t) is a window function that encodes the finite pulse extent of the waveform and has a value equal to unity when its argument is between 0 and 1 and zero elsewhere. Note that the parameter a is equal to B/T, or the ramp slope divided by the time length of the pulse.

A pulse delayed in time by an amount $\tau = R/c$ is given by

$$s(t+\tau) = \frac{A}{\sqrt{T}} rect\left(\frac{1}{2} - \frac{t+\tau}{T}\right) e^{2\pi ja\left(\frac{T}{2}-t-\tau\right)^2} e^{2\pi jf_c(t+\tau)} e^{j\phi} \tag{5.16}$$

Convolving the delayed pulse and a reference pulse gives

$$\begin{aligned} s(t)s^*(t+\tau) &= \frac{A}{T} rect\left(\frac{1}{2} - \frac{t+\tau}{T}\right) rect\left(\frac{1}{2} - \frac{t}{T}\right) e^{2\pi ja\left[\left(\frac{T}{2}-t\right)^2 - \left(\frac{T}{2}-t-\tau\right)^2\right]} e^{2\pi jf_c[t-t-\tau]} \\ &= \frac{A}{T} rect\left(\frac{1}{2} - \frac{t+\tau}{T}\right) rect\left(\frac{1}{2} - \frac{t}{T}\right) e^{2\pi ja\left[\left(\frac{T}{2}\right)^2 + t^2 - tT - \left(\frac{T}{2}\right)^2 - t^2 + tT + \tau T - 2t\tau - \tau^2\right]} e^{-2\pi jf_c\tau} \\ &= \frac{A}{T} rect\left(\frac{1}{2} - \frac{t+\tau}{T}\right) rect\left(\frac{1}{2} - \frac{t}{T}\right) e^{-2\pi ja(2t\tau - \tau T + \tau^2)} e^{-2\pi jf_c\tau} \end{aligned} \tag{5.17}$$

Note that, after removal of the carrier frequency term, the phase is given by

$$\varphi(t,\tau) = 2\pi a(2t\tau - \tau T + \tau^2) \tag{5.18}$$

And frequency of the matched filter output is the time derivative of phase

$$\frac{d\varphi}{dt} = 4\pi a\tau \tag{5.19}$$

Because the delay variable τ represents the delay due to the time of the two-way propagation to the target, the target range is immediate from the output of the matched filter.

5.1.3 Frequency-Modulated Continuous Wave (FMCW)

The FMCW waveform combines the CW waveform with an LFM waveform to achieve a waveform that has the timing properties of a pulsed waveform but that maintains 100 percent duty. The waveform consists of a series of chirped LFM segments over the duration of the waveform. The chirps can be alternated between up- and down-chirps or can be all one variety of chirped LFM. As with LFM, the time bandwidth product of each LFM segment determines the range resolution of the waveform. For any given range cell, the compressed pulses are separated by the duration T of the LFM pulse. Thus the equivalent PRF is $1/T$. The Doppler resolution is determined by the duration of the train of the LFM sequences that make up the waveform. Figure 5.6 shows the FMCW train of LFM sequences and the effect of Doppler on the return waveform. Let N be the number of LFM segments in an FMCW waveform; then range and Doppler ambiguities are set by the following expressions:

$$R_{\text{amb}} = \frac{cT}{2} \tag{5.20}$$

$$f_{\text{Doppamb}} = \frac{1}{T} \tag{5.21}$$

However, the velocity ambiguity is wavelength or frequency dependent, and for LFM each range cell corresponds to a different frequency. Let λ_c denote the wavelength of the carrier frequency of the LFM chirp.

$$V_{\text{amb}} = \frac{\lambda_c}{2T} \tag{5.22}$$

The primary advantage of FMCW is that it is easily implemented using a homodyne receiver, just as CW Doppler waveforms are easily implemented using a

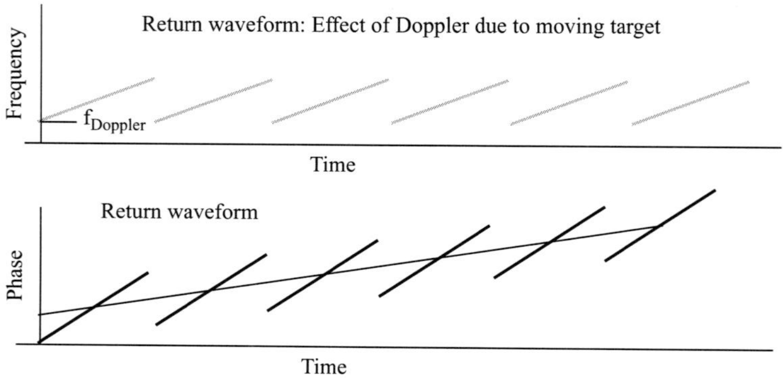

Figure 5.6. Transmit and Receive FMCW Waveforms

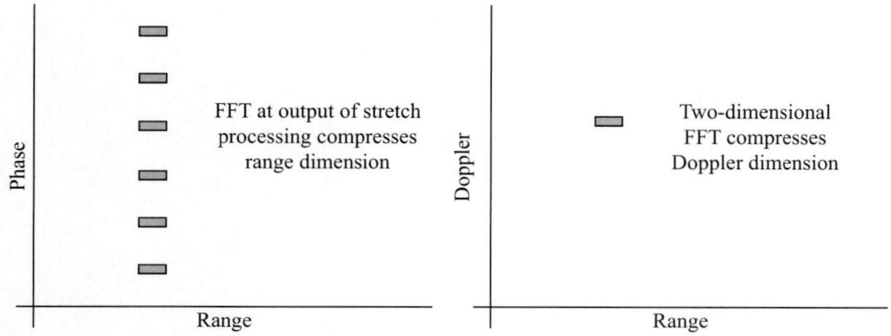

Figure 5.7. Creation of the Range-Doppler Map for FMCW Waveforms: One-Dimensional FFT (left) Compresses Output of Stretch Processing in Range Dimension; Second One-Dimensional FFT (right) Compresses in the Doppler Dimension

homodyne architecture. The homodyne receiver mixes a reference signal with the return signal in analog to an intermediate frequency that is significantly lower in frequency than the carrier of the transmitted waveform. This can be accomplished because the FWCW and CW waveforms operate at 100 percent duty and thus are always available as reference waveforms when the transmitted signal returns to the radar receive antenna. For FMCW, the analog mixing is equivalent to stretch processing, which results in a tone, and the Fourier transform of the tone establishes the range to the target. Figure 5.7 shows the creation of the range-Doppler map for an FMCW waveform using two one-dimensional FFT transforms. As a result, all of the angle-of-arrival processing results for LFM that was derived in Section 5.1.2 are applicable for FMCW.

FMCW waveforms are used in homodyne radars to reduce cost and complexity. The FMCW waveform is usually applied to short-range systems due to the 100 percent duty. Longer-range radars need significant sensitivity that can be delivered with higher peak power over short pulses at lower duty. Military applications include perimeter surveillance and force protection radars for intrusion detection. A popular application for FMCW waveform is sports tracking. For sports radars, the range is typically short, and it is necessary to measure range, angle, and Doppler accurately. Thus FMCW radar interferometers are a popular choice for tracking golf balls, baseballs, tennis balls, and other objects, such as the golf club swing dynamics.

5.1.4 Nonlinear Frequency Modulation

We now develop a process that generates NLFM waveforms. Consider a signal $u(t)$ modulated by the function $\varphi(t)$, and let $U(f)$ be the Fourier transform of $u(t)$. Then we have

$$u(t) = g(t)e^{i\varphi(t)} \tag{5.23}$$

$$\begin{aligned}
U(f) &\equiv |U(f)|e^{i\Phi(f)} \\
&= \int_{-\infty}^{\infty} g(t)e^{i\varphi(t)}e^{-2\pi ift}dt \\
&= \int_{-\infty}^{\infty} g(t)e^{i(-2\pi ft+\varphi(t))}dt \\
&\approx g(c)e^{if(c)+i\frac{\pi}{4}}\sqrt{\frac{2\pi}{|\varphi''(c)|}}
\end{aligned} \tag{5.24}$$

If we apply the principle of stationary phase (see Appendix C) to the function $u(t)$, we have

$$\begin{aligned}
u(t) &= \frac{1}{2\pi}\int_{-\infty}^{\infty} U(f)e^{2\pi ft}df \\
&= \frac{1}{2\pi}\int_{-\infty}^{\infty} |U(f)|e^{i\Phi(f)}e^{2\pi ift}df \\
&= \frac{1}{2\pi}\int_{-\infty}^{\infty} |U(f)|e^{i(2\pi ft+\Phi(t))}df \\
&\approx \frac{1}{2\pi}U(f)e^{itc+i\frac{\pi}{4}}\sqrt{\frac{2\pi}{|\Phi''(f)|}}
\end{aligned} \tag{5.25}$$

Assuming that Φ is real, we have

$$\begin{aligned}
\Phi''(f) &= \frac{|U(f)|^2}{2\pi|u(t)|^2} \\
&= \frac{|U(f)|^2}{2\pi g^2(t)}
\end{aligned} \tag{5.26}$$

To determine a desired behavior for an NLFM signal, we choose a function V that approximates U, and we set $g(t)$ equal to a constant:

$$\Phi''(f) = V(f) \tag{5.27}$$

The group delay $T(f)$ is determined by

$$T(f) = \frac{1}{2\pi}\Phi'(f) = \frac{1}{2\pi}\int_{0}^{f}\Phi''(z)dz = \frac{1}{2\pi}\int_{0}^{f}V(z)dz \tag{5.28}$$

The instantaneous frequency versus time $f(t)$ is the inverse of $T(f)$.

$$f(t) = T^{-1}(f) \tag{5.29}$$

And finally the phase function $\varphi(t)$ is defined by

$$\varphi(t) = 2\pi \int_0^t f(x)dx \tag{5.30}$$

The NLFM time waveform $u(t)$ then becomes

$$u(t) = g(t)e^{i\varphi(t)} \tag{5.31}$$

where $g(t)$ is a constant, and φ is defined by the preceding process.

5.1.4.1 Example 1: Gaussian Spectrum

In this example, we set $V(f)$ to be a Gaussian density function.

$$V(f) = \frac{\gamma_B}{\sqrt{2\pi\sigma^2}} e^{-\frac{f^2}{2\sigma^2}} \qquad -\frac{B}{2} \leq f \leq \frac{B}{2} \tag{5.32}$$

$$\begin{aligned} T(f) &= \frac{1}{2\pi} \int_{-\frac{B}{2}}^{f} \frac{\gamma_B}{\sqrt{2\pi\sigma^2}} e^{-\frac{x^2}{2\sigma^2}} dx \\ &= \frac{\gamma_B}{4\pi} \left(erf\left(\frac{f}{\sqrt{2}\sigma} \right) + erf\left(\frac{B/2}{\sqrt{2}\sigma} \right) \right) \end{aligned} \tag{5.33}$$

where $erf(x) = \frac{2}{\sqrt{\pi}} \int_0^x e^{t^2} dt$ and $\gamma_B = erf\left(\frac{B/2}{\sqrt{2}\sigma} \right)$.

The challenge is to invert $T(f)$ with sufficient accuracy to achieve the desired frequency response defined by $V(f)$. To achieve the inversion $T(f)$, we first compute values (f, T), and then we reflect through the 45° diagonal line through the reversal (T, f). These reflected values define $f(t)$. Now $\varphi(t)$ is computed by integrating $f(t)$ numerically using the nonuniformly spaced time interval T. The discrete $f(t)$ values are fit with a polynomial of high degree $p(t)$. Finally, $u(t)$ is defined using $p(t)$.

$$u(t) = e^{ip(t)} \tag{5.34}$$

Figures 5.8 and 5.9 show the function $f(t)$, $\varphi(t)$, real($u(t)$), and the autocorrelation of u when $\sigma = 0.05$ and time is 318 units.

Figure 5.10 shows the Gaussian density function $V(f)$ with $\sigma = 0.05$ and the achieved spectrum $U(f)$. The difference is due to the fact that the time was inflated to achieve a larger bandwidth and thus larger time-bandwidth product in order to

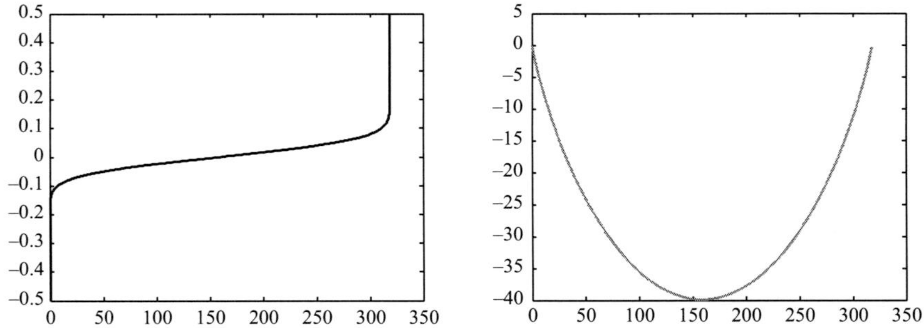

Figure 5.8. The Function f(t) (left) and the Phase Functions φ(t) and p(t) (right)

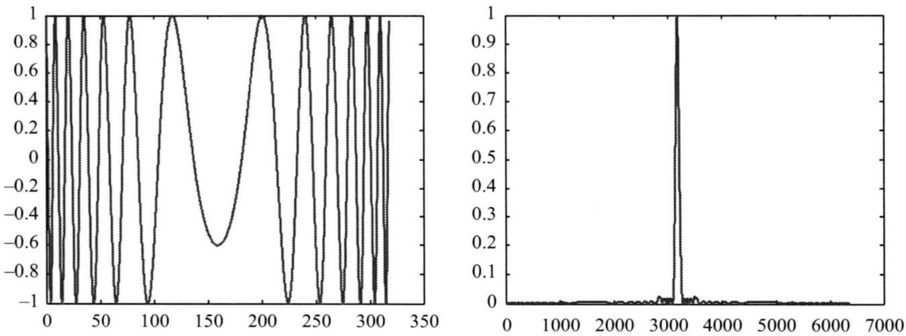

Figure 5.9. The Real Part of the Function u(t) (left) and the Autocorrelation of u (right)

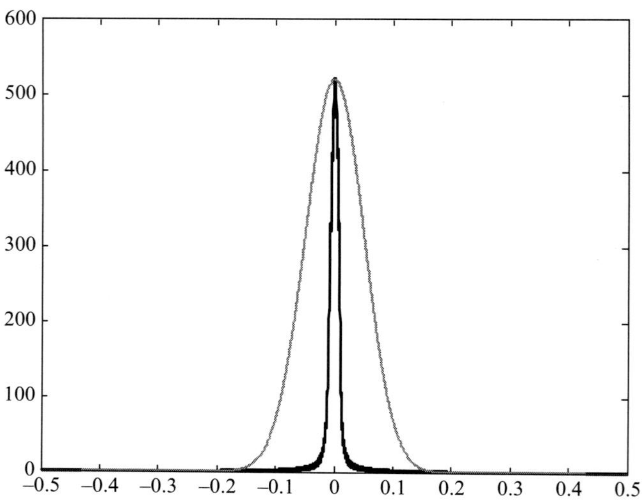

Figure 5.10. The Desired Spectrum V(f) (light gray) and the Achieved Spectrum U(f) (dark gray)

achieve relatively low time sidelobes on the right plot in Figure 5.9. The overall time sidelobe product is on the order of 1000, thus achieving nearly −30 dB sidelobes below the peak.

5.1.4.2 Example 2: Fourier Series Approximation

Byron Keel at GTRI has discovered the following NLFM relationship:

$$f(t) = 2\pi\frac{\beta}{\tau}t + \pi\beta\sum_{n=1}^{N} K_n \sin\left(2\pi\frac{nt}{\tau}\right) \qquad -\frac{\tau}{2} \le t \le \frac{\tau}{2} \tag{5.35}$$

$$\varphi(t) = \pi\frac{\beta}{\tau}t^2 - \pi\beta\tau\sum_{n=1}^{N}\frac{K_n}{\pi n}\cos\left(2\pi\frac{nt}{\tau}\right) \qquad -\frac{\tau}{2} \le t \le \frac{\tau}{2} \tag{5.36}$$

where β is the waveform bandwidth, τ is the pulse width, and K_n are the coefficients used to generate the NLFM waveform. The coefficients in Table 5.3 are

Table 5.3. NLFM Coefficients

n	K_n
1	−0.11417607723306
2	0.03960138311910
3	−0.02048549632323
4	0.01253307329411
5	−0.00840992355201
6	0.00598620805378
7	−0.00444406248014
8	0.00340066531153
9	−0.00265806073847
10	0.00210938646003
11	−0.00169902707554
12	0.00139108276338
13	−0.00115689275330
14	0.00096693003633
15	−0.00080241178523
16	0.00066411304862
17	−0.00056409896538
18	0.00049971940882
19	−0.00043663940977
20	0.00036352600042
21	−0.00030225236729
22	0.00025435747083
23	−0.00023125401709
24	0.00022214432810
25	−0.00019499398675
26	0.00015746842818
27	−0.00011571222127
28	0.00010580212284
29	−0.00011517925950
30	0.00011232412077

derived when $V(f)$ (previously defined) is the Fourier transform of a square wave with a 40-dB Taylor weighting.

Again the NLFM waveform is defined by

$$u(t) = e^{i\varphi(t)} \qquad -\frac{\tau}{2} \leq t \leq \frac{\tau}{2} \tag{5.37}$$

5.2 Phase Coding

In this section we introduce phase-coded waveforms that have low autocorrelation and cross-correlation properties. In Appendix D, maximal length sequences are developed that form the foundation for pseudorandom noise (PRN) codes discussed in Appendix D and, for the modulated PRN codes, achieve RMS sidelobes levels of $-20\log(N)$ [14], where N is the code length but the cross-correlation performance among PRN codes of equal length is relatively high compared to other classes, such as Kasami codes.

Modulated PRN codes are introduced where the modulation operators form an Abelian group (operators that commute). The cross-correlation values from any two codes in these PRN modulated code groups are relatively low (less than the square root of the code length). In addition, a set of modulated PRN codes is defined with codes that are essentially orthogonal (cross-correlation values between codes in the code group are predominantly small [$\ll$1]). Although the essentially orthogonal codes have autocorrelation ambiguities, these codes have application in a multiple-code environment where cross-correlation interference is the dominant factor.

Code division multiple access (CDMA) waveforms, or phase-coded waveforms, offer a trade between achieving orthogonality and effective SNR performance. Phase-coded waveforms essentially are phase step functions where phase takes on discrete values such as binary codes where the phase alternates between values of $+1$ and -1 randomly. As a result, the second derivative of phase is not defined as the jumps in phase. However, these jumps are defining the energy density. As such, the ambiguity function for these waveforms are thumbtack functions.

5.2.1 Pseudorandom Noise Codes (Kasami Codes)

By encoding pulses with Kasami codes, the jammer cannot easily capture the signal pulse due to the low cross correlation between Kasami codes. The separation of the signal pulse from the jammer pulse is achieved through the cross-correlation suppression of the orthogonal biphase waveforms. Figure 5.11 presents an example of the auto- and cross-correlation of two orthogonal Kasami-coded waveforms. As can be seen, the cross-correlation interference is suppressed by the square root of the number of bits in the code.

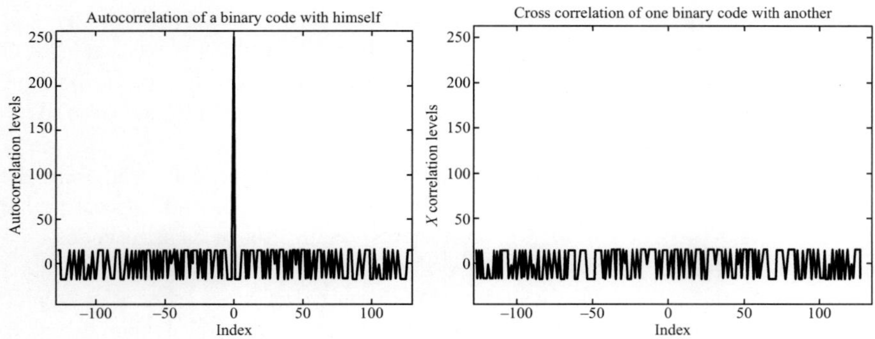

Figure 5.11. Autocorrelation of a 255-Bit Kasami Biphase-Modulated Code (left) Cross Correlation of Two Orthogonal 255-Bit Kasami Biphase-Modulated Codes (right)

5.2.1.1 Properties of General Binary Codes

Let $\theta_{u,v}$ denote the cross correlation between the codes u and v, and let $\theta_{u,v}(k)$ denote the kth element of this vector. It is a general property of PRN binary sequences of odd length that

$$\sum_{k=0}^{N-1} \theta_{u,v}(k) = 1 \tag{5.38}$$

$$\sum_{k=0}^{N-1} |\theta_{u,v}(k)|^2 = N^2 + N - 1 \tag{5.39}$$

To use such codes in radar applications, we want the in-phase values $\theta_{u,v}(0)$ to be large in magnitude compared with any out-of-phase values $\theta_{u,v}(k)$, $0 < k < N$, and $\theta_{u,v}(k)$, $0 \leq k < N$. For this reason, our goal is to find a set of binary sequences (codes) for which the maximum magnitude of any out-of-phase term is as small as possible. From (5.39), we can see that a theoretical lower bound for such a minimum cross-correlation value exists.

For instance, one might hope that two codes u and v exist for which $\theta_{u,v}(0) = N$ and that $|\theta_{u,v}(k)| = 1$ for all k different from zero. Using (5.39) one can see that this can never happen with binary codes because $\Sigma|\theta_{u,v}(k)|^2 = N$ (if u is different from v) and not $N^2 + N - 1$, as it must.

5.2.1.2 Kasami Codes

With a little consideration, we can see that the best situation we can hope for is that $|\theta_{u,v}(k)| = \sqrt{N + 1 + 1/N}$ for all k different from zero. This situation is almost achieved by the Kasami codes, which have out-of-phase values $|\theta_{u,v}(k)| = \sqrt{N+1}$. It is not possible to construct Kasami codes of arbitrary length; however, an explicit construction exists starting from any PRN code of length $N = 4^n - 1$, for positive integers n. Further examination shows $\sqrt{N+1}$ to be the closest integer value to the theoretical minimum $\sqrt{N+1+1/N}$ previously computed.

A set of Kasami codes with these properties is produced by shuffling the entries of the base PRN code systematically. Another key point of interest is that any PRN code of length $N = 4^n - 1$ has a corresponding set of Kasami codes containing $\sqrt{N+1}$ codes in the set.

Kasami codes can produce waveforms with really low cross correlation, which is important to mitigate a coherent jammer that is creating multiple-signal replica waveforms.

5.2.1.3 Quadra-Phase Codes

Quadra-phase codes use four-phase states as opposed to the two-phase states used in binary phase codes. These additional phase states allow for lower cross-correlation performance at the expense of being Doppler intolerant.

5.2.2 Group Modulation of PRN Codes

The compression performance of binary phase codes is sensitive to Doppler modulation due to the changes in received waveform phase that result in filter mismatch. In this section, we use waveform modulation to create a family of waveforms that possess the properties of a mathematical group. A mathematical group is defined as a set G and an operation $*$ such that the following properties are satisfied,

1. For all $g_1, g_2 \in G$, $g_1 * g_2 \in G$ (closure).
2. There exists $e \in G$ such that for all $g \in G$, we have $g * e = e * g = g$ (identity).
3. For all $g \in G$, there exists $g^{-1} \in G$ such that $g * g^{-1} = g^{-1} * g = e$ (inverse).

A group G is an Abelian group if the following is satisfied:

4. For all $g_1, g_2 \in G$, $g_1 * g_2 = g_2 * g_1$ (commutative).

For a code x of length L, the set of modulations $\exp(\pi jkx)$ form a group under the operation of Hadamard product denoted by the MATLAB® notation asterisk (.*). Figure 5.12 shows examples of group modulations that are elements of the group of modulations. Using these structured modulations, a group of waveforms can be created from maximum length sequence codes that have low cross-correlation properties with excellent autocorrelation sidelobes. We begin by introducing formalism that develops the general theory of group modulation.

A suite of nearly orthogonal codes can be derived using group operations on pseudorandom noise waveforms. Consider an operator R_θ that has the following properties:

$$R_\theta R_\phi = R_{\theta+\phi} = R_\phi R_\theta \tag{5.40}$$

$$R_\theta R_{-\theta} = I \tag{5.41}$$

$$R_\theta^* = R_{-\theta} \tag{5.42}$$

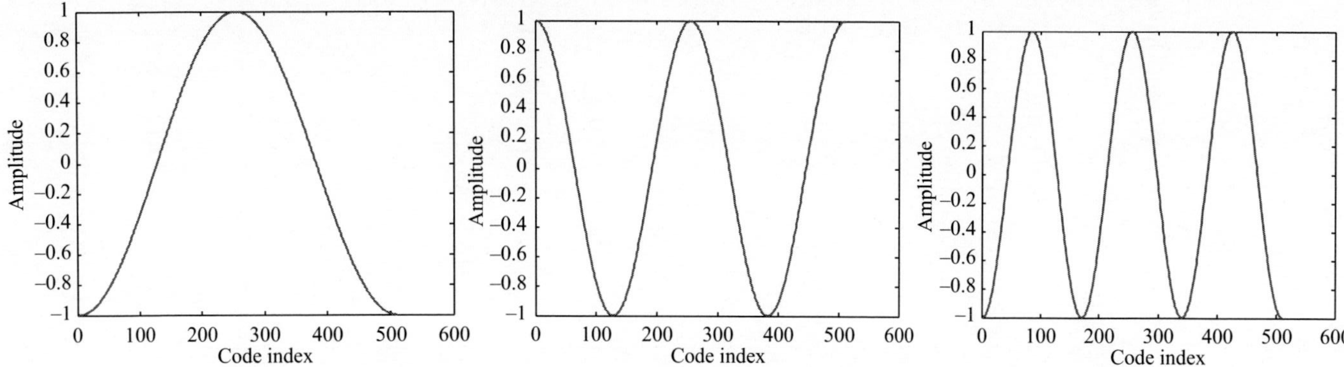

Figure 5.12. Example of Group Modulations (1/2 period, 1 period, and 3/2 period)

where I is the identity operator. The operators R_θ operate on the elements of a waveform and form an Abelian group, and we now investigate properties of group operators on PRN waveforms. In particular, we consider the following special subgroup of R_θ, which is defined from a unique θ.

$$R_{m\theta}R_{n\theta} = R_{(m+n)\theta} = R_{n\theta}R_{m\theta} \tag{5.43}$$

$$R_{n\theta}R_{-n\theta} = I \tag{5.44}$$

Let M denote the circular shift matrix defined in Appendix D, and consider a PRN sequence s and the following group action on s.

$$corr(s, R_\theta s) \equiv \langle s, R_\theta s\rangle = \langle R_\theta^* s, s\rangle \tag{5.45}$$

$$\begin{aligned} A_{i,k} &= corr_{i,k}(s, R_\theta s) == \langle \exp(j\pi M^i c), \exp(j\pi M^k(c+\theta))\rangle \\ &= \sum \exp(j\pi(M^i c \oplus_2 M^k c)).^*\exp(j\pi\theta) \\ &= \sum \exp(j\pi((M^i \oplus_2 M^k)c)).^*\exp(j\pi\theta) \\ &= \langle M^{i+k}s, R_\theta\rangle \\ &= \langle \exp(j\pi c'), \exp(j\pi\theta)\rangle \\ &= \sum u_i.^*(\cos(\pi\theta(i)) + j\sin(\pi\theta(i))) \end{aligned} \tag{5.46}$$

where $\oplus_2$ denotes addition modulo 2, and c' is the $i + k$ shifted version of c. Computing the magnitude squared, we have

$$\begin{aligned} |A_{i,k}|^2 &= \left(\sum_i u_i.^*\cos(\pi\theta(i))\right)^2 + \left(\sum_i u_i.^*\sin(\pi\theta(i))\right)^2 \\ &= \sum_i u_i^2\cos^2(\pi\theta(i)) + \sum_k M^k u_i \cos(\pi\theta(i))\cos(\pi\theta(k)) \\ &\quad + \sum_i u_i^2\sin^2(\pi\theta(i)) + \sum_k M^k u_i \sin(\pi\theta(i))\sin(\pi\theta(k)) \\ &= N + \sum_i\sum_k M^k u_i\cos(\pi\theta(i) - \pi\theta(k)) \\ &= N + \sum_k\sum_i M^k u_i\cos(\pi\theta(i) - \pi\theta(k)) \end{aligned} \tag{5.47}$$

Now define θ as follows:

$$\theta(k) = k\theta_0 \tag{5.48}$$

where θ_0 is defined such that, for an integer m, $(N+1)\theta_0 = 2m$. For a PRN sequence length of $N = 2^q - 1$, we have

$$\begin{aligned} \cos(\pi\theta_0(N-k)) - \cos(\pi\theta_0 k) &= \cos\left(\frac{2\pi m}{N+1}(N-k)\right) - \cos\left(\frac{2\pi m}{N+1}k\right) \\ &= \cos\left(2\pi m\frac{N}{N+1} - \frac{2\pi m}{N+1}k\right) - \cos\left(\frac{2\pi m}{N+1}k\right) \\ &\to \cos\left(\frac{2\pi m}{N}k\right) - \cos\left(\frac{2\pi m}{N}k\right) = 0 \quad \text{as } N \to \infty \end{aligned} \tag{5.49}$$

And the magnitude of the cross correlation can be expressed as follows:

$$
\begin{aligned}
|A_{i,k}|^2 &= N + \sum_i \sum_k M^k u_i \cos((i-k)\pi\theta_0) \\
&= N + \sum_k \sum_m M^k u_{k+m} \cos(m\pi\theta_0) \\
&= N + \sum_m \pm\cos(m\pi\theta_0) = N \pm 1 \pm \varepsilon
\end{aligned} \tag{5.50}
$$

where $\varepsilon \ll 1$ is a small deviation (see Appendix D), and the magnitude is

$$
|A_{i,k}| = \sqrt{N \pm 1} = \begin{cases} \sqrt{2^q - 2} \pm \varepsilon & \text{if} \sum_i u_i = -1 \\ 2^{q/2} \pm \varepsilon & \text{if} \sum_i u_i = 1 \end{cases} \tag{5.51}
$$

5.2.2.1 Example: $N=3$

We assume a three-element code $[u_1 \quad u_2 \quad u_3]$, and we modulate the code with the modulation $\exp(2\pi jk/3) = 1, 2, 3$. Then we have

$$
\begin{aligned}
\cos(3) &\equiv \cos\left(3\frac{2\pi}{3}\right) = \cos\left(0\frac{2\pi}{3}\right) \equiv \cos(0) \\
\cos(1) &\equiv \cos\left(1\frac{2\pi}{3}\right) = \cos\left(2\frac{4\pi}{3}\right) \equiv \cos(2)
\end{aligned}
$$

And we can compute

$$
\begin{aligned}
&(u_1\cos(1) + u_2\cos(2) + u_3\cos(3))^2 + (u_1\sin(1) + u_2\sin(2) + u_3\sin(3))^2 \\
&\quad = {u_1}^2 + {u_2}^2 + {u_3}^2 + 2u_1u_2\cos(1)\cos(2) + 2u_1u_3\cos(1)\cos(3) \\
&\qquad + 2u_2u_3\cos(2)\cos(3) + 2u_1u_2\sin(1)\sin(2) + 2u_1u_3\sin(1)\sin(3) \\
&\qquad + 2u_2u_3\sin(2)\sin(3) \\
&\quad = {u_1}^2 + {u_2}^2 + {u_3}^2 + 2u_1u_2\cos(1-2) + 2u_1u_3\cos(1-3) + 2u_2u_3\cos(2-3) \\
&\quad = {u_1}^2 + {u_2}^2 + {u_3}^2 + u_1u_2\cos(1) + u_2u_3\cos(1) + u_3u_1\cos(1) \\
&\qquad + u_1u_3\cos(2) + u_2u_1\cos(2) + u_3u_2\cos(2) \\
&\quad = {u_1}^2 + {u_2}^2 + {u_3}^2 + (u_1u_2 + u_1u_3 + u_2u_3)\cos(1) + (u_1u_3 + u_2u_1 + u_3u_2)\cos(2) \\
&\quad = 3 - 1^*(\cos(1) + \cos(2)) = 3 - 2^*\cos(1) = 2
\end{aligned}
$$

5.2.2.2 Example: Modulation on a PRN Code

Let $N = 2^q - 1$ and define θ_0 as follows

$$
\theta_0(m) = \frac{2m}{N} \quad -(2^{q/2} - 1) \le m \le (2^{q/2} - 1) \tag{5.52}
$$

The angle $\pi\theta_0$ is a fundamental harmonic consisting of a single period of phase rotation over the PRN code length N. As such, the function $\theta(k) = k\theta_0$ satisfies (5.52). For a given PRN code u with length $N = 2^q - 1$, the following code group can be defined:

$$G_u = \{R_\theta u \quad \theta = k\theta_0, \text{ where } k \in Z\} \tag{5.53}$$

Using these results, we see that the magnitude of the cross correlation between any two codes in the code group G_u is

$$corr(R_{i\theta}u, R_{k\theta}u) = \sqrt{N \pm 1} = \begin{cases} \sqrt{2^q - 2} & if \sum_i u_i = -1 \\ 2^{q/2} & if \sum_i u_i = 1 \end{cases} \tag{5.54}$$

Figure 5.13 shows the cross correlation between two codes in G_u ($k = 1$ and $k = 2$) where $N = 1023$. Notice that magnitude is centered at 32, which is the square root of 1024. For the other preceding case, the magnitude of the cross correlation is $\sqrt{N-1} = \sqrt{1022} < 32$. When $m = 0$ in (5.50), it can be seen that the magnitude of cross-correlation values is $+1$, which can be observed in Figure 5.13 for the correlation lag equal to 2046.

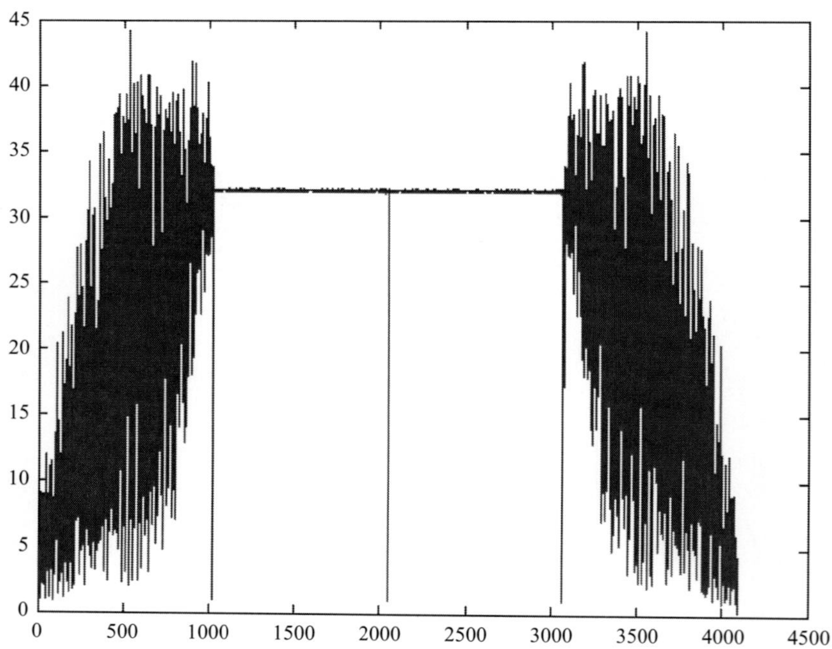

Figure 5.13. Cross-Correlation Magnitude Between Two Codes in G_u ($k = 1$ and $k = 2$)

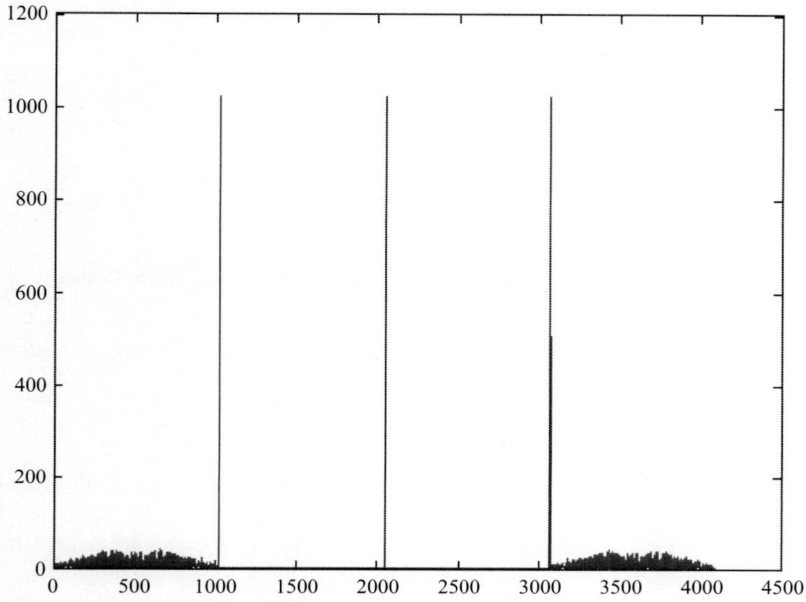

Figure 5.14. Autocorrelation of the Fundamental Code (k = 1) in G_u

5.2.2.3 Application

The code group G_u opens up a class of codes that achieve the cross-correlation lower bound of the square root of the code length or slightly less but that achieve maximal code length autocorrelation. Figure 5.14 shows the magnitude of autocorrelation of the fundamental code ($k = 1$) in G_u. Thus the code group G_u can provide codes with optimal auto- and cross-correlation performance that are easily implemented into a radar waveform. Essentially, a waveform generator must implement a basic PRN code that is modulated by a Doppler frequency defined by the code group.

5.2.2.4 Multiple Codes

We compare how multiple codes increase the interference floor due to cross-correlation interference. Figure 5.15 shows a comparison of the matched filter output for Kasami codes and four modulated PRN codes. The interference level from the three interfering codes is 130 for the Kasami and 97 for the modulated PRN codes. Basically, the interference level increases proportionately as the number of codes increases. For Kasami codes, the maximum interference level is approximately $(N_c + 1)2^{q/2} + 1$, and for Modulated PRN codes, the maximum interference level is $2^q N_c + 1$ where N_c is the number of interfering codes. Also notice the peak gain loss from 1023 to 990 for four Kasami codes. This loss does not occur for the four modulated PRN codes.

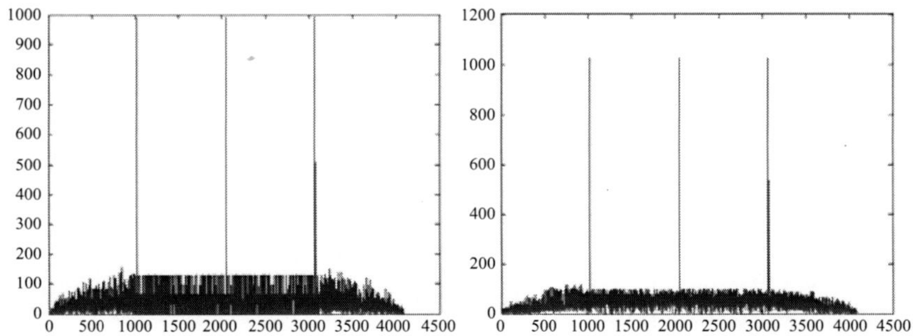

Figure 5.15. Comparison of Kasami (left) and Modulated PRN (right) Matched Filter Output for Four Waveforms

5.2.3 Essentially Orthogonal Waveforms

Based on the generation of the code group G_u, waveforms can be designed to provide lower cross-correlation values than those for codes in G_u. The property that $\cos(N-k)=\cos(k)$ implies that the code length needs to be an even number to reduce the error.

Consider the code $w=[u \quad u]$ where u is repeated, has length $N=2^q-1$, and its associated code group is G_w. Figure 5.16 shows that the magnitude of the autocorrelation sidelobes for the fundamental code ($k=1$) attains a constant value of 2 except for ambiguities. Analysis shows that the modulation codes need have an

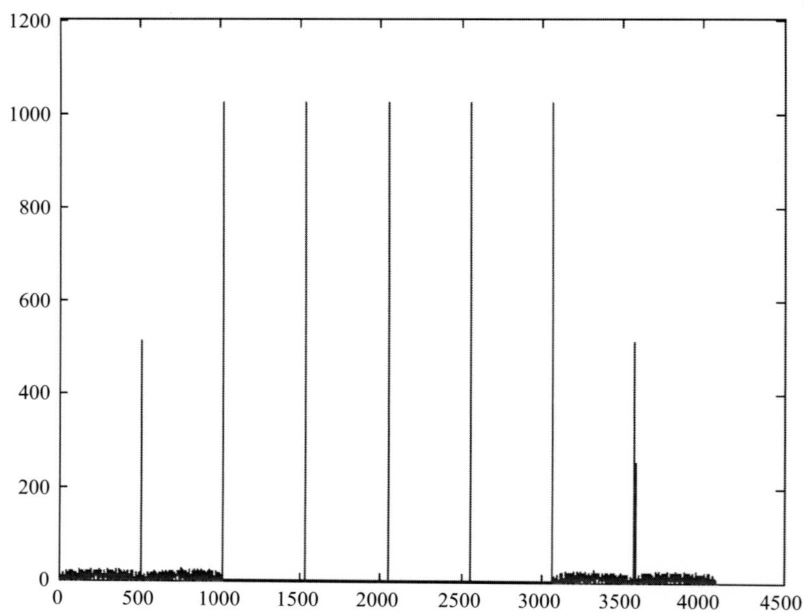

Figure 5.16. Autocorrelation Magnitude of the Fundamental Code ($k=1$) in G_w

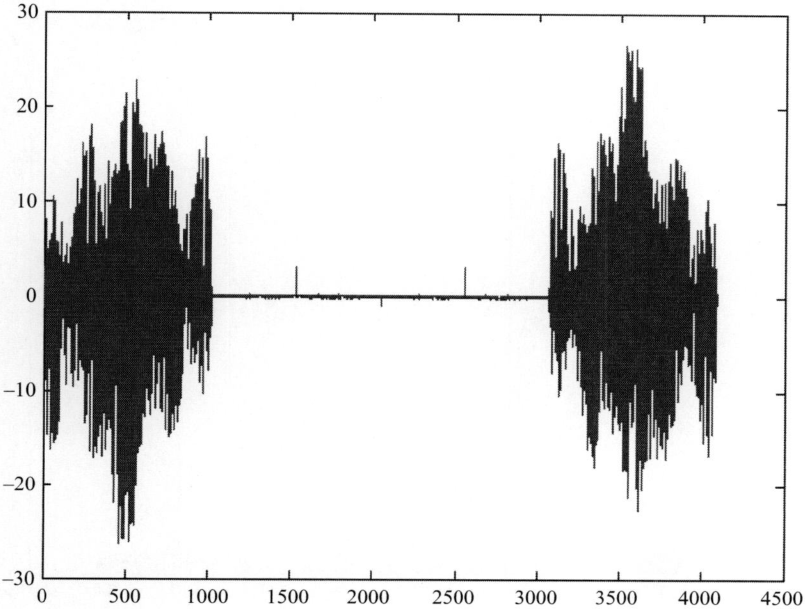

Figure 5.17. Real Part of the Cross Correlation Between Two Codes in G_w (k = 1 and k = 2)

odd modulation difference between the interfering codes and the primary code (the code used for the matched filter). Figure 5.17 shows that the cross correlation between two codes in G_w ($k = 1$ and $k = 2$) is essentially zero since $2 - 1 = 1$, which is odd. The addition of multiple codes that have odd modulation difference with the primary code does not affect the time sidelobes. For example, if $k = 1$ is the primary code, then adding an even modulation code does not affect the sidelobes because the difference in any even modulation and $k = 1$ results in an odd modulation.

5.2.3.1 Doppler Sensitivity

One issue with modulated PRN codes is the sensitivity to Doppler in a dynamic target environment. Any target moving radially with respect to the radar induces additional phase rotation other than the rotation induced by the code group. Doppler filter banks can be implemented that estimate the target Doppler, and this estimated Doppler can be derotated for the return signal. However, any residual Doppler degrades autocorrelation and cross-correlation performance. Figure 5.18 show the effect for a 0.1 single period rotation for residual Doppler with four modulated Doppler waveforms from code group G_w. The average interference level is around the value 10 due to the increase in interference level resulting from the residual Doppler. Thus it is imperative that the residual Doppler be small relative to the rotation group rotation in order to minimize multiple signal interference. However, even with a residual Doppler rotation, the combined interference is less than interference from the four Kasami codes.

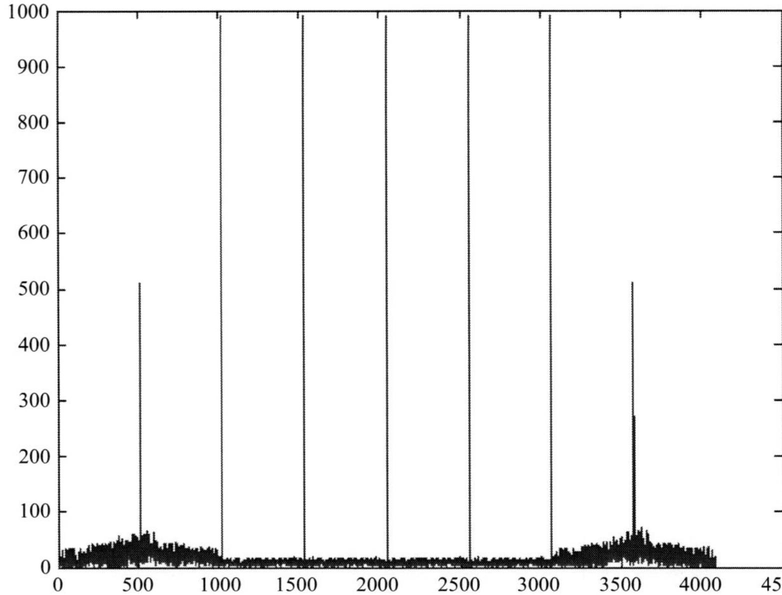

Figure 5.18. Magnitude of the Matched Filter Output from Four Codes in G_w with a 0.1 Period Residual Doppler Rotation

5.2.4 Optimized Multiphase Waveforms

Using multiphase waveforms that can take advantage of multiple phase states to achieve desired performance has advantages over binary phase codes. Waveforms can be computed using optimal search techniques that implement dynamic programming methods. Consider the following two examples that illustrate multiphase waveform design using a quadratic penalty constrained optimal search algorithm. Consider waveforms with 15 phase states. Let s_x and s_y denote two multiphase waveforms.

$$s_x = \begin{bmatrix} e^{jx_1} & e^{jx_2} & \cdots & e^{jx_{15}} \end{bmatrix} \quad \text{and} \quad s_y = \begin{bmatrix} e^{jy_1} & e^{jy_2} & \cdots & e^{jy_{15}} \end{bmatrix}$$

In the first example, Figure 5.19 shows two 15-bit waveforms that were determined through an optimal search routine that attempted to minimize the autocorrelation sidelobe energy and cross-correlation energy subject to constraining the cross correlation to have a 34-dB null. Note that the achieved null is actually 34 dB down from the peak and that the peak autocorrelation sidelobes and the peak cross-correlation sidelobes are about -12 dB down from the peak.

In the next example, we consider the design of two waveforms where we intentionally place nulls in autocorrelation sidelobes. Figure 5.20 shows the autocorrelation performance for 15-bit multiphase waveforms with constraints in the 7–8 and 22–23 lags. As a result, any discrete clutter located at those range sidelobe positions is reduced by an additional 15 dB. The waveform design algorithm is a

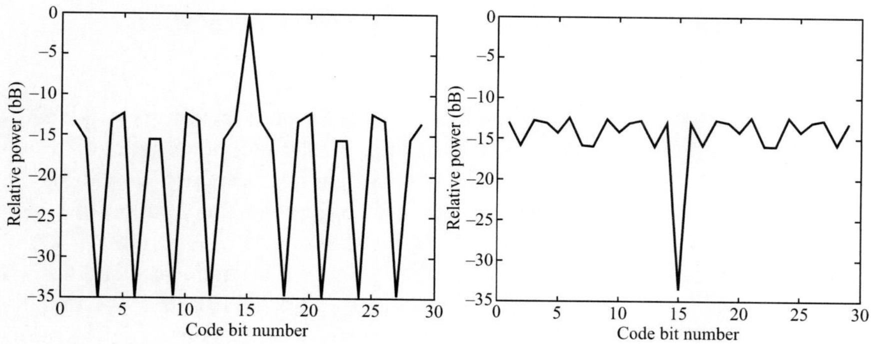

Figure 5.19. Two Multiphase Optimized Waveforms: Autocorrelation of First and Second Waveforms (left); Cross Correlation of Waveforms with 34-dB Null at Correlation Number 15 (right)

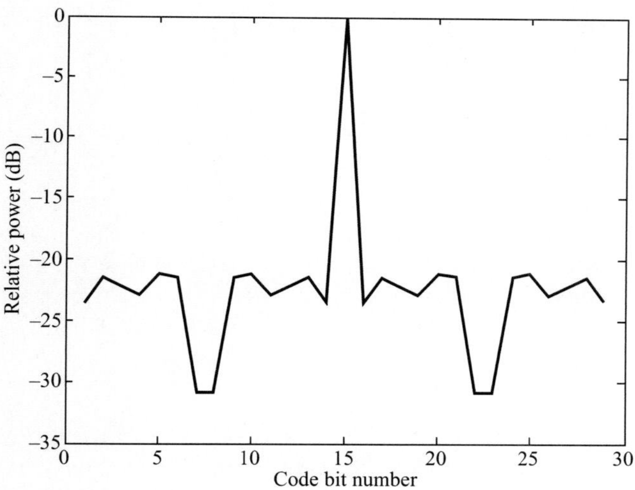

Figure 5.20. PRA Optimal Waveform Design with Low Correlation Constraints at the 7–8 and 22–23 Correlation Lags

constrained optimization algorithm that minimizes correlation sidelobes while maintaining the sidelobe constraints. In the case of Figure 5.20, the correlation sidelobe constraints were chosen to achieve low sidelobes at the selected correlation lags.

As opposed to multiphase matched filters, mismatched filters offer the potential to achieve both low autocorrelation sidelobes and low cross-correlation performance. The primary reason that mismatched filters can achieve lower time sidelobes is that the transmitted code and the ideal filter code can be completely different, hence mismatched. These extra degrees of freedom enable improved correlation and autocorrelation performance over standard binary codes.

5.3 Bounds on Autocorrelation and Cross-Correlation Performance

An important consideration in developing a class of radar waveforms is the autocorrelation of the individual and cross correlation of the individual pairs of waveforms. The ideal autocorrelation and cross-correlation response would be an impulse response where the values for the linear or circular convolution of waveforms is zero except for the zero lag autocorrelation.

For binary phase-coded waveforms, achieving this impulse condition is not possible, and a significant body of work by Welch [10] in 1974 initially developed lower bounds for binary sequences. Consider unit vectors $S=\{X_1, X_2, \ldots, X_M\}$ in $\mathbb{C}^N$, where $M>N$; then Welch [10] was able to develop lower bounds on the maximal cross correlation:

$$c_{\max} = \max_{i \neq j} |\langle X_i, X_j \rangle| \tag{5.55}$$

Specifically, Welch showed that, for all integers k,

$$c_{\max}^{2k} \geq \frac{1}{M-1}\left[\frac{M}{\binom{N+k-1}{k}} - 1\right] \tag{5.56}$$

A geometric proof for the Welch bound was presented in [11], which is found below for the case $k=1$. We now apply this Welch bound to the autocorrelation and cross correlation for a set of binary codes for $k=1$. Let S be a set of vectors $X=(x_1, x_2, \ldots, x_M)$, and define the lth cyclic shift of X by

$$X^l = (x_l \ \ \ldots \ \ x_M \ \ x_1 \ \ \ldots \ \ x_{l-1}) \tag{5.57}$$

Now define the set of vectors C that contains the set S along with all of the cyclic shifts of the vectors contained in S. If there are M members in S, then there are $M \cdot N$ number of vectors in C. Now apply the inequality in (5.56) derived for unit vectors to the set of binary vectors C that take on values of $+1$ and -1 for the case $k=1$:

$$\frac{c_{\max}^2}{N^2} \geq \frac{M-1}{M \cdot N - 1} \ \text{(circular convolution)} \tag{5.58}$$

This set C contains all of the autocorrelations and cross correlations for the vectors in the set S, and (5.56) establishes a lower bound for periodic correlation or circular convolution. Welch also established the lower bound for aperiodic correlation or linear convolution.

$$\frac{c_{\max}^2}{N^2} \geq \frac{M-1}{M \cdot (2N-1) - 1} \ \text{(linear convolution)} \tag{5.59}$$

Considering the asymptotic behavior of (5.58), we see that, as the code length is increased, we have

$$|c_{\max}| \geq \sqrt{\frac{N^2(M-1)}{M \cdot N - 1}} \rightarrow \sqrt{N} \text{ for large } N \tag{5.60}$$

And for (5.59),

$$|c_{\max}| \geq \sqrt{\frac{N^2(M-1)}{M \cdot (2N-1) - 1}} \rightarrow \sqrt{\frac{N}{2}} \text{ for large } N \tag{5.61}$$

Sarwate and Pursley [12] developed bounds that incorporate the trade between autocorrelation and cross-correlation performance. Define the peak cross-correlation magnitude for a set of vectors S:

$$\theta_c = \max_{i \neq j} \left\{ |\langle X_i^l, X_j^m \rangle| : \ X_i, X_j \in S, \ 0 \leq l, \ m \leq N-1 \right\} \tag{5.62}$$

Define the peak out-of-phase autocorrelation as

$$\theta_a = \max_{l \neq 0} \{ |\langle X_i^l, X_i^0 \rangle| : \ X_i \in S, \ 0 \leq l \leq N-1 \} \tag{5.63}$$

Thus Sarwate and Pursley [12] derived the following bounds:

$$\frac{\theta_c^2}{N} + \frac{N-1}{N(M-1)} \left(\frac{\theta_a^2}{N} \right) \geq 1 \text{ (circular convolution)} \tag{5.64}$$

If, for example, the goal is to achieve zero cross correlation, then (5.64) becomes

$$\theta_a^2 \geq \frac{N^2(M-1)}{N-1} \rightarrow N \text{ for large } N \tag{5.65}$$

However, (5.64) shows that there is a balance between autocorrelation and cross-correlation performance. If the cross-correlation performance is increased, then autocorrelation performance can be decreased and vice versa.

Sawarte and Pursley [12] also derived similar bounds for the linear convolution case.

$$\frac{(2N-1)}{N} \frac{\theta_c^2}{N} + \frac{2(N-1)}{N(M-1)} \left(\frac{\theta_a^2}{N} \right) \geq 1 \text{ (linear convolution)} \tag{5.66}$$

5.3.1 Correlation and Cross Correlation of Random Binary Phase Sequences

Let X and Y be two random binary phase vectors of length M. Without loss of generality, we will assume the M is an even integer. We compute the probability that the cross correlation will attain certain values. Define the lth cyclic shift of X by

$$X^l = (x_l \quad \ldots \quad x_M \quad x_1 \quad \ldots \quad x_{l-1}) \tag{5.67}$$

The *l-k* cross correlation of X and Y is defined by

$$\langle X^l, Y^k \rangle = \sum_{i=1}^{M} x_{(l+i)_M} y_{(k+i)_M} \tag{5.68}$$

For a binary phase vector of even length M, only even-valued correlation values are possible, which can be seen in (5.68). For any correlation, (5.68) results in a sum of q values equal to $+1$ and $M-q$ values equal to -1. The correlation is then given by

$$\langle X^l, Y^k \rangle = \sum_{i=1}^{M} x_{(l+i)_M} y_{(k+i)_M} = \sum_{i=1}^{q} 1 - \sum_{i=1}^{M-q} 1 = 2q - M \quad q \geq M/2 \tag{5.69}$$

Because we assumed M to be an even number, the cross correlation is even valued. And we compute

$$\mathrm{Prob}\left(\langle X^l, Y^k \rangle = M - 2q\right) = \begin{cases} \binom{M}{M-q} p_{+1}^{M-q} p_{-1}^{q} + \binom{M}{M-q} p_{-1}^{M-q} p_{+1}^{q} & \text{for } q = 1, \ldots, \frac{M}{2} - 1 \\ \binom{M}{M-q} p_{+1}^{M-q} p_{-1}^{q} & \text{for } q = \frac{M}{2} \end{cases} \tag{5.70}$$

where p_{+1} is the probability of choosing $+1$ and p_{-1} is the probability of choosing -1.

Figure 5.21 shows a plot of (5.70) for $M = 32$, and Figure 5.22 shows a histogram of correlation values for 10,000 Monte Carlo runs. Note the agreement between the values determined by (5.70) and the Monte Carlo simulation.

Figure 5.23 presents a plot for the autocorrelation values for 31,000 Monte Carlo runs, which shows the complexity in estimating the autocorrelation values. Also note that the Welch bound for this waveform is approximately equal to the square root of $M = 32$. So we see that in both Figure 5.21 and Figure 5.22, the mean of the distributions is reasonably close to the Welch bound.

5.3.2 Derivation of the Welch Bound for $k = 1$

Let $\{X_1, \; X_2, \; \ldots, \; X_M\}$ be a set of unit vectors in a Hilbert space H that span a subspace V of dimension N. Define the operator F: $V \rightarrow \mathbb{C}^M$ by

$$F(Y) = [\langle X_1, Y \rangle \quad \langle X_2, Y \rangle \quad \ldots \quad \langle X_M, Y \rangle] \tag{5.71}$$

And define the adjoint operator F^*: $\mathbb{C}^M \rightarrow V$ by

$$F^*(c_1, \; c_2, \; \ldots, \; c_M) = \sum_{i=1}^{M} c_i X_i \tag{5.72}$$

The frame operator T: $V \rightarrow V$ is defined as

$$T = F^* F(Y) = \sum_{i=1}^{M} \langle X_i, Y \rangle X_i \tag{5.73}$$

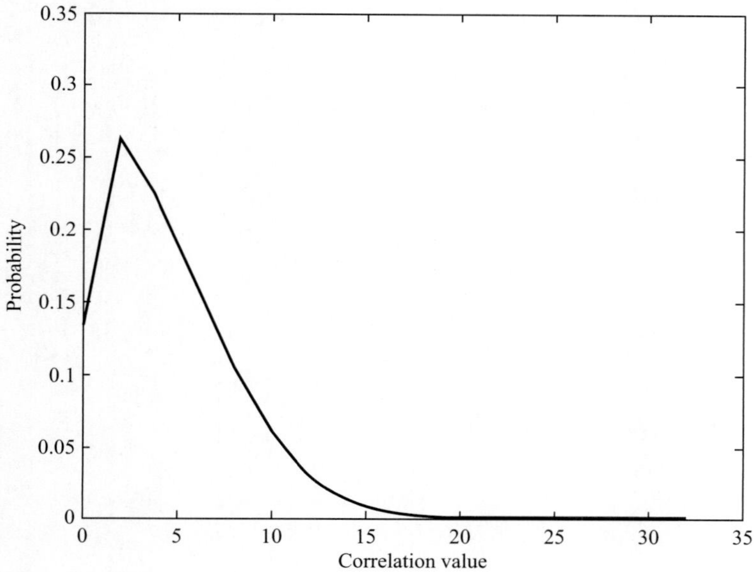

Figure 5.21. Probability of Correlation Values for M = 32

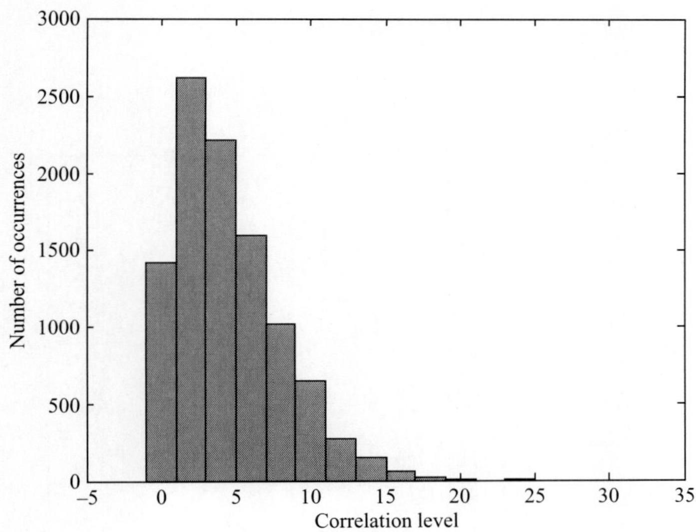

Figure 5.22. Histogram of Correlation Values for M = 32 (10,000 Monte Carlo Runs)

And the Grammian operator T^*: $\mathbb{C}^M \rightarrow \mathbb{C}^M$ is defined by

$$\begin{aligned} T^* &= FF^*(c_1, \quad c_2, \quad \ldots, \quad c_M) \\ &= \left[\left\langle X_1, \sum_{i=1}^{M} c_i X_i \right\rangle \quad \left\langle X_2, \sum_{i=1}^{M} c_i X_i \right\rangle \quad \ldots \quad \left\langle X_M, Y \sum_{i=1}^{M} c_i X_i \right\rangle \right] \\ &= \left[\sum_{i=1}^{M} c_i \langle X_1, X_i \rangle \quad \sum_{i=1}^{M} c_i \langle X_2, X_i \rangle \quad \ldots \quad \sum_{i=1}^{M} c_i \langle X_M, X_i \rangle \right] \\ &= \begin{bmatrix} \langle X_1, X_1 \rangle & \langle X_1, X_2 \rangle & \cdots & \langle X_1, X_M \rangle \\ \langle X_2, X_1 \rangle & \langle X_2, X_2 \rangle & & \langle X_2, X_M \rangle \\ \vdots & & \ddots & \vdots \\ \langle X_M, X_i \rangle & \langle X_M, X_2 \rangle & \cdots & \langle X_M, X_M \rangle \end{bmatrix} \begin{bmatrix} c_1 \\ c_2 \\ \vdots \\ c_M \end{bmatrix} \\ &\equiv G \begin{bmatrix} c_1 \\ c_2 \\ \vdots \\ c_M \end{bmatrix} \end{aligned} \tag{5.74}$$

and

$$\|T^*\| = \|G\| = \left(\sum_{i=1}^{M} \sum_{j=1}^{M} |\langle X_i, X_j \rangle|^2 \right)^{\frac{1}{2}} \tag{5.75}$$

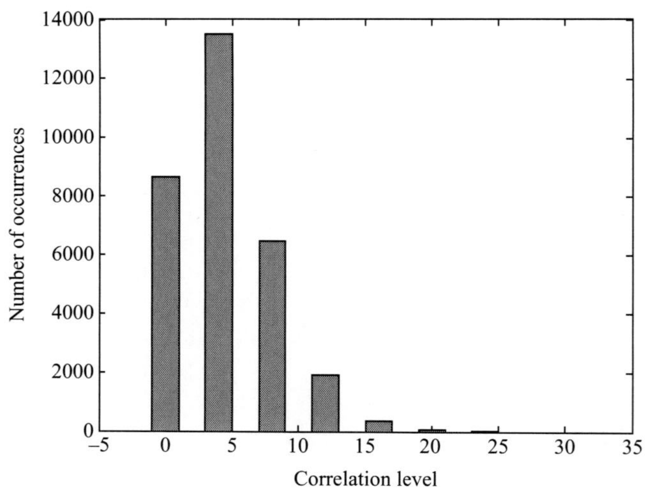

Figure 5.23. Histogram of Autocorrelation for M = 32 (31,000 Monte Carlo Runs)

The rank of T^* is the rank of G, which is equal to N. Let $\lambda_1, \lambda_2, \ldots, \lambda_N$ be the eigenvalues of G; then applying the Cauchy-Schwartz inequality

$$\|G\|^2 = \sum_{i=1}^{N} |\lambda_i|^2 \geq \frac{1}{N}\left(\sum_{i=1}^{N} |\lambda_i|\right)^2 = \frac{|tr(G)|^2}{\dim(V)} \tag{5.76}$$

Hence,

$$\sum_{i=1}^{M}\sum_{j=1}^{M} |\langle X_i, X_j\rangle|^2 \geq \frac{M^2}{N} \tag{5.77}$$

(5.77) is because the X_i have unit norm, the inequality in (5.77) is equivalent to

$$\sum_{i\neq j} |\langle X_i, X_j\rangle|^2 \geq \frac{M^2}{N} - M \tag{5.78}$$

(5.77) is because the $M(M-1)$ terms in the sum are all non-negative, the maximum value of the terms must be at least as large as their average value, and

$$c_{\max}^2 \geq \frac{1}{M(M-1)}\sum_{i\neq j} |\langle X_i, X_j\rangle|^2 \geq \frac{1}{M(M-1)}\left(\frac{M^2}{N} - M\right)$$
$$= \frac{1}{M-1}\left(\frac{M}{N} - 1\right) \tag{5.79}$$

Note that, in the derivation of (5.79), the bound in (5.78) was established. The inequality in (5.78) establishes the lower bound for root mean square correlation values for a set of unit vectors.

5.4 Chaotic Waveforms

Chaotic waveforms arose from using nonlinear dynamical systems to create a waveform that has random properties. Numerous chaotic processes can be used to generate a coded signal, and within each process numerous codes can be generated. As a result, these chaotic codes have applications for netted radars where the number of radars is large and the signals among the various radars in the network need to be separated. These chaotic waveforms can be implemented into distributed apertures that can be configured into interferometers.

Several nonlinear dynamical systems can be used to generate a chaotic waveform [13]. For example, the Lorenz system is system of nonlinear equations first used as an atmospheric model that is sensitive to its initial conditions. With small variations in initial conditions, the random-like behavior of the system changes.

$$\begin{aligned}\dot{x}_1 &= 10(x_2 - x_1)\\ \dot{x}_2 &= 25x_1 - x_1(x_3 - x_2) \quad \text{(Lorenz attractor)}\\ \dot{x}_3 &= x_1x_2 - 8/3x_3\end{aligned} \tag{5.80}$$

The Lorenz system oscillates randomly between two attractors in phase space to create a chaotic phase sequence. Other nonlinear systems oscillate between attractors with unique behavior to create chaotic sequences. The primary advantage for these waveforms is that multiple waveforms can be generated with varying initial conditions to produce low cross-correlation sequences.

The nonlinear systems can generate continuous-phase, binary-phase, or poly-phase sequences. The following MATLAB® algorithm generates a binary phase–coded waveform that has chaotic phase switching properties:

```
z=2*rand;
r=2.1;
q=2;
for k=1:600
     z=mod(z,1);
     z=r*z;
     zc(k)=z;
end
zc=zc*q;
xx=[];
yy=[];
zz=zeros(1,100);
N=zeros(1,100);
for kz=1:100
     zz(kz)=zc(kz+500)+.5;
     N(kz)=round(zz(kz));
end
x=sin(2*pi*i/N);
xcode=2*(round(.5*x+.5)-.5)
```

The initial value for z and the parameters r and q are chaotic input parameters. By setting $q=2$ and varying the parameter r, the correlation sidelobe behavior can be changed. Table 5.4 summarizes the mean correlation values for this chaotic waveform for various values of r. Note that, for $r=1.1$, a minimum mean level is achieved that is less than the Welch bound. As a result, it is apparent from Table 5.4 that correlation performance is a function of the input parameter q and that a minimum correlation performance is achievable. Thus various correlation values can be achieved by changing the chaotic input parameter r.

Table 5.4. Mean Correlation Values for Chaotic Waveform ($M=32$)

	Autocorrelation	**Cross Correlation**	**Welch Bound**
$r=4$	6.2480	6.8166	5.6569
$r=2$	5.9240	5.8045	5.6569
$r=1.1$	4.7840	5.3412	5.6569
$r=0.6$	13.5860	16.2951	5.6569

References

1. N. Levanon and E. Mozeson, *Radar Signals*, IEEE Press, John Wiley & Sons, Hoboken, NJ, USA, 2002.
2. B. M. Keel, J. A. Saffold, M. R. Walbridge, and J. Chadwick, "Non-linear stepped chirp waveforms with sub-pulse processing for range sidelobe suppression," *Proceedings of the SPIE, The International Society for Optical Engineering Conference on Radar Sensor Technology III*, Orlando, FA, USA, vol. 3395, pp. 87–98, Apr. 16, 1998.
3. B. M. Keel, "Brief summary of a nonlinear frequency modulated waveform," Georgia Tech Research Institute Technical Memorandum, July 17, 2002.
4. L. R. Varshney and D. Thomas, "Sidelobe reduction for matched filter range processing," *IEEE Radar Conference*, 2003.
5. J. R. Klauderer, A. C. Price, S. Darlington, and W. L. Albersheim, "The theory and design of chirp radars," *Bell System Tech. J.*, vol. 39, pp. 745–808, Jul. 1996.
6. P. Z. Peebles, *Radar Principles*, John Wiley & Sons, New York, 1998.
7. W. Renaud et al., "Improved active sonar performance using Costas codes," Queens University, Costas Array Symposium 2004.
8. K. Drakakis, "A review of Costas arrays," *JAM*, vol. 2006, ID 26385, pp. 1–32.
9. M. Soumekh "SAR-ECCM using phase-perturbed LFM chirp signals and DRFM repeat jammer penalization," *IEEE Trans. AES*, vol. 42, no. 1, Jan. 2006.
10. L. R. Welch, "Lower bounds on the maximum cross correlation of signals," *IEEE Trans. IT*, vol. 20, no. 3, pp. 397–399, May 1974.
11. S. Datta, S. Howard, and D. Cochran, "Geometry of the Welch bounds," September 2009 http://arxiv.org/pdf/0909.0206.pdf.
12. D. V. Sawarte and M. B. Pursley, "Cross correlation properties of pseudorandom and related sequences," *Proc. IEEE*, vol., 68, no. 5, pp. 583–619, May 1980.
13. M. Wicks et al., *Principles of Waveform Diversity and Design (Part II)*, Chapters 55–59, SciTech Publishing, Inc., Rayleigh, NC, USA, 2010.
14. F. J. Harris, "On the use of windows for harmonic analysis with the discrete Fourier transform," *Proc. IEEE*, vol. 66, pp. 61–83, Jan. 1978.
15. R. B. Ash, *Information Theory*, Dover Publications, New York, 1965.

Chapter 6

The Radar Interferometer

In their classic book, Barton and Ward [2] characterized angle-of-arrival errors for various antenna types, including interferometers. In this chapter, we focus on the effect of additive random noise on angle-of-arrival accuracy for several types of interferometers: monopulse interferometer, digital interferometer, orthogonal interferometer, amplitude interferometer, and bistatic interferometer. In each case, angle-of-arrival is computed as a difference in measurements and the error due to additive random noise affects each interferometer type differently. The basic angle accuracy equations are derived for each interferometer type for the case that noise is independent and identically distributed (IID) and for the case when it is not IID. For the monopulse and digital interferometer, the general case is considered where the noise error is neither independent nor identically distributed. These basic accuracy equations relate radar design parameters to angle accuracy and can be used to design interferometer architectures. For this chapter, we derive the accuracy equations assuming an interferometer that uses two antennas to measure angle in one dimension. The angle precision relationships presented in this chapter are derived from principles developed in Chapters 2, 3, and 4. In Chapter 7, we will consider interferometers with more than two antennas that are designed to estimate the two-dimensional angle-of-arrival.

Consider a simple interferometer consisting of two antennas separated by a distance D as shown in Figure 6.1. The voltages at the antennas are denoted ε_1 and ε_2 for a plane-wave signal impinging at an angle θ. The standard deviation of the angle error will be computed for two kinds of interferometers: an interferometer using monopulse to compute angle-of-arrival and an interferometer using digitized phase difference. We will refer to the two approaches as monopulse interferometer and digital interferometer. The primary difference between the two approaches is that the monopulse interferometer coherently sums the energy from the two antennas, thus creating a radiation pattern that contains grating lobes, whereas the digital interferometer digitizes the phase at each antenna and determines angle-of-arrival using phase difference without coherently summing the energy from the antennas.

6.1 Monopulse Interferometry

The monopulse angle estimation techniques discussed in Section 4.2 can be implemented using the separated interferometer antennas. In this section, we derive

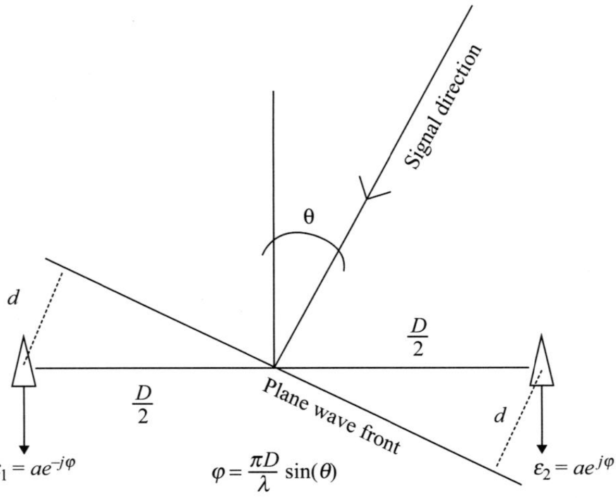

Figure 6.1. Interferometer Configuration

the basic accuracy equation for monopulse interferometry and later show that monopulse interferometry accuracy is equivalent to digital interferometer accuracy. We also derive results for the case of correlated error that is not identically distributed.

6.1.1 Monopulse Interferometer Phase Sensitivity

We first develop an expression for the monopulse interferometer sensitivity to phase error that ultimately defines the monopulse slope parameter first introduced in Section 4.2. The relative phase at each antenna is

$$\varphi = \frac{2\pi}{\lambda}\frac{D}{2}\sin(\theta) \tag{6.1}$$

and the voltage at each antenna is thus

$$\varepsilon_1 = ae^{j\varphi} + n_1 \quad \varepsilon_2 = ae^{j\varphi} + n_2 \tag{6.2}$$

where n_1 and n_2 are complex noise terms given by

$$n_1 = n_{1I} + jn_{1Q} \quad n_2 = n_{2I} + jn_{2Q} \tag{6.3}$$

with standard deviations of

$$\sigma_{iI}^2 = \sigma_{iQ}^2 = \frac{\sigma_{n1}^2}{2} = \frac{\sigma_{n2}^2}{2} \tag{6.4}$$

Assuming $\varphi \ll 1$, the monopulse ratio (difference/sum) is defined by

$$\frac{d}{s} = \mathrm{Im}\left(\frac{\varepsilon_2 - \varepsilon_1}{\varepsilon_2 + \varepsilon_1}\right) = \mathrm{Im}\frac{j\sin(\phi) + \dfrac{n_1 - n_2}{2a}}{\cos(\phi) + \dfrac{n_1 + n_2}{2a}}$$
$$\approx \tan(\varphi) + \frac{n_{\Delta Q}}{a\cos(\varphi)}$$
$$\approx \varphi + \frac{n_{\Delta Q}}{a} \tag{6.5}$$

where $n_\Delta = \frac{n_1 - n_2}{2}$ and $n_{\Delta Q} = \mathrm{Im}\left(\frac{n_1 - n_2}{2}\right)$

$$\sigma_{d/s} = \frac{\sigma_{n_{\Delta Q}}}{a} = \frac{\sqrt{\sigma_{1Q}^2 + \sigma_{2Q}^2}}{2a} = \frac{\sigma_{nQ}}{a\sqrt{2}} \tag{6.6}$$

On the other hand, for the interferometer we have

$$\varphi = \tan^{-1}\left(\frac{\mathrm{Im}(\varepsilon_2)}{\mathrm{Re}(\varepsilon_2)}\right) = \tan^{-1}\left(\frac{a\sin(\varphi) + n_{2Q}}{a\cos(\varphi) + n_{2I}}\right)$$
$$\approx \varphi + \frac{n_{2Q}\cos(\varphi) - n_{2I}\sin(\varphi)}{a} \tag{6.7}$$

$$\sigma_\varphi = \frac{\sqrt{\sigma_{2Q}^2\cos^2(\varphi) + \sigma_{2I}^2\sin^2(\varphi)}}{a} = \frac{\sigma_n}{a\sqrt{2}} \tag{6.8}$$

$$\sigma_{\Delta\varphi} = \sqrt{2}\sigma_\varphi = \frac{\sigma_n}{a} \tag{6.9}$$

and comparing the slope parameters,

$$\frac{\sigma_{\Delta\varphi}}{\sigma_{d/s}} = 2 \tag{6.10}$$

Thus the digital interferometer slope is twice the monopulse interferometer slope parameter.

6.1.2 Monopulse Beamwidth

For an interferometer, we need to derive an expression for the beamwidth of the combined antennas in order to relate angle accuracy to the beamwidth. Because an interferometer consists of two or more separate antennas, the beamwidth will be the 3-dB width of the mainlobe, θ_3, for the combined antennas.

$$s = \varepsilon_1 + \varepsilon_2 = 2\cos(\varphi) = 2\cos\left[\frac{2\pi}{\lambda}\frac{D}{2}\sin(\theta)\right] \tag{6.11}$$

$$\cos\left(\frac{\pi}{4}\right) = \frac{1}{\sqrt{2}} \tag{6.12}$$

$$2\frac{\pi}{4} = \frac{2\pi}{\lambda}\frac{D}{2}\sin(\theta_3) \approx \frac{2\pi}{\lambda}\frac{D}{2}\theta_3 \tag{6.13}$$

$$\theta_3 = \frac{\lambda}{2D} \tag{6.14}$$

6.1.3 Monopulse Interferometer Angle Error

We now relate the monopulse beamwidth to an expression for monopulse angle accuracy.

$$\frac{d}{s} = \tan(\varphi) \tag{6.15}$$

$$\sigma_{d/_s} = \sigma_\varphi = \frac{2\pi}{\lambda}\frac{D}{2}\cos(\theta)\sigma_\theta \tag{6.16}$$

$$\begin{aligned}\sigma_\theta^M &= \frac{\lambda}{\pi D\cos(\theta)}\sigma_{d/_s} = \frac{\lambda}{\pi D\cos(\theta)}\frac{\sigma_{n\Delta}}{a\sqrt{2}} = \frac{\lambda}{\pi D\cos(\theta)}\frac{1}{\sqrt{2a^2/\sigma_{n\Delta}^2}}\\ &= \frac{\theta_3}{\frac{\pi}{2}\cos(\theta)\sqrt{2S/N_\Delta}}\\ &= \frac{\theta_3}{\frac{\pi}{2}\cos(\theta)\sqrt{2SNR_\Delta}}\end{aligned} \tag{6.17}$$

where SNR_Δ is the signal-to-noise in the difference channel.

The angle error calculated in (6.17) for a monopulse interferometer assumes that the target is being tracked in the null of the difference beam. However, this may not be the case, especially if the interferometer antennas cannot be scanned. If the target is located off the monopulse null, then an additional angle bias error can occur due to both correlated noise between the sum and difference channels and differences between the actual off-null monopulse slope and the estimated slope.

Now consider the case when the errors are correlated and not independent. Referring to the analysis for IID errors in (6.7), we have

$$\begin{aligned}\frac{d}{s} &= \mathrm{Im}\left(\frac{\varepsilon_2 - \varepsilon_1}{\varepsilon_2 + \varepsilon_1}\right) = \mathrm{Im}\,\frac{j\sin(\varphi) + \dfrac{n_1 - n_2}{2a}}{\cos(\varphi) + \dfrac{n_1 + n_2}{2a}}\\ &\approx \tan(\varphi) + \frac{n_{\Delta Q}}{a\cos(\varphi)}\\ &\approx \varphi + \frac{n_{\Delta Q}}{a}\end{aligned} \tag{6.18}$$

$$n_{\Delta Q} = \mathrm{Im}\left(\frac{n_1 - n_2}{2}\right) \tag{6.19}$$

We now assume that n_1 and n_2 are not IID random variables. In this case,

$$\sigma_{d/s} = \frac{\sigma_{n_{\Delta Q}}}{a} = \frac{\sqrt{\sigma_{1Q}^2 + 2E(n_{1Q}n_{2Q}) + \sigma_{2Q}^2}}{2a}$$

$$= \frac{\sqrt{\sigma_{1Q}\sigma_{2Q}}}{2a}\sqrt{\frac{\sigma_{1Q}}{\sigma_{2Q}} - 2\frac{E(n_{1Q}n_{2Q})}{\sigma_{1Q}\sigma_{2Q}} + \frac{\sigma_{2Q}}{\sigma_{1Q}}}$$

$$= \frac{\sqrt{\sigma_{1Q}\sigma_{2Q}}}{2a}\sqrt{\frac{\sigma_{1Q}}{\sigma_{2Q}} - 2\rho + \frac{\sigma_{2Q}}{\sigma_{1Q}}}$$

$$= \sqrt{\frac{1}{4a^2/\sigma_{1Q}\sigma_{2Q}}\left(\frac{\sigma_{1Q}}{\sigma_{2Q}} - 2\rho + \frac{\sigma_{2Q}}{\sigma_{1Q}}\right)}$$

$$= \sqrt{\frac{1}{2\sqrt{2SNR_{1Q}}\sqrt{2SNR_{2Q}}}\left(\frac{\sqrt{SNR_{2Q}}}{\sqrt{SNR_{1Q}}} + \frac{\sqrt{SNR_{1Q}}}{\sqrt{SNR_{2Q}}} - 2\rho\right)} \tag{6.20}$$

$$\sigma_\theta^M = \frac{\lambda}{\pi D\cos(\theta)}\sigma_{d/s}$$

$$= \frac{\lambda}{\pi D\cos(\theta)}\sqrt{\frac{1}{2\sqrt{2SNR_{1Q}}\sqrt{2SNR_{2Q}}}\left(\frac{\sqrt{SNR_{2Q}}}{\sqrt{SNR_{1Q}}} + \frac{\sqrt{SNR_{1Q}}}{\sqrt{SNR_{2Q}}} - 2\rho\right)}$$

$$= \frac{\theta_3}{\frac{\pi}{2}D\cos(\theta)}\sqrt{\frac{1}{2\sqrt{2SNR_{1Q}}\sqrt{2SNR_{2Q}}}\left(\frac{\sqrt{SNR_{2Q}}}{\sqrt{SNR_{1Q}}} + \frac{\sqrt{SNR_{1Q}}}{\sqrt{SNR_{2Q}}} - 2\rho\right)} \tag{6.21}$$

6.1.4 Off-Axis Monopulse Error

In monopulse tracking using a phased array, the array beam is usually steered in the direction of the target in an effort to keep the target in the monopulse null. In many cases, the target can be placed in the monopulse null and kept there. However, in some cases, the target cannot be positioned near the null axis, and thus off-axis monopulse tracking is required. For angles-of-arrival located off of the null axis, we consider $\varphi \gg 0$ and derive the following:

$$\frac{d}{s} = \mathrm{Im}\left(\frac{\varepsilon_2 - \varepsilon_1}{\varepsilon_2 + \varepsilon_1}\right) = \mathrm{Im}\left(\frac{j\sin(\varphi) + \frac{n_1 - n_2}{2a}}{\cos(\varphi) + \frac{n_1 + n_2}{2a}}\right)$$

$$\approx \tan(\varphi) + \frac{n_{\Delta Q}}{a\cos(\varphi)} + \frac{\sin(\varphi)n_{SQ}}{a\cos^2(\varphi)} \tag{6.22}$$

$$\sigma_{d/s} = \sqrt{\left(\frac{\sigma_{n_{\Delta Q}}}{a\cos(\varphi)}\right)^2 + \left(\frac{\sin(\varphi)\sigma_{n_{SQ}}}{a\cos^2(\varphi)}\right)^2} \tag{6.23}$$

$$\sigma_\theta^M = \frac{\lambda}{\pi D\cos(\theta)}\sigma_{d/s} = \frac{\lambda}{\pi D\cos(\theta)}\sqrt{\left(\frac{\sigma_{n_{\Delta Q}}}{a\cos(\varphi)}\right)^2 + \left(\frac{\sin(\phi)\sigma_{n_{SQ}}}{a\cos^2(\varphi)}\right)^2} \tag{6.24}$$

$$\sigma_\theta^M = \frac{\lambda}{\pi D\cos(\theta)}\sigma_{d/s} = \frac{\lambda}{\pi D\cos(\varphi)\cos(\theta)}\sqrt{\frac{1}{\sqrt{2SNR_\Delta}} + \frac{\tan^2(\varphi)}{\sqrt{2SNR_S}}} \tag{6.25}$$

where SNR_S is the signal-to-noise in the sum channel and SNR_Δ is previously defined as the signal-to-noise in the difference channel. For off-axis tracking, the signal-to-noise in the sum and difference channels can be approximately equal ($SNR = SNR_\Delta = SNR_S$), which results in the following simplification:

$$\sigma_\theta^M = \frac{\lambda}{\pi D\cos(\varphi)\cos(\theta)\sqrt{2SNR}} \tag{6.26}$$

where

$$n_{\Delta Q} = \mathrm{Im}\left(\frac{n_1 - n_2}{2}\right) \quad \text{and} \quad n_{SQ} = \mathrm{Im}\left(\frac{n_1 + n_2}{2}\right) \tag{6.27}$$

For $\varphi \gg 0$, we see that the monopulse slope is inversely proportional to $\cos^2(\varphi)$.

The preceding analysis assumed that the noise terms n_1 and n_2 are mean zero independent and identically distributed. This assumption assures that $n_{\Delta Q}$ and n_{SQ} are mean zero independent and identically distributed. When n_1 and n_2 are not mean zero independent and identically distributed random variables, the noise terms $n_{\Delta Q}$ and n_{SQ} are correlated.

$$E(n_{\Delta Q}n_{SQ}) \neq 0 \tag{6.28}$$

In this case, (6.24) becomes

$$\begin{aligned}
\sigma_\theta^M &= \frac{\lambda}{\pi D\cos(\theta)}\sigma_{d/s} \\
&= \frac{\lambda}{\pi D\cos(\theta)}\sqrt{\left(\frac{\sigma_{n_{\Delta Q}}}{a\cos(\varphi)}\right)^2 + 2E(n_{\Delta Q}n_{SQ})\frac{\sin(\varphi)}{a\cos^3(\varphi)} + \left(\frac{\sin(\varphi)\sigma_{n_{SQ}}}{a\cos^2(\varphi)}\right)^2} \\
&= \frac{\lambda}{\pi D\cos(\theta)}\sqrt{\left(\frac{1}{\frac{a}{\sigma_{n_{\Delta Q}}}\cos(\varphi)}\right)^2 + 2\frac{E(n_{\Delta Q}n_{SQ})}{\sigma_{n_{\Delta Q}}\sigma_{n_{SQ}}}\frac{\sigma_{n_{\Delta Q}}\sigma_{n_{SQ}}\sin(\varphi)}{a^2\cos^3(\varphi)} + \left(\frac{\sin(\varphi)}{\frac{a}{\sigma_{n_{SQ}}}\cos^2(\varphi)}\right)^2} \\
&= \frac{\lambda}{\pi D\cos(\theta)}\sqrt{\left(\frac{1}{\frac{a}{\sigma_{n_{\Delta Q}}}\cos(\varphi)}\right)^2 + 2\rho\frac{\sin(\varphi)}{\frac{a}{\sigma_{n_{\Delta Q}}}\frac{a}{\sigma_{n_{SQ}}}a^2\cos^3(\varphi)} + \left(\frac{\sin(\varphi)}{\frac{a}{\sigma_{n_{SQ}}}\cos^2(\varphi)}\right)^2} \\
&= \frac{\lambda}{\pi D\cos(\varphi)\cos(\theta)}\sqrt{\left(\frac{1}{\sqrt{2SNR_{\Delta Q}}}\right)^2 + 2\rho\frac{\tan(\varphi)}{\sqrt{2SNR_{\Delta Q}}\sqrt{2SNR_{SQ}}} + \left(\frac{\tan(\varphi)}{\sqrt{2SNR_{SQ}}}\right)^2}
\end{aligned} \tag{6.29}$$

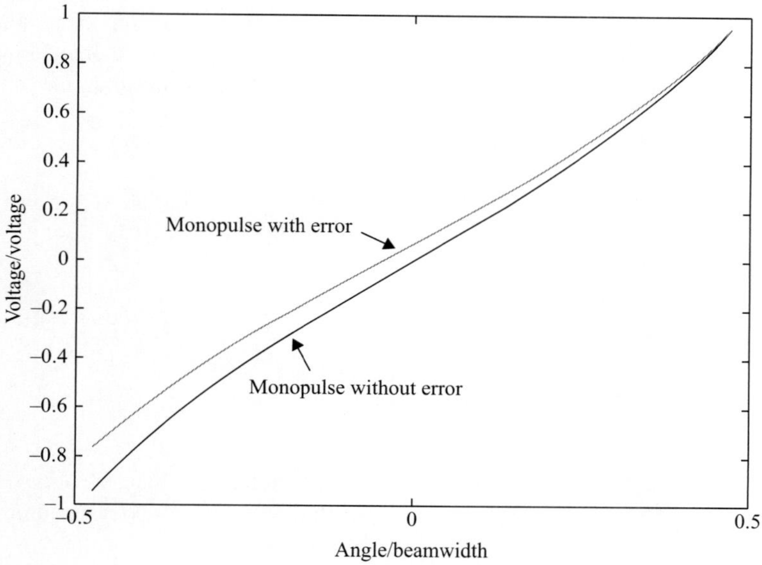

Figure 6.2. Effect of Off-Axis Monopulse Slope with and without Correlated Error

The off-axis monopulse angle error for correlated error is expressed as follows:

$$\sigma_\theta^M = \frac{\lambda}{\pi D\cos(\varphi)\cos(\theta)}\sqrt{\frac{1}{2SNR_\Delta} + 2\rho\frac{\tan(\varphi)}{\sqrt{2SNR_\Delta}\sqrt{2SNR_S}} + \frac{\tan^2(\varphi)}{2SNR_S}} \tag{6.30}$$

where ρ is the correlation coefficient defined by

$$\rho = \frac{E(n_{\Delta Q}n_{SQ})}{\sigma_{\Delta Q}\sigma_{SQ}} \tag{6.31}$$

Equation (6.30) shows the effect of correlation on monopulse angle estimation for off-axis target tracking. For on-axis tracking, $\varphi \ll 1$ and (6.30) show that correlation in the sum and difference error has less of an effect. Figure 6.2 shows the effect of correlation on monopulse angle estimation. Note that the correlated error degrades the monopulse slope and also induces a bias in the angle estimate. It may be possible to eliminate or minimize the bias by calibrating the monopulse curve, but the degradation in angle estimation due to decreased monopulse effects cannot be mitigated.

6.2 Digital Interferometer Angle Error

We now make a distinction between a digital interferometer and an analog interferometer. The digital interferometer processes the digital output from each antenna to determine angle-of-arrival, whereas the analog interferometer combines the output from each antenna channel at RF or IF. Thus, the digital interferometer determines the phase of the signal at each antenna and determines angle:

$$\Delta\varphi = \varphi_1 - \varphi_2 = \frac{2\pi}{\lambda} D \sin(\theta) \tag{6.32}$$

The angle accuracy for a digital interferometer can be determined as follows:

$$\sigma_{\Delta\varphi} = \frac{2\pi}{\lambda} D \cos(\theta)\sigma_\theta \tag{6.33}$$

$$\begin{aligned} \sigma_\theta^I = \frac{\lambda}{2\pi D \cos(\theta)}\sigma_{\Delta\varphi} &= \frac{\lambda}{2\pi D \cos(\theta)\sqrt{{}^S/_N}} = \frac{\theta_3}{\pi \cos(\theta)\sqrt{{}^S/_N}} \\ &= \frac{\lambda}{2\pi D \cos(\theta)\sqrt{SNR}} \end{aligned} \tag{6.34}$$

Because $N_\Delta = \frac{N}{2}$, it follows from the preceding equations that the expressions for monopulse interferometer accuracy and digital interferometer accuracy are equivalent. Hence,

$$\sigma_\theta^M = \sigma_\theta^I \tag{6.35}$$

Figure 6.3 shows the slope and null depth for both monopulse and digital interferometry for two values of SNR. Notice that the digital interferometer slope is twice the slope of monopulse interferometer, as shown in the preceding analysis.

In the preceding angle error derivation, the phase errors were assumed to be independent and identically distributed with zero mean. We now investigate the effects of correlated phase errors on angle estimation. As before, consider an interferometer with phases at two antennas, defined as follows:

$$\varphi_1 = \frac{2\pi}{\lambda}\frac{D}{2}\sin(\theta) \tag{6.36}$$

$$\varphi_2 = -\frac{2\pi}{\lambda}\frac{D}{2}\sin(\theta) \tag{6.37}$$

$$\Delta\varphi = \varphi_1 - \varphi_2 = \frac{2\pi}{\lambda} D \sin(\theta) \tag{6.38}$$

$$\sigma_{\Delta\varphi} = \frac{2\pi D}{\lambda}\cos(\theta)\sigma_\theta \tag{6.39}$$

$$\sigma_\theta = \frac{\lambda}{2\pi D \cos(\theta)}\sigma_{\Delta\varphi} \tag{6.40}$$

6.2.1 *Correlated and Nonidentically Distributed Error Effects*

We now compute $\sigma_{\Delta\varphi}$ when phase noise is correlated and not identically distributed. First define a correlation coefficient ρ:

$$\rho = \frac{E(\varphi_1\varphi_2)}{\sigma_{\varphi_1}\sigma_{\varphi_2}} = \frac{E(\eta_1\eta_2)}{\sigma_{\eta_1}\sigma_{\eta_2}} \tag{6.41}$$

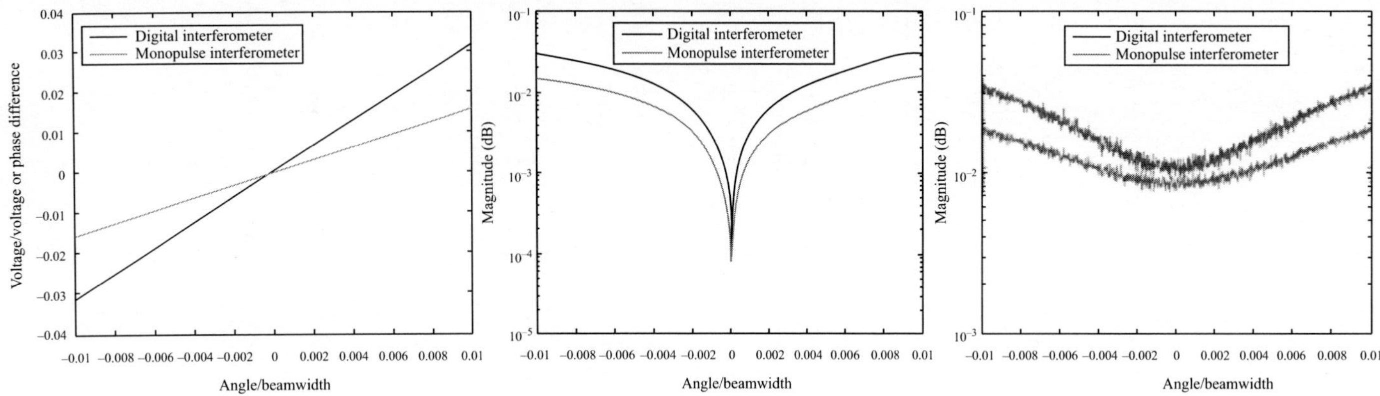

Figure 6.3. Monopulse and Phase Difference Slope (left), Null Depth for 60-dB SNR (center), and Null Depth for 20-dB SNR (right)

Assume that φ_1 and φ_2 are not independent ($\rho \neq 0$) and not identically distributed. We derive the general expression for angle-of-arrival error.

$$\begin{aligned}
\sigma_{\Delta\varphi}^2 &= E\left((\varphi_1 - \varphi_2)^2\right) \\
&= E(\varphi_1\varphi_1) - E(\varphi_1\varphi_2) - E(\varphi_2\varphi_1) + E(\varphi_2\varphi_2) \\
&= \sigma_{\varphi_1}^2 - 2E(\varphi_1\varphi_2) + \sigma_{\varphi_2}^2 \\
&= \sigma_{\varphi_1}\sigma_{\varphi_2}\left(\frac{\sigma_{\varphi_1}}{\sigma_{\varphi_2}} - 2\rho + \frac{\sigma_{\varphi_2}}{\sigma_{\varphi_1}}\right) \\
&= \frac{1}{\sqrt{2SNR_1}\sqrt{2SNR_2}}\left(\frac{\sqrt{SNR_2}}{\sqrt{SNR_1}} + \frac{\sqrt{SNR_1}}{\sqrt{SNR_2}} - 2\rho\right)
\end{aligned} \tag{6.42}$$

$$\begin{aligned}
\sigma_\theta &= \frac{\lambda}{2\pi D\cos(\theta)}\sigma_{\Delta\varphi} \\
&= \frac{\lambda}{2\pi D\cos(\theta)}\sqrt{\frac{1}{\sqrt{2SNR_1}\sqrt{2SNR_2}}\left(\frac{\sqrt{SNR_2}}{\sqrt{SNR_1}} + \frac{\sqrt{SNR_1}}{\sqrt{SNR_2}} - 2\rho\right)}
\end{aligned} \tag{6.43}$$

Assume that φ_1 and φ_2 are identically distributed but not independent (i.e., $\sigma_\varphi = \sigma_{\varphi_1} = \sigma_{\varphi_2}$). Then (6.43) reduces to the following:

$$\sigma_\theta = \frac{\lambda\sqrt{1-\rho}}{2\pi D\cos(\theta)\sqrt{SNR}} \tag{6.44}$$

Note that (6.44) shows that the correlation represented by the correlation coefficient ρ affects the angle-of-arrival accuracy. For a positive ρ, we see that accuracy is improved. For example, if $\rho = 1$, the phase errors η are perfectly correlated, which means that the two phase errors are identical. In that case, the phase error difference is zero. However, when the phase errors are negatively correlated, the angle accuracy is degraded. For example if $\rho = -1$, then the angle accuracy is degraded by the square root of 2 because, in the phase difference, the errors add perfectly.

6.2.2 Impact of Baseline Errors

In all the preceding equations that estimate angle accuracy, the baseline length term D appears in the denominator. Determining the baseline D is not always an accurate process because the baseline is defined as the distance between antenna phase centers, and antenna phase centers can change due to scan angle and environmental factors, such as temperature. Consider (6.45):

$$\sigma_\theta = \frac{\lambda}{2\pi D\cos(\theta)\sqrt{SNR}} \tag{6.45}$$

Small errors in the baseline D impact the angle accuracy. Referring to the expression for phase difference,

$$\Delta\varphi = \frac{2\pi}{\lambda}D\sin(\theta) \tag{6.46}$$

Computing the impact of baseline error, we have

$$\sigma_\theta = \sqrt{\left(\frac{\lambda}{2\pi D\cos(\theta)\sqrt{SNR}}\right)^2 + \sigma_D^2\tan^2(\theta)} \tag{6.47}$$

Notice that, as the angle θ approaches 90°, the impact of baseline errors becomes greater and that, at 0°, the impact of baseline error is zero. At 0° angle, the delta phase is zero regardless of the baseline.

6.3 Transmit Interferometry

A conventional interferometer (CI) is an array with spatially diverse receive antennas, like the design shown in Figure 6.4 (left), which consists of three receive arrays located at the vertices of an equilateral triangle and one transmit array located at the center of the triangle. The angle accuracy for a CI is achieved through the spatial diversity of the receive antennas. We now consider alternative interferometer configurations that use transmit spatial diversity. A configuration for an unconventional interferometer (UCI) is shown in Figure 6.4 (center), and a configuration for the orthogonal interferometer (OI) is shown in Figure 6.4 (right). The UCI architecture is really the inversion of the UC architecture where transit and receive are reversed. The angle accuracy CRLB for both are thus identical. The example OI architecture has three coherent transmit/receive arrays located at the vertices of an equilateral triangle. For both the UCI and the OI configurations, each array (transmit only or transmit/receive) is transmitting an orthogonal waveform, and each receive array is capable of receiving all three waveforms. For the configuration to operate as an interferometer, all waveforms must be coherent and sufficiently orthogonal to have low cross-correlation properties.

The orthogonal interferometer falls into a class of multiple in or multiple out (MIMO) radar technologies that use multiple orthogonal waveforms for enhanced performance. MIMO radar has proven to be useful in providing clutter multipath mitigation in severe multipath environments such as urban canyons. A specific example of multipath mitigation using the orthogonal interferometer is presented in Section 8.1.1.

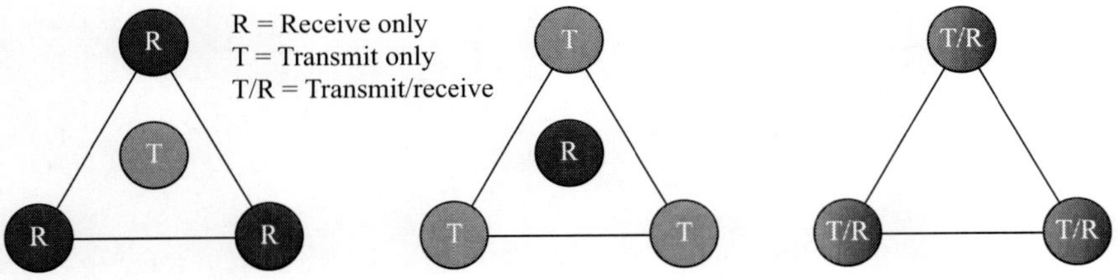

Figure 6.4. Conventional Interferometer (left); Unconventional Interferometer (right); Orthogonal Interferometer Antenna Architecture

The principal advantage of the orthogonal interferometer (OI) is that each array can process three coherent waveforms simultaneously due to their low cross-correlation properties. Thus the OI array contains three separate CI arrays (transmit out of one array and receive out of all three arrays) and three separate UCI arrays (transmit out of three arrays and receive out of one array). This unique multi-architecture allows for multiple ways of computing angle-of-arrival estimation that can be used to calibrate the OI antenna. The angle ambiguity for the OI is half that for the CI or UCI, making angle ambiguity resolution more difficult for OI. The individual CI and UCI interferometer can assist with angle unwrapping for the OI array by first unwrapping the CI and UCI arrays.

In an OI configuration, the three waveforms can be integrated to provide an additional 4.7 dB of sensitivity in each channel. However, when each T/R array for the OI system is smaller in aperture than the CI transmit array, the orthogonal interferometer can experience an SNR loss at each array relative to the CI system due to the loss in transmit power and gain. Thus array sizing for the OI system must consider power and weight constraints, as well as overall system performance.

For the case of both the monopulse interferometer and the digital interferometer using a CI architecture, the target is assumed to be illuminated by an independent illumination source. This source could be a separate antenna, or one of the interferometer antennas could be used for both transmit and receive. Thus, the outbound phase from the transmitter to the target is the same for each measurement of phase at the two antennas.

$$\varphi = \varphi_{outbound} + \varphi_{inbound} \tag{6.48}$$

Angle-of-arrival is determined by the difference of the inbound phase values for the interferometer antennas. In contrast, the orthogonal interferometer assumes that each antenna has both transmit and receive capability and can operate independently. The phase at each antenna now measures the total inbound and outbound phase difference. The phase term has an extra factor of 2 due to this two-way-path phase measurement. This factor of 2 creates twice as many angle ambiguities for OI as the number of ambiguities for CI or UCI.

$$\varphi = \frac{4\pi}{\lambda}\frac{D}{2}\sin(\theta) \tag{6.49}$$

As a result, the angle error expression now becomes

$$\sigma_\theta^{OI} = \frac{\lambda}{4\pi D\cos(\theta)}\sigma_{\Delta\phi} = \frac{\lambda}{4\pi D\cos(\theta)\sqrt{SNR}} = \frac{\theta_3}{2\pi D\cos(\theta)\sqrt{SNR}} \tag{6.50}$$

$$\sigma_\theta^{OI} = \frac{\lambda}{4\pi D\cos(\theta)}\sigma_{\Delta\phi} = \frac{\lambda}{4\pi D\cos(\theta)\sqrt{SNR}} \tag{6.51}$$

The utility of the orthogonal interferometer lies in using the extra factor of 2 in the denominator, as compared with the digital interferometer, to enhance performance. The two antennas can operate independently by transmitting waveforms

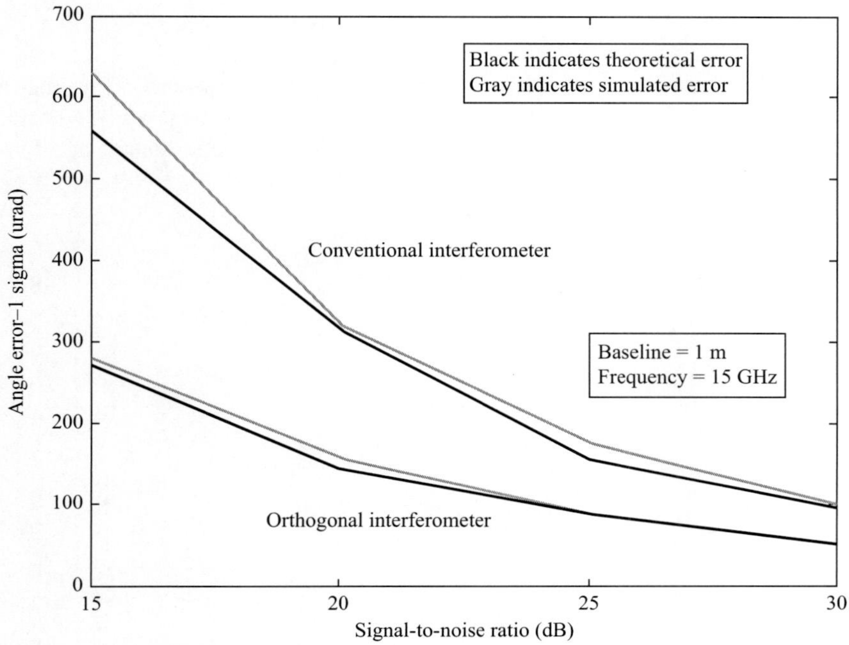

Figure 6.5. Angle Error for Conventional Interferometer Versus Orthogonal Interferometer for Equivalent SNR

that are orthogonal in time, frequency, or phase. Each domain has its advantages and disadvantages, but all three domains offer the potential for the orthogonal interferometer to improve performance over either the monopulse or digital interferometer. However, the penalty for this improved angle accuracy is that the unambiguous angle interval is half that of the conventional interferometer (see 7.3.1).

Figure 6.5 shows the reduction in angle error with an orthogonal interferometer compared to the conventional interferometer or, equivalently, the unconventional interferometer when the *SNR* at each array is equivalent for each architecture.

6.3.1 Correlated and Nonidentically Distributed Error Effects

Similar to the derivation for the digital interferometer, the effects due to errors that are correlated and not identically distributed result in the following expression for the orthogonal interferometer:

$$\sigma_\theta = \frac{\lambda}{4\pi D\cos(\theta)}\sigma_{\Delta\varphi} = \frac{\lambda}{4\pi D\cos(\theta)}\sqrt{\frac{1}{\sqrt{2SNR_1}\sqrt{2SNR_2}}\left(\frac{\sqrt{SNR_2}}{\sqrt{SNR_1}}+\frac{\sqrt{SNR_1}}{\sqrt{SNR_2}}-2\rho\right)} \quad (6.52)$$

6.4 Cramer-Rao Lower Bound Analysis

We compute the Cramer-Rao lower bound [3,4] for angle-of-arrival estimation using an interferometer and extend the analysis to an N element array with uniformly spaced elements. First, consider a simple interferometer consisting of two antennas separated by a distance D, as shown in Figure 6.1.

The relative phase at each antenna is

$$\varphi_{\pm} = \pm\frac{2\pi}{\lambda}\frac{D}{2}\sin(\theta) + \eta_{\pm} \tag{6.53}$$

where

$$\eta_{\pm} \sim N\left(0, \sigma_{\varphi}^2\right) \tag{6.54}$$

Let $g_{\pm}(\theta) = \pm\frac{2\pi}{\lambda}\frac{D}{2}\sin(\theta)$.

Let the joint distribution (f) for $\varphi_{\pm}$ be defined as follows:

$$f(\varphi_+, \varphi_-; \theta) = f(\varphi_+; \theta) f(\varphi_-; \theta) \tag{6.55}$$

where

$$f(\varphi_{\pm}; \theta) = \frac{1}{\sqrt{2\pi\sigma_{\varphi}^2}} e^{-\frac{(\varphi_{\pm} - g_{\pm}(\theta))^2}{2\sigma_{\varphi}^2}} \tag{6.56}$$

Now the CRLB can be computed as follows:

$$\begin{aligned}
J(\theta) &= E_{\theta}\left(\frac{\partial}{\partial\theta}\ln(f(\varphi_+, \varphi_-; \theta))\right)^2 \\
&= \frac{1}{\sigma_{\varphi}^4} E_{\theta}\left((\varphi_+ - g_+(\theta))g'_+(\theta) + (\varphi_- - g_-(\theta))g'_-(\theta)\right)^2 \\
&= \frac{1}{\sigma_{\varphi}^4}\left(E_{\theta}(\varphi_+ - g_+(\theta))^2 g'_+(\theta)^2 + E_{\theta}(\varphi_- - g_-(\theta))^2 g'_-(\theta)^2\right) \\
&= \frac{1}{\sigma_{\varphi}^4}\left(g'_+(\theta)^2\sigma_{\varphi}^2 + g'_-(\theta)^2\sigma_{\varphi}^2\right) \\
&= \frac{1}{\sigma_{\varphi}^2}\left(g'_+(\theta)^2 + g'_-(\theta)^2\right) \\
&= \frac{2\left(\frac{2\pi}{\lambda}\frac{D}{2}\cos(\theta)\right)^2}{\sigma_{\varphi}^2} = \frac{2\pi^2 D^2\cos^2(\theta)}{\lambda^2\sigma_{\varphi}^2}
\end{aligned} \tag{6.57}$$

and

$$\frac{1}{J(\theta)} = \frac{\lambda^2\sigma_{\varphi}^2}{2\pi^2 D^2\cos^2(\theta)} = \frac{\lambda^2\frac{1}{2SNR}}{2\pi^2 D^2\cos^2(\theta)} = \frac{\lambda^2}{4\pi^2 D^2\cos^2(\theta)SNR} \tag{6.58}$$

Therefore,

$$\sigma_\theta^2 \geq \frac{\lambda^2}{4\pi^2 D^2 \cos^2(\theta) SNR} \tag{6.59}$$

or

$$\sigma_\theta \geq \frac{\lambda}{2\pi D \cos(\theta)\sqrt{SNR}} \tag{6.60}$$

Note the agreement in the CLRB with the error computed for the digital interferometer in Section 6.2. This agreement indicates that, when the signal is corrupted by purely white Gaussian noise, the estimate of angle-of-arrival using an interferometer achieves the CLRB.

Now we extend the above derivation of the CLRB for an interferometer to an N-element array with uniformly spaced elements separated by a distance D. For the kth element in the array, we have

$$g_\pm(\theta, k) = \pm\frac{2\pi}{\lambda}\frac{(2k-1)D}{2}\sin(\theta) \tag{6.61}$$

Without loss of generality, we assume that N is even (if N were odd, then we set the $(N+1)/2$ center element to the zero position, such that the remaining $N-1$ elements only factor into the computation of the CRLB). As a result, we can write

$$\begin{aligned}
f\left(\varphi_+^1, \ldots \varphi_+^{N/2}, \varphi_-^1, \ldots \varphi_-^{N/2}; \theta\right) &= \prod_{i=1}^{N/2} f\left(\varphi_+^i, \theta\right) f\left(\varphi_-^i, \theta\right) \\
&= \left(\frac{1}{\sqrt{2\pi\sigma_\varphi^2}}\right)^N e^{-\frac{\sum_{i=1}^{N/2}(\varphi_+ - g_+(\theta))^2 + \sum_{i=1}^{N/2}(\varphi_- - g_-(\theta))^2}{2\sigma_\varphi^2}} \\
&= \left(\frac{1}{\sqrt{2\pi\sigma_\varphi^2}}\right)^N e^{-\frac{\sum_{i=1}^{N/2}(\varphi_+ - g_+(\theta))^2 + \sum_{i=1}^{N/2}(\varphi_- - g_-(\theta))^2}{2\sigma_\varphi^2}}
\end{aligned} \tag{6.62}$$

Using the following identity,

$$\sum_{i=1}^{N/2}(2i-1)^2 = \frac{N}{6}\left(N^2-1\right) \tag{6.63}$$

we have

$$J(\theta) = \frac{2}{\sigma_\varphi^2}\left(\frac{2\pi}{\lambda}\frac{D}{2}\cos(\theta)\right)^2\left(\frac{N}{6}(N^2-1)\right) \tag{6.64}$$

$$\sigma_\theta^2 \geq \frac{1}{J(\theta)} = \frac{6\lambda^2(N-1)}{4\pi^2L^2\cos^2(\theta)N(N+1)SNR} \tag{6.65}$$

where $L=(N-1)D$ is the array length. Note that if $N=2$, the result agrees with the case for the interferometer previously derived. For $D=\lambda/2$, the CRLB simplifies to

$$\sigma_\theta^2 \geq \frac{6}{\pi^2\cos^2(\theta)N(N^2-1)SNR} \tag{6.66}$$

6.5 Amplitude Interferometer

We consider the case for amplitude relative to the known amplitude at boresight. This case assumes that amplitude is sufficiently calibrated to measure amplitude at boresight accurately.

$$\varphi_\pm = \pm\frac{2\pi}{\lambda}\frac{D}{2}\sin(\theta) \tag{6.67}$$

$$\varepsilon_1 = ae^{j\varphi_+} \tag{6.68}$$

$$\varepsilon_2 = ae^{j\varphi_-} \tag{6.69}$$

$$\begin{aligned} A^2 &= (\varepsilon_1+\varepsilon_2)(\varepsilon_1+\varepsilon_2)^* = 2a^2(1+\cos(\varphi_+-\varphi_-)) \\ &= 4a^2\cos^2(\varphi_+-\varphi_-) \end{aligned} \tag{6.70}$$

$$A = 2a\cos(2\varphi) = 2a\cos\left(\frac{2\pi}{\lambda}D\sin(\theta)\right) \tag{6.71}$$

Let $A_0=2a$, which is the amplitude at boresight for the two interferometer antennas.

$$\sin(\theta) = \frac{\lambda}{2\pi D}\cos^{-1}\left(\frac{A}{A_0}\right) \tag{6.72}$$

$$\tilde{A} = \cos^{-1}\left(\frac{A}{A_0}\right) = \varphi_+ - \varphi_- \tag{6.73}$$

$$A = 2a\cos(\varphi_+ - \varphi_-) \tag{6.74}$$

$$\sigma_A = 2a\sin(\varphi_+-\varphi_-)\sqrt{\sigma_{\varphi_+}^2+\sigma_{\varphi_-}^2} = 2a\sin(\varphi_+-\varphi_-)\sqrt{2}\sigma_\varphi \tag{6.75}$$

$$\sigma_A = 2a\sin\left(\frac{2\pi}{\lambda}D\sin(\theta)\right)\frac{\sqrt{2}}{\sqrt{2SNR}} \tag{6.76}$$

Also,

$$A = 2a\cos\left(\frac{2\pi}{\lambda}D\sin(\theta)\right) \tag{6.77}$$

$$\sigma_A = 2a\sin\left(\frac{2\pi D}{\lambda}\sin(\theta)\right)\frac{2\pi D}{\lambda}\cos(\theta)\sigma_\theta \tag{6.78}$$

Therefore, equating (6.76) and (6.78),

$$\sigma_\theta = \frac{\lambda}{2\pi D\cos(\theta)}\frac{\sqrt{2}}{\sqrt{2SNR}} \tag{6.79}$$

$$\sigma_\theta = \frac{\lambda}{2\pi D\cos(\theta)\sqrt{SNR}} \tag{6.80}$$

6.6 Bistatic Interferometer

Most of the discussion about interferometry has assumed that the transmitter antenna and receive antennas are either colocated or are nearly so. However, the transmit and receive antennas do not need to be located anywhere near each other. The primary difficulty in bistatic radar is achieving coherence over long distances, but, for interferometer angle measurements, coherence between the transmitter and receiver is not required because angle is determined by relative phase among antennas and a homodyne receiver is adequate. Bistatic range measurements require a common clock or clocks that have been aligned to sufficient precision in addition to knowledge of the time of transmit.

An orthogonal interferometer can also operate as a bistatic radar where the transmit arrays are physically separated in distance from the receive arrays. In this case, the transmit signals from each antenna must be transmitted simultaneously because transmit timing errors translate into phase error bias. The timing requirement is also a function of wavelength because a timing bias is equivalent to the fraction of a wavelength that determines the phase bias. For example, if the wavelength is 3 cm and the transmit timing bias is 1 ps, then the phase bias is about 1/100 of a wavelength, or about 3.6°. This error can be eliminated through time alignment if the bias is constant. Angle estimation is accomplished using the time difference of arrival between two or more bistatic receivers.

6.7 Differential Interferometry

In some applications, the differential angle measurement between two targets is of interest. One such application is determining the relative distance between two geostationary satellites, as discussed in Chapter 7. Another application is in beam rider guidance, where an interceptor is guided to engage a target by nulling the relative angle. The fundamental assumption in differential interferometry is that the differential angle $\Delta\theta$ between two targets is small. If the differential angle is smaller than the antenna beamwidth for the interferometer antennas and both targets are contained within the beam, then correlated system errors can cancel out to improve the accuracy of the differential angle measurement. For example, if the interferometer antennas consist of phased arrays that require phase shifters or time delay units to point beams, then, when two targets are located within a beam, the errors associated with the beam-forming network are highly correlated for both targets. Also, any angle dependent alignment and calibration will be cancelled when two targets are located in the same beam.

Assume that a one-dimensional interferometer measures the differential phase measurement for two targets and that each measurement is biased by an error ε_c. Then the differential phase measurement for each target is given by

$$\Delta\varphi^1 = \varphi_1^1 - \varphi_2^1 + \varepsilon_c \tag{6.81}$$

$$\Delta\varphi^2 = \varphi_1^2 - \varphi_2^2 + \varepsilon_c \tag{6.82}$$

where $\Delta\varphi_i^j$ is the phase associated with the jth target at the ith antenna. The angle-of-arrival for each target is given by

$$\sin(\theta_1) = \frac{\lambda}{2\pi D}\left(\Delta\varphi^1 + \varepsilon_c + 2\pi N_1\right) \tag{6.83}$$

$$\sin(\theta_2) = \frac{\lambda}{2\pi D}\left(\Delta\varphi^2 + \varepsilon_c + 2\pi N_2\right) \tag{6.84}$$

where θ_i is the angle for the ith target, and N_i is the integer ambiguity for the ith target. Using a first-order Taylor series approximation, we have

$$\sin(\theta_2) = \sin(\theta_1 - \Delta\theta) \approx \sin(\theta_1) - \cos(\theta_1)\Delta\theta \tag{6.85}$$

Or, solving for differential angle $\Delta\theta$,

$$\begin{aligned} \cos(\theta_1)\Delta\theta &= \sin(\theta_1) - \sin(\theta_2) \\ &= \frac{\lambda}{2\pi D}\left(\Delta\varphi^1 + \varepsilon_c + 2\pi N_1\right) - \frac{\lambda}{2\pi D}\left(\Delta\varphi^2 + \varepsilon_c + 2\pi N_2\right) \end{aligned} \tag{6.86}$$

and

$$\Delta\theta = \frac{\lambda}{2\pi D\cos(\theta_1)}\left(\left(\Delta\varphi^1 + \varepsilon_c + 2\pi N_1\right) - \left(\Delta\varphi^2 + \varepsilon_c + 2\pi N_2\right)\right) \tag{6.87}$$

$$\Delta\theta = \frac{\lambda(\Delta\varphi^1 - \Delta\varphi^2 + 2\pi(N_1 - N_2))}{2\pi D\cos(\theta_1)} \tag{6.88}$$

So the differential angle can be estimated from differential phase measurements without estimating each angle separately. The key point is that the common error cancels in the estimate of differential angle. Thus, in estimating differential angle, it is beneficial to have both targets located within a single beam in order to improve the angle accuracy by cancelling correlated systematic biases. The angle precision for the differential measurement in (6.88) is

$$\sigma_{\Delta\theta} = \frac{\lambda 2\sigma_{\varphi}}{2\pi D\cos(\theta)} = \frac{\lambda\sqrt{2}}{\pi D\cos(\theta)\sqrt{SNR}} \tag{6.89}$$

If we refer to Section 6.2, we see that, for a single target, the angle error is given by

$$\sigma_{\theta} = \frac{\lambda}{2\pi D\cos(\theta)\sqrt{SNR}} \tag{6.90}$$

If we compute the differential angle by estimating each angle instead of using differential phase, we have

$$\Delta\theta = \theta_1 - \theta_2 \tag{6.91}$$

Then the differential angle precision is given by

$$\sigma_{\Delta\theta} = \frac{\lambda\sqrt{2}}{2\pi D\cos(\theta)\sqrt{SNR}} \tag{6.92}$$

Thus, the angle precision using both differential phase, (6.88), and using direct angle estimation, (6.90), is the same. However, the correlated bias errors cancel directly using differential phase but not exactly using (6.91) due to the nonlinear expression for angle-of-arrival that uses the inverse sine function. Although both techniques provide correlated error cancellation, the technique defined by (6.88) yields complete cancellation.

6.8 Synthetic Aperture Radar Interferometry

Synthetic aperture radar (SAR) interferometry is an application that uses unwrapped phase measurements relative to the phase of a reference point to create 3-D images as opposed to the conventional SAR 2-D (down-range and cross-range) images. Because SAR interferometry uses differential phase measurements, the resolution is determined by baseline separation and operating frequency, as in radar interferometers discussed in this chapter, and the accuracy is a function of SNR. Consider the geometry defined in Figure 6.6, where a satellite passes over an area of interest in two orbits.

We let P denote the reference point on the ground; we assume that we know the ranges R_1 and R_2 to the reference point from each of the orbit positions. Interferometric SAR attempts to measure the height of the point P' where P' is within the same range cell as P. In other words, the points P and P' cannot be resolved in range but can be resolved using interferometry. The SAR image for the point P has complex voltage for a signal transmitted and received at both orbit 1 and orbit 2 positions.

$$x_1 = a_1 e^{j\varphi_1} \tag{6.93}$$

$$x_2 = a_2 e^{j\varphi_2} \tag{6.94}$$

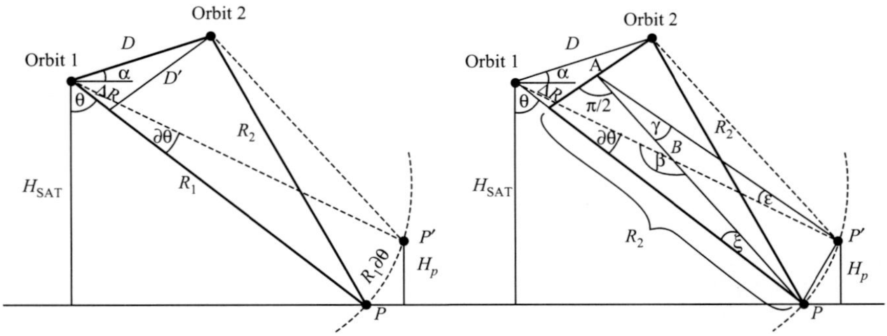

Figure 6.6. Synthetic Aperture Radar Interferometer Geometry

And the complex interferogram is defined by complex multiplication

$$v = x_1 x_2^* = a_1 a_2 e^{j(\varphi_1 - \varphi_2)} \tag{6.95}$$

6.8.1 SAR Interferometry Using Differentials

The interferometric phase for the point P can be written as

$$\varphi_{1P} = \frac{4\pi R_1}{\lambda} \tag{6.96}$$

$$\varphi_{2P} = \frac{4\pi R_2}{\lambda} \tag{6.97}$$

And the phase difference is defined by

$$\phi_P = \varphi_{1P} - \varphi_{2P} = \frac{4\pi(R_1 - R_2)}{\lambda} = \frac{4\pi \Delta R}{\lambda} \tag{6.98}$$

To understand how the interferometer can resolve P', we first take the derivative of (6.98):

$$\partial\phi_P = \frac{4\pi}{\lambda}\partial\Delta R \tag{6.99}$$

Referring to Figure 6.6, we can see that we can define ΔR in terms of the geometric angles θ and α.

$$\Delta R = D\sin(\theta - \alpha) \tag{6.100}$$

Taking the derivative we have,

$$\partial\Delta R = D\cos(\theta_0 - \alpha)\partial\theta \tag{6.101}$$

Substituting (6.101) into (6.99),

$$\partial\phi_P = \frac{4\pi}{\lambda} D\cos(\theta_0 - \alpha)\partial\theta \tag{6.102}$$

The key is to realize that H_P is the change in H_{SAT} as a function of the change in θ.

$$H_{\text{SAT}} = R_1 \cos(\theta) \tag{6.103}$$

and the change in H_{SAT} is given by

$$\partial H_{\text{SAT}} = H_P = R_{1P} \sin\left(\theta_P^0\right)\partial\theta \tag{6.104}$$

Substituting (6.102) into (6.104),

$$H_P = \frac{\lambda R_{1P} \sin\left(\theta_p^0\right)}{4\pi D'_\perp}\partial\phi_P \tag{6.105}$$

where

$$D'_\perp = D\cos\left(\theta_P^0 - \alpha\right) \tag{6.106}$$

6.8.2 SAR Interferometry Using Angle-of-Arrival

We now derive (6.105) using angle-of-arrival in a similar fashion to derivations used in other interferometer techniques presented in this chapter. We refer to the angles definitions in Figure 6.6 (right), and, in particular, we use the two triangles ΔS_1BP and $\Delta ABP'$ with common angle β to show

$$\gamma + \varepsilon = \pi - \beta = \partial\theta + \xi \tag{6.107}$$

Now the key is to realize that

$$\lim_{\delta\theta\to 0} P' = P \tag{6.108}$$

$$\lim_{\delta\theta\to 0} \angle\varepsilon = \angle\xi \tag{6.109}$$

And from (6.107) we have

$$\lim_{\delta\theta\to 0} \gamma = \partial\theta \tag{6.110}$$

Define the phases for the point P' relative to orbit 1 and orbit 2 locations,

$$R'_1 = |S_1 - P'| \tag{6.111}$$

$$R'_2 = |S_2 - P'| \tag{6.112}$$

and

$$\varphi'_1 = \frac{4\pi R'_1}{\lambda} \tag{6.113}$$

The phases for the point P' are defined as follows,

$$\varphi'_1 = \frac{4\pi R'_1}{\lambda} \tag{6.114}$$

$$\varphi_2' = \frac{4\pi R_2'}{\lambda} \tag{6.115}$$

and

$$\Delta\varphi' = \frac{4\pi \Delta R'}{\lambda} \tag{6.116}$$

The change in phase is defined as the difference between the phase difference for P' and the phase difference for P. We use the projected interferometer defined by D' to compute the angle-of-arrival, which in this case is the angle γ. But substituting (6.110), we have

$$\partial\varphi = \Delta\varphi' - \frac{4\pi\Delta R}{\lambda} = \left(\varphi_1' - \frac{4\pi\Delta R}{\lambda}\right) - \varphi_2' = \frac{4\pi D'}{\lambda}\sin(\gamma) = \frac{4\pi D'}{\lambda}\sin(\partial\theta) \tag{6.117}$$

Using the geometry in Figure 6.6 (right), we can solve for the height H_P.

$$H_P = 2R_1\sin(\theta)\tan\left(\frac{\partial\theta}{2}\right) = 2R_1\sin(\theta)\tan\left(\frac{1}{2}a\sin\left(\frac{\lambda\partial\varphi}{4\pi D'}\right)\right) \tag{6.118}$$

Or, using the small angle approximation,

$$H_P = \frac{R_1\sin(\theta)\lambda\partial\varphi}{4\pi D'} \tag{6.119}$$

Note that both approaches lead to the same derivation for the H_P in (6.105) and (6.119). The second derivation, using angle-of-arrival, clearly defines the phase differences that are used to compute the angle-of-arrival including the phase differences for the reference point P. Because the location of point P, relative to the satellite locations, is presumed to be known, the only phase error contribution is from the measured phase for the point P', which is made perfectly clear in the second derivation. Whereas the first derivation relates phase to range, the second derivation relates phase to angle-of-arrival, but both approaches lead to the same result when the reference phase is taken into account.

6.8.3 SAR Interferometry Height Error

Assuming that the reference phase is perfectly known, we can determine the error for estimating the height H_P from (6.105) and (6.119). The accuracy in estimating the height H_P can be derived in a way similar to how accuracy is derived for an interferometer in previous sections [5–6].

$$\sigma_{H_P} = \frac{\lambda R_{1P}|\sin(\theta_p^0)|}{4\pi D'_\perp}\sigma_{\partial\phi_P} = \frac{\lambda R_{1P}|\sin(\theta_p^0)|}{4\pi D'_\perp\sqrt{SNR}} \tag{6.120}$$

where we have shown previously that, when $\partial\varphi$ is noise limited,

$$\sigma_{\partial\varphi} = \frac{1}{\sqrt{SNR}} \tag{6.121}$$

The result for height error in (6.120) assumes that there is no error in the reference phase. When there is error in the reference phase and the error is correlated due to atmospheric effects, the expression for height error must take into account the additional phase error and the correlation in the error. In this case, we are interested in determining the error in the difference between two points P and Q. We define the differential height as

$$\Delta H = H_P - H_Q \tag{6.122}$$

And (6.120) becomes [1]

$$\sigma_{\Delta H} = \sqrt{\left(\frac{\lambda R_{1P}|\sin(\theta_p^0)|}{4\pi D'_\perp}\right)^2 \sigma_{\varphi_P}^2 + \left(\frac{\lambda R_{1P}|\sin(\theta_q^0)|}{4\pi D'_\perp}\right)^2 \sigma_{\varphi_Q}^2 - 2\frac{\lambda R_{1P} R_{1Q}\sin(\theta_p^0)\sin(\theta_q^0)}{4\pi D'_\perp}\sigma_{\varphi_P \varphi_Q}} \tag{6.123}$$

Assuming that ρ is the correlation coefficient, we have

$$\sigma_{\Delta H} = \sqrt{\left(\frac{\lambda R_{1P}|\sin(\theta_p^0)|}{4\pi D'_\perp}\right)^2 \sigma_{\varphi_P}^2 + \left(\frac{\lambda R_{1P}|\sin(\theta_q^0)|}{4\pi D'_\perp}\right)^2 \sigma_{\varphi_Q}^2 - 2\rho\frac{\lambda R_{1P} R_{1Q}\sin(\theta_p^0)\sin(\theta_q^0)}{4\pi D'_\perp}\sigma_{\varphi_P}\sigma_{\varphi_Q}} \tag{6.124}$$

In Chapter 7, we will discuss the generation of the reference point P, and the differential phase for P is derived based on the geometry. The differential phase for the point Q is measured by the interferometer determined by the two different orbits of a satellite, and the differential phase error can be noise limited but will also be affected by both tropospheric and ionospheric refraction. However, for the reference point P, the differential phase error is unlikely to be noise limited because it is not measured by the radar. Instead, the differential phase error associated with P is erred by propagation effects such as time delays resulting from atmospheric refraction. Because the second orbit occurs at a later time than the first orbit, the correlation in these errors is related to the stability and stationarity of error sources such as atmospheric refraction and dispersion.

6.9 Cramer Rao Lower Bound for Time-of-Arrival

The Cramer-Rao lower bound (CRLB) can be computed for estimating the absolute time delay between a transmitted pulse and a receive pulse [1,2].

We will follow the derivation as provided in [1] where details that were omitted have been included. The CRLB for the variance for estimating a parameter θ is

$$\sigma_\theta^2 \geq \frac{1}{E_y\left(\left(\frac{\partial}{\partial\theta}\ln(p(y,\theta))\right)^2\right)} \tag{6.125}$$

where $p(y,\theta)$ is the probability density for the random variable y that depends on the parameter θ.

Consider the following radar signal model $s(t)$ for a pulse that is transmitted at time t and $s(t - t_d)$ for the same pulse received at a time t_d later.

$$s(t) = a(t)e^{\omega t} \tag{6.126}$$

$$s(t - t_d) = a(t - t_d)e^{\omega(t - t_d)} \tag{6.127}$$

where $a(t)$ is the envelope function that includes pulse modulation such as linear frequency modulation or phase modulation, and ω is the carrier frequency.

Typically, the peak of the convolution $(s * s)(t_d)$ determines an estimate of the time delay. The time delay is therefore determined solely by the envelope function $a(t)$ and not by the carrier ω.

$$\begin{aligned} c(t_d) = |(s * s)(t_d)| &= \left|\int_{-\infty}^{\infty} s(t)s^*(t - t_d)dt\right| \\ &= \left|\int_{-\infty}^{\infty} a(t)e^{\omega t}a^*(t - t_d)e^{-(t-t_d)}dt\right| \\ &= \left|e^{\omega t_d}\int_{-\infty}^{\infty} a(t)a^*(t - t_d)dt\right| \\ &= \left|\int_{-\infty}^{\infty} a(t)a^*(t - t_d)dt\right| \\ &= |(a * a)(t_d)| \end{aligned} \tag{6.128}$$

When the carrier is involved in time delay or range estimation, the process is typically called "phase derived range," and the CRLB includes additional terms involving the carrier frequency ω. This discussion considers only time delay estimation that uses the envelope function.

Assume that the pulse waveform is sampled numerous times over the duration of the waveform s_k. Then the probability density function for a signal s that depends on the parameter θ is given as follows:

$$p(y,\theta) = \prod_k \frac{1}{\sqrt{2\pi\sigma^2}}e^{-\frac{(y - s_k(\theta))^2}{2\sigma^2}} \tag{6.129}$$

Computing the CRLB for estimating the parameter θ, we have

$$
\begin{aligned}
E_y\left(\left(\frac{\partial}{\partial\theta}\ln(p(y,\theta))\right)^2\right) &= \sum_k E\left(\frac{y-a_k}{\sigma^2}\frac{da_k}{d\theta}\right)^2 \\
&= \sum_k \left(\frac{da_k}{d\theta}\right)^2 E\left(\frac{y-a_k}{\sigma^2}\right)^2 \\
&= \sum_k \frac{1}{\sigma^4}\left(\frac{da_k}{d\theta}\right)^2 E(y-a_k)^2 \\
&= \sum_k \frac{1}{\sigma^2}\left(\frac{da_k}{d\theta}\right)^2 \qquad E(y-a_k)^2=\sigma^2 \\
&= \sum_k \frac{2\Delta t}{\mathcal{N}}\left(\frac{da_k}{d\theta}\right)^2 \qquad \frac{\sigma^2}{\Delta t}\to\frac{\mathcal{N}}{2} \\
&\to \frac{2}{\mathcal{N}}\int_{-\infty}^{\infty}\left(\frac{da}{d\theta}\right)^2 dt
\end{aligned}
\tag{6.130}
$$

where $\mathcal{N}$ is the noise power of the continuous noise process $n(t)$. The relationship between the continuous noise power spectrum and the discrete variance is given by

$$
\frac{\sigma^2}{\Delta t}\to\frac{\mathcal{N}}{2} \tag{6.131}
$$

This relationship is derived in Appendix E in more detail. Applying (6.131) to $\theta = t_d$, we have

$$
\begin{aligned}
\int_{-\infty}^{\infty}\left(\frac{da(t-t_d)}{dt_d}\right)^2 dt &= \int_{-\infty}^{\infty}\left(\frac{da(\xi)}{d\xi}\right)^2 d\xi \qquad \xi = t-t_d,\ \frac{ds}{dt} = -\frac{ds}{dt_d} \\
&= \int_{-\infty}^{\infty}\omega^2 A^2(\omega)d\omega \qquad \text{Parseval's theorem} \\
&= \int_{-\infty}^{\infty}\omega^2 A^2(\omega)d\omega \frac{\int_{-\infty}^{\infty}A^2(\omega)d\omega}{\int_{-\infty}^{\infty}A^2(\omega)d\omega} \\
&= B^2\int_{-\infty}^{\infty}A^2(\omega)d\omega \qquad \text{Definition of bandwidth}
\end{aligned}
\tag{6.132}
$$

where the mean bandwidth B is defined by

$$B^2 = \frac{\int_{-\infty}^{\infty} \omega^2 A^2(\omega) d\omega}{\int_{-\infty}^{\infty} A^2(\omega) d\omega} \tag{6.133}$$

Rearranging terms for the variance of time delay, the CRLB becomes

$$\begin{aligned} \sigma_{t_d}^2 &= \frac{1}{B^2} \frac{N}{2 \int_{-\infty}^{\infty} |A(\omega)|^2 d\omega} \\ &= \frac{1}{2B^2 SNR} \end{aligned} \tag{6.134}$$

$$\sigma_{t_d} = \frac{1}{B\sqrt{2SNR}} \tag{6.135}$$

For an interferometer using time delay to estimate angle-of-arrival, we have

$$\begin{aligned} \Delta\varphi &= \varphi_1 - \varphi_2 = 2\pi f t_1 - 2\pi f t_2 \\ &= \frac{2\pi c}{\lambda} \Delta t \end{aligned} \tag{6.136}$$

and

$$\Delta\varphi = \varphi_1 - \varphi_2 = \frac{2\pi D}{\lambda} \sin(\theta) \tag{6.137}$$

$$\Delta t = \frac{D}{c} \sin(\theta) \tag{6.138}$$

$$\sqrt{2}\sigma_t = \frac{D}{c} \cos(\theta)\sigma_\theta \tag{6.139}$$

$$\sigma_\theta = \frac{c\sqrt{2}}{D\cos(\theta)} \sigma_t = \frac{c}{BD\cos(\theta)\sqrt{SNR}} \tag{6.140}$$

To determine range for a two-way signal propagation, we have

$$\text{Range} = \frac{ct_d}{2} \tag{6.141}$$

where t_d is the time delay resulting from the two-way signal propagation. Range accuracy is determined by multiplying both sides of (6.135) by one-half the speed of light c; we have the CRLB for range estimation using the envelope function.

$$\sigma_{Range} = \frac{c\sigma_{t_d}}{2} = \frac{c}{2B\sqrt{2SNR}} \tag{6.142}$$

To simplify (6.142) and provide an intuitive understanding of the derivation, we derive a CRLB estimate for a particular class of waveforms in Appendix F.

6.10 Coherent Phase Trilateration

Coherent phase trilateration (CPT) is a process that accurately locates the position of a transmitting or receiving signal source applying highly accurate range difference measurements to multiple distributed receivers or distributed transmitters with known positions. The range difference measurements are determined using unwrapped phase difference between the sources and the distributed transmitter or receiver elements. The global positioning system (GPS) is an example of a CPT system using multiple distributed transmitting sources to locate GPS receivers. In this discussion, we focus on implementing CPT with multiple distributed receivers to locate either a transmitting source (one-way signal) or a target illuminated by a separate transmitter (bistatic signal). However, the results of this section can also be applied to multiple distributed transmitters. Using phase measurements determined by multiple distributed receiver or transmitters to locate targets is somewhat analogous to interferometers that accurately measure angle-of-arrival with the exception that phase ambiguities occur in range for CPT as opposed to angle for interferometers.

With N number of receivers there are $N - 1$ range difference measurements and provided that the geometry of the receivers relative to the signal source provides uniform coverage over 360°, the position error is inversely proportional to the square root of the total number of measurements.

For $i = 1, 2, \ldots, N$, let $r_i = (x_i \quad y_i \quad z_i)$ denote the position vector of the ith radar, and let $p_T = (x_T \quad y_T \quad z_T)$ denote the target position. If R_i denotes the range measurement from the ith radar to the target, then

$$R_i = \|r_i - p_T\| \quad i = 1, 2, \ldots, N \tag{6.143}$$

determines N equations with three unknowns. The solution to the system of equations defined by (6.143) yields the solution for the target position. Because the system is nonlinear, a Newton-Raphson convergence technique is required to find the solution. The difference between the conventional trilateration technique and the coherent phase trilateration technique is that, in the former technique, range is determined by range gate splitting (RGS), and the latter range is determined by phase.

The accuracy of the range measurements using range gate splitting is determined by the instantaneous bandwidth B and the signal-to-noise ratio SNR.

$$\sigma_R^{RGS} = \frac{c}{2B\sqrt{2SNR}} \tag{6.144}$$

For CPT, the range is determined by unwrapping the phase measurements, whereas the radar unwrapped phase is determined by

$$\varphi_i^w = \left[\frac{2\pi}{\lambda} R_i + \pi\right]_{2\pi} - \pi \tag{6.145}$$

and, unwrapping the phase, we have,

$$\varphi_i = \frac{2\pi R_i}{\lambda} = \varphi_i^w + 2\pi M_i + \varphi_{\text{noise}} \tag{6.146}$$

$$R_i = \frac{\lambda}{2\pi}\left(\varphi_i^w + 2\pi M_i + \varphi_{\text{noise}}\right) \tag{6.147}$$

where the superscript w denotes the wrapped phase measurements, and φ_{noise} is the receiver phase noise. Because range is now determined using phase, the accuracy of the CPT range measurement is

$$\sigma_R^{CPT} = \frac{\lambda}{2\pi\sqrt{2SNR}} \tag{6.148}$$

However, the CPT accuracy defined by (6.148) is usually smaller than the accuracy defined by (6.144). If we assume that the bandwidth of the radar signal is a fraction γ of the signal frequency where $0 < \gamma < 0.5$, then the ratio of the one-sigma accuracies becomes

$$\frac{\sigma_R^{RGS}}{\sigma_R^{CPT}} = \frac{\pi c}{\lambda B} = \frac{\pi f}{\gamma f} = \frac{\pi}{\gamma} > 2\pi \tag{6.149}$$

For most radar signals, $\gamma < 0.1$, and the ratio defined in (6.149) shows that using CPT to determine the target position yields accuracies that are more accurate by an order of magnitude or greater compared to using RGS.

One of the primary issues in using CPT is that the phases must be unwrapped in order to define the range magnitude for each radar. In other words, the integers M_i must be computed. In general, unwrapping the phase cannot be accomplished using CPT alone. Another technique is required. One such possible technique is to use the RGS technique to unwrap phase and then apply the unwrapped phase to implement CPT for increased accuracy. Of course, the RGS technique must use sufficient bandwidth to ensure that RGS measurement accuracy is sufficiently less than one phase ambiguity defined by the CPT phase unwrapping. Other method to unwrap phase involves frequency agility or frequency hopping and using two or three closely spaced receivers and multiple interferometers to triangulate in order to determine target position for the purpose of unwrapping range phase ambiguities.

Another way to deal with phase ambiguities is to use range (phase) differencing where (6.143) becomes

$$R_{i,j} = \|r_i - p_T\| - \|r_j - p_T\| \quad i,j = 1,\ 2,\ \ldots,\ N \quad i \neq j \tag{6.150}$$

Note that, when using range or phase differencing, the minimum number of radars required is four, whereas, when using absolute range, the minimum number is three. By using range differencing, the number of ambiguities is significantly reduced because the range difference has a fewer number of wavelengths than absolute range. Figure 6.7 shows the result of using CPT range differencing with four distributed radars. Figure 6.8 shows the position error projected into range and cross-range for 1000 Monte Carlo runs. The distribution of black points is determined using RGS, and the small white spot within the black dots is the distribution of 1000 points determined by the CPT technique. Notice the increased accuracy of CPT over RGS for the radar parameters used. Here $SNR = 20$ dB, the radar

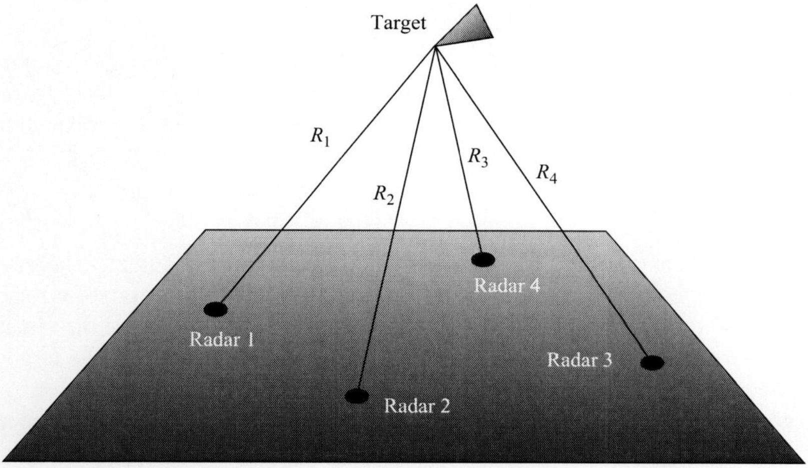

Figure 6.7. CPT Range Differencing Using Four Radars

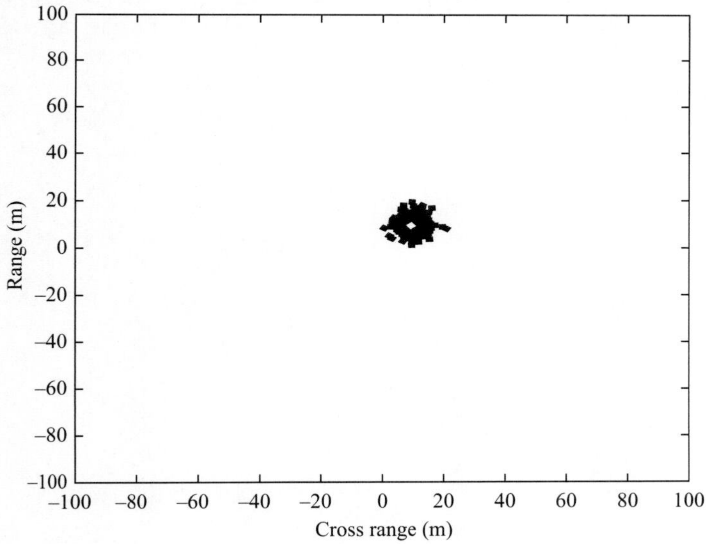

Figure 6.8. The Distribution of Points for Trilateration Using RGS (black dots) and CPT (white dots in center of black dots)

operating frequency is 2 GHZ, and the radar bandwidth chosen is 2 MHz. Thus, the parameter $\gamma = 0.001$. As a result, (6.149) shows that CPT should be more accurate by a factor of 3142 times compared with using RGS ranging. The result in Figure 6.8 shows a ratio of 3038, which shows reasonable agreement with the theory. In Appendix G, MATLAB® code is presented for a 2-D solution that implements a Newton-Raphson algorithm for solving (6.150) for either the CPT or RGS range determination methods.

6.10.1 Geometric Dilution of Precision

Geometric dilution of precision (GDOP) refers to how the configuration geometry impacts the positioning error. Figure 6.9 shows two situations that impact trilateration accuracy. If R_1 and R_2 are the covariance estimates of two radar measurements, then the resultant covariance R for the combined measurements is given by

$$R = \left(R_1^{-1} + R_2^{-1}\right)^{-1} \tag{6.151}$$

Figure 6.9 (left) shows an ideal situation where the range measurement errors are evenly distributed about the target. In this case, the position estimate errors are approximately the same for all coordinates. Figure 6.9 (right) shows the situation where the radars are closely separated with respect to the location of the target. In this case, the position error is amplified transverse to the radial direction. The geometry showed in Figure 6.9 (left) is ideal, whereas, when the radars are located close together relative to the target as in Figure 6.9 (right), the resulting error is degraded because the radars are primarily dispersed along a single dimension. The issues resulting from GDOP can present challenges in the processing and position determination. A measure of GDOP is the ratio of the major and minor axes for the resulting ellipse (black ellipses in Figure 6.9). The GDOP for the radar geometry for Figure 6.9 (left) is 1, whereas the GDOP for the Figure 6.9 (right) is 3. When the GDOP measure is exceedingly large (>10), the resulting position error can be significant in one dimension, making the resulting estimate unreliable.

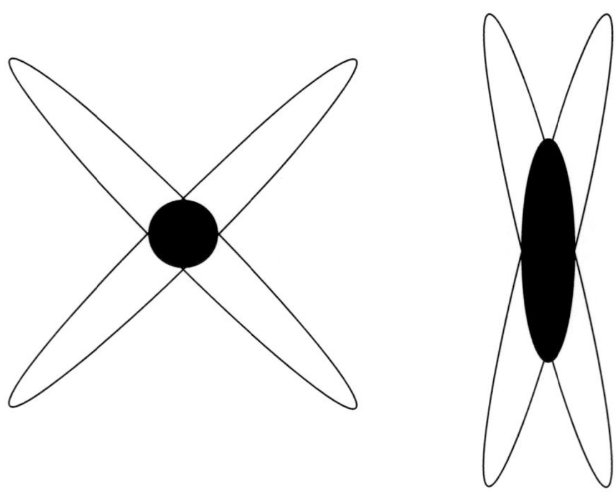

Figure 6.9. The Result of Combining the Position Estimates Derived from Two Radars (black line ellipses) Where the Major Ellipse Axis Is Angle and the Minor Axis Is Range: Example of When the Receivers Provide a Good Geometry for an Equally Distributed Resultant Ellipse (left, solid black); Where the Receivers Are Closely Separated, Causing the Resultant Errors (right, solid black) to Be Elongated in the Radial Direction

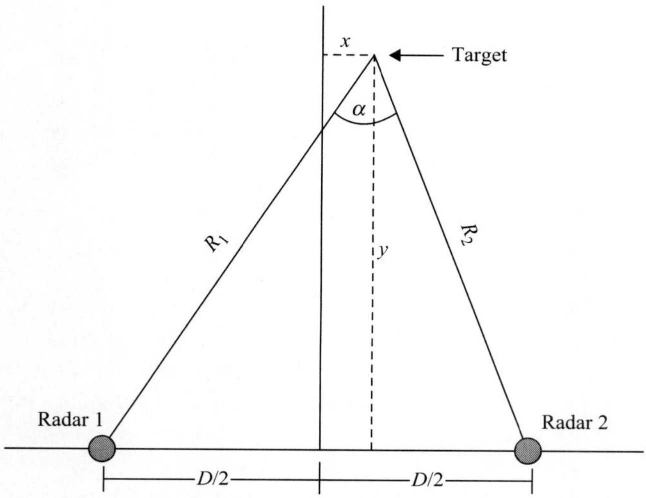

Figure 6.10. Two-Dimensional Geometry for Trilateration

To better understand the impact of GDOP, consider the 2-D geometry defined in Figure 6.10. For $D < \sqrt{R_1^2 + R_2^2}$ we can approximate the angle α defined in Figure 6.10 by

$$\sin(\alpha) \sim \frac{D}{2R_1R_2}\sqrt{2(R_1^2 + R_2^2) - D^2} \tag{6.152}$$

Furthermore, using basic trigonometry, the error in estimating x is derived to be

$$\sigma_x = \frac{\sqrt{R_1^2 + R_2^2}}{D}\sigma_R \tag{6.153}$$

or relating the error in x to α, we have,

$$\begin{aligned}\sigma_x &= \frac{\sqrt{R_1^2 + R_2^2}}{2R_1R_2\sin(\alpha)}\left(\sqrt{2(R_1^2 + R_2^2) - D^2}\right)\sigma_R \\ &\approx \frac{R_1^2 + R_2^2}{\sqrt{2}R_1R_2\sin(\alpha)}\sigma_R\end{aligned} \tag{6.154}$$

It turns out the error in estimating y can be approximated by

$$\sigma_y = \frac{\sigma_R}{\sqrt{2}\cos\left(\frac{\alpha}{2}\right)} \tag{6.155}$$

Notice that, when $D \ll \sqrt{R_1^2 + R_2^2}$, the error in estimating x grows significantly due to lack of observability in estimating x due to the geometry. For this simplified geometry, we defined GDOP as the ratio of the error in x to the error in y.

$$GDOP = \frac{\sigma_x}{\sigma_y} = \frac{\left(R_1^2 + R_2^2\right)\cos\left(\frac{\alpha}{2}\right)}{\sqrt{2}R_1R_2\,\sin(\alpha)} = \frac{R_1^2 + R_2^2}{2\sqrt{2}R_1R_2\,\sin\left(\frac{\alpha}{2}\right)} \tag{6.156}$$

which is a valid approximation for $0 < \alpha \leq \pi/2$. If we further assume that R_1 and R_2 are nearly the same, then

$$GDOP = \frac{1}{\sqrt{2}\,\sin\left(\frac{\alpha}{2}\right)} \tag{6.157}$$

and GDOP = 1 when $\alpha = 90°$ or when the x and y errors are exactly the same. The geometry in this case is an isosceles right triangle with legs equal to the ranges $R_1 = R_2$ and making a 45° angle with both the x and y axes. Thus the observability in x and y is the same. Also, as α approaches zero, GDOP approaches infinity due to lack of observability to x.

6.11 Summary of Interferometer Angle Precision

Table 6.1 summarizes the angle precision equations derived in this chapter.

Table 6.1. Angle Precision for Various Interferometer Types

Interferometer Type	Angle Precision
Monopulse:	
On-axis IID	$\sigma_\theta^M = \dfrac{\theta_3}{\frac{\pi}{2}\cos(\theta)\sqrt{2SNR_\Delta}}$
On axis non-IID	$\sigma_\theta^M = \dfrac{\theta_3}{\frac{\pi}{2}\cos\,(\theta)}\sqrt{\dfrac{1}{2\sqrt{2SNR_{1Q}}\sqrt{2SNR_{2Q}}}\left(\dfrac{\sqrt{SNR_{2Q}}}{\sqrt{SNR_{1Q}}} + \dfrac{\sqrt{SNR_{1Q}}}{\sqrt{SNR_{2Q}}} - 2\rho\right)}$
Off-axis IID	$\sigma_\theta^M = \frac{\lambda}{\pi D\,\cos(\theta)}\sigma_{d/s} = \frac{\lambda}{\pi D\,\cos(\varphi)\cos(\theta)}\sqrt{\frac{1}{\sqrt{2SNR_\Delta}} + \frac{\tan^2(\varphi)}{\sqrt{2SNR_S}}}$
Off-axis Non-IID $\varphi \gg 0$	$\sigma_\theta^M = \frac{\lambda}{\pi D\,\cos(\varphi)\cos(\theta)}\sqrt{\frac{1}{2SNR_\Delta} + 2\rho\frac{\tan(\varphi)}{\sqrt{2SNR_\Delta}\sqrt{2SNR_S}} + \frac{\tan^2(\varphi)}{2SNR_S}}$
Digital:	
IID	$\sigma_\theta^I = \frac{\lambda}{2\pi D\,\cos(\theta)}\sigma_{\Delta\phi} = \frac{\lambda}{2\pi D\,\cos(\theta)\sqrt{SNR}}$
Non-IID	$\sigma_\theta = \frac{\lambda}{2\pi D\,\cos(\theta)}\sqrt{\frac{1}{\sqrt{2SNR_1}\sqrt{2SNR_2}}\left(\frac{\sqrt{SNR_2}}{\sqrt{SNR_1}} + \frac{\sqrt{SNR_1}}{\sqrt{SNR_2}} - 2\rho\right)}$
Orthogonal:	
IID	$\sigma_\theta^{OI} = \frac{\lambda}{4\pi D\,\cos(\theta)}\sigma_{\Delta\phi} = \frac{\lambda}{4\pi D\,\cos(\theta)\sqrt{SNR}}$
Non-IID	$\sigma_\theta = \frac{\lambda}{4\pi D\,\cos(\theta)}\sqrt{\frac{1}{\sqrt{2SNR_1}\sqrt{2SNR_2}}\left(\frac{\sqrt{SNR_2}}{\sqrt{SNR_1}} + \frac{\sqrt{SNR_1}}{\sqrt{SNR_2}} - 2\rho\right)}$
Amplitude:	Same as digital interferometer
Bistatic:	Same as digital interferometer with SNR defined as the bistatic signal-to-noise ratio

References

1. P. Z. Peebles, *Radar Principles*, John Wiley & Sons, New York, 1998.
2. D. K. Barton and H. R. Ward, *Handbook of Radar Measurements*, Artech House, Dedham, MA, USA, 1984.
3. A. Papoulis and S. U. Pillai, *Probability, Random Variables, and Stochastic Processes*, McGraw-Hill, New York, 2002.
4. H. L. Van Trees, *Detection, Estimation, and Modulation Theory, Part 1*, John Wiley & Sons, New York, 1968.
5. R. F. Hannsen, *Radar Interferometry*, Kluwer Academic Publishers, Dordrecht, Netherlands, 2001.
6. E. Ridriguez and J. M Martin, "Theory and design of interferometric synthetic aperture radars," *IEE Proceedings-F*, vol. 139, no. 2, pp. 147–159, 1992.

Chapter 7

Interferometer Signal Processing

In this chapter, the fundamental signal processing required to implement an interferometer is introduced. The processing required for an interferometer is similar to the processing required for radars in general with the exception that angle estimation is implemented using phase comparison. Because phase is defined only over an interval [0 2π], the phase can wrap over many 2π intervals, creating an ambiguity in the phase measurement. For the interferometer, that relative phase ambiguity between two antennas must be resolved in order to estimate angle-of-arrival. In this chapter, we discuss methods to resolve interferometer phase ambiguities. In particular, all receive interferometer antennas or arrays must be calibrated to ensure that relative phase is consistent across the radar field of regard. The orthogonal interferometer has unique processing requirements due to its implementation of pseudo-orthogonal waveforms. Calibration of an orthogonal interferometer requires that all antennas be calibrated for transmit as well as receive. The derivation of angle-of-arrival from relative phase measurements is presented for various far-field conditions. The first-order derivation assumes that the constant phase contours that emanate from a target are essentially linear across the interferometer distributed antennas. This linear far-field condition is satisfied when the target is sufficiently far in range from the interferometer but depends on the interferometer baseline and radar operating frequency [1–3]. Second-order derivations are also presented for quadratic phase behavior across the interferometer antenna.

7.1 Basic Interferometer Processing

The processing for a basic RF interferometer is essentially the same as that for most radars, with the exception that interferometer angle processing uses a direct phase comparison; also, due to the spacing among interferometer antennas, the processing must also resolve ambiguities due to the multiple 2π rotations in phase that occur because of the spacing. In general, the RF interferometer utilizes a suite of pulse-Doppler or continuous wave (CW) waveforms that are required to perform search, track, and fire control functions. The latter function may require an imaging waveform where the radar is guiding an interceptor to hit a specific aim point.

For the pulse-Doppler application, the transmitter and receiver are cohered using a common local oscillator. After the waveform is transmitted, interacts with a

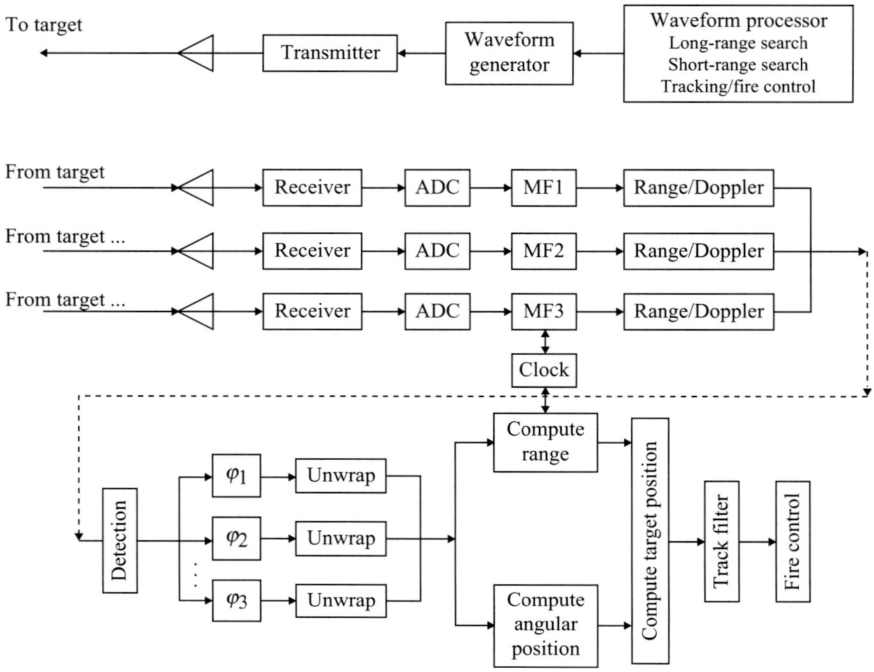

Figure 7.1. Conventional Interferometer Processing Flow Diagram

target, and returns to the receive antennas, the signal is received and mixed down to either IF or baseband, where it is digitized and passed through a matched filter at each receive antenna channel. Figure 7.1 shows this processing chain for three receive channels. Note that, after detection, the phases are computed in each channel and unwrapped to resolve the angle ambiguities. Various methods to accomplish unwrapping or angle ambiguity resolution are discussed later in the chapter. Range is computed using time-of-arrival and a clock that is common to both the receiver and transmitter. Angle is computed using phase differencing algorithms (presented later in this chapter). The measured range and angles are passed to a Kalman filter, which further enhances the target position estimate by assuming certain dynamic behavior.

Another waveform that is commonly used in RF interferometry is the continuous wave (CW) waveform [1–3]. This application is popular in sports and other targeted applications, such as dismount detection and through-the-wall sensing. The primary difference between a pulse-Doppler interferometer and a CW interferometer is that the pulse-Doppler system uses a heterodyne receiver, whereas the CW system uses a homodyne receiver [4]. Thus the transmit signal is directly mixed with the receive signal in all interferometer channels in order to create a baseband signal.

7.2 Orthogonal Interferometer Processing

The orthogonal interferometer (OI) concept relies on being able to transmit three coherent pseudo-orthogonal codes (one from each transmitter array) and process those codes at receivers either colocated with the transmitters or located on the missile interceptor. For OI, waveforms must be transmitted and processed for search, track, and guidance. Figure 7.2 shows the signal flow required for OI signal generation and processing. Typically, both a long-range and a short-range search waveform are implemented in an objective OI design. Figure 7.3 shows the signal processing flow required for each waveform. The basic signal processing steps are the matched filter, Doppler processing, and angle ambiguity resolution.

7.3 Angle Ambiguity Resolution

Because the phase center separation between interferometer arrays is several wavelengths, the determination of angle-of-arrival using phase difference is ambiguous by the number of wavelengths. The expression for angle-of-arrival is

$$\sin(\theta) = \frac{\lambda(\Delta\varphi + 2\pi N)}{2\pi D} \tag{7.1}$$

where D is the array phase center separation, and N is the ambiguity number. The measured wrapped phase difference is determined by

$$\Delta\varphi_{\text{wrap}} = W(\Delta\varphi) = [\Delta\varphi + \pi]_{2\pi} - \pi = \text{mod}\{\Delta\varphi + \pi : 2\pi\} - \pi \tag{7.2}$$

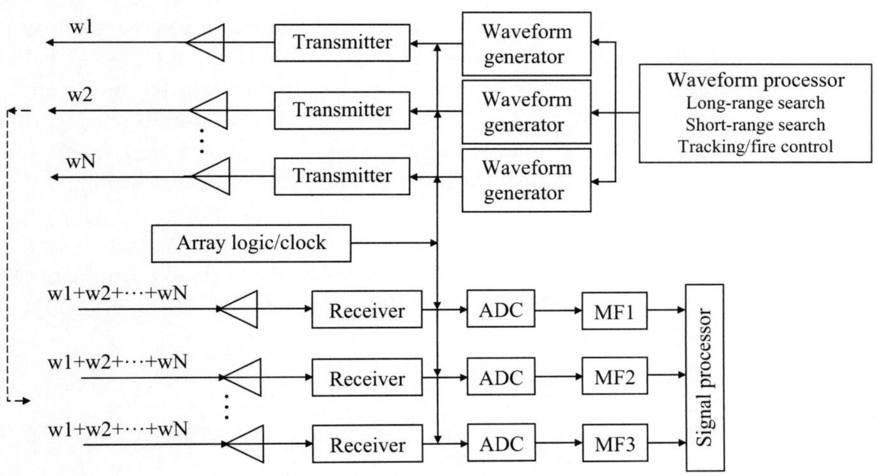

Figure 7.2. Orthogonal Interferometer Signal Flow Diagram

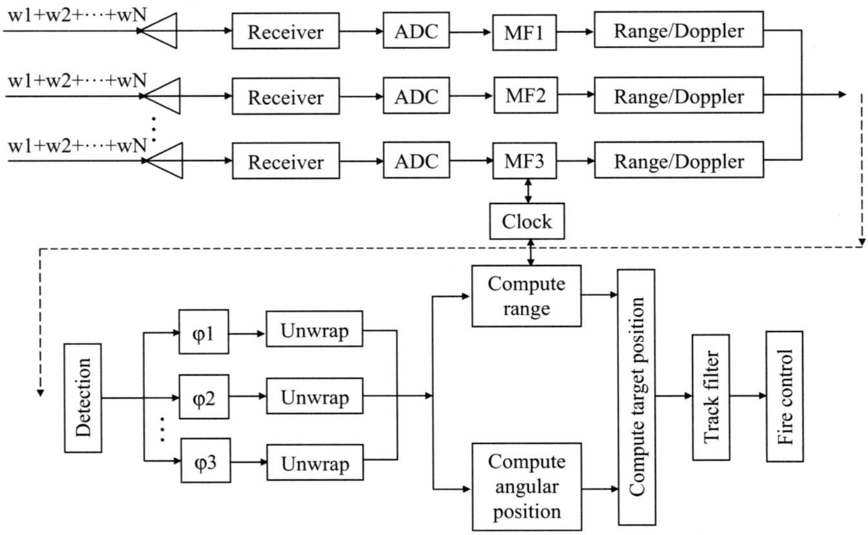

Figure 7.3. Orthogonal Interferometer Signal Processing Diagram

with

$$\Delta\varphi = \frac{4\pi\Delta R}{\lambda} + \varphi_{\text{noise}} = \Delta\varphi_{\text{wrap}} + 2\pi N + \varphi_{\text{noise}} \tag{7.3}$$

where W is the phase wrapping operator, $\Delta R = R_1 - R_2$, and $\varphi_{\text{noise}} \in [-\pi, \pi)$ is additive phase noise. The phase unwrap problem is to determine the number N, given measurements of $\Delta\varphi_{\text{wrap}}$. As stated, the phase unwrap problem is an ill posed inverse problem that has no solution, but other measurements can be taken into account that can lead to a well posed inverse. In this section, we discuss using unambiguous angle measurements to resolve the ambiguous interferometer angle measurements as expressed in (7.1). For a two-antenna interferometer, the requirement to resolve the ambiguity is that the phase satisfies a smoothness criterion that means the phase gradient must behave linearly, as shown in Figure 7.4.

The discrete jumps in phase in Figure 7.4 occur as a result of phase wrappings that exceed $2\pi k$ for some integer k. It is possible to unwrap phase only if phase has this nearly linear behavior. Note the similarity in the positive slope portions of the graphs. Mathematically, we define the following operator.

$$D^2\left(\varphi^i\right) = \varphi^{i-1} - 2\varphi^i + \varphi^{i+1} \tag{7.4}$$

In theory, differential phase is nearly linear, provided $D^2(\Delta\varphi^i) = 0$ except at a small number of points (mathematically defined as a set of measure zero). In practice, the operation D^2 is not identically zero due to noise and phase errors, but

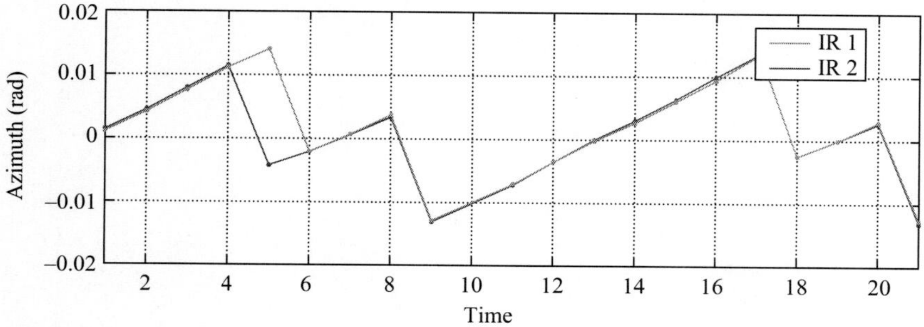

Figure 7.4. Nearly Linear Phase Behavior of Phase as a Function of Time for Two Interferometer Measurements

we can still define the concept of nearly linear phase. For $\varepsilon < 1$, let Z_0 and Z_1 be defined as follows:

$$Z_0 = \left\{ i : \left| D^2 \left(\Delta\varphi^i_{\text{unwrap}} \right) \right| < \varepsilon \right\} \tag{7.5}$$

$$Z_1 = \left\{ i : \left| D^2 \left(\Delta\varphi^i_{\text{unwrap}} \right) \right| > \varepsilon \right\} \tag{7.6}$$

Then the nearly linear criterion corresponds to the size of the set Z_0, being significantly larger than the size of the set Z_1.

$$\frac{\langle Z_0 \rangle}{\langle Z_1 \rangle} \gg 1 \tag{7.7}$$

where $\langle \cdot \rangle$ denotes the size of the set. Also, the larger the ratio of the set sizes defined in (7.7), the lower the number of phase wrappings that must be dealt with. In two dimensions, this criterion is defined as a two-dimensional phase derivative (phase difference in two orthogonal spatial directions) equal to zero. The two-dimensional phase criterion is discussed in Section 7.7.

The ratio of the two set sizes defined in (7.7) is a function of the operating frequency, interferometer baseline, ADC sample rate, and *SNR*. When the requirement in (7.7) is not satisfied, changing or modifying any of these parameters can result in satisfying (7.7). As a result, (7.7) imposes a requirement on the design parameters previously defined. For the remainder of this section, we assume that (7.7) is satisfied so that ambiguity resolution is possible.

7.3.1 Nyquist Sampling for a Spatial Array

In an antenna array, the individual antennas or elements represent digital samples in space. Spatial sampling, like time sampling, has a Nyquist sampling rate related to the spatial frequency of the signal opposed to the temporal frequency. The spatial Nyquist interval is $\lambda/2$, where λ is the signal wavelength. Figure 7.5 shows the antenna patterns for arrays with antenna spacing both greater and smaller than $\lambda/2$.

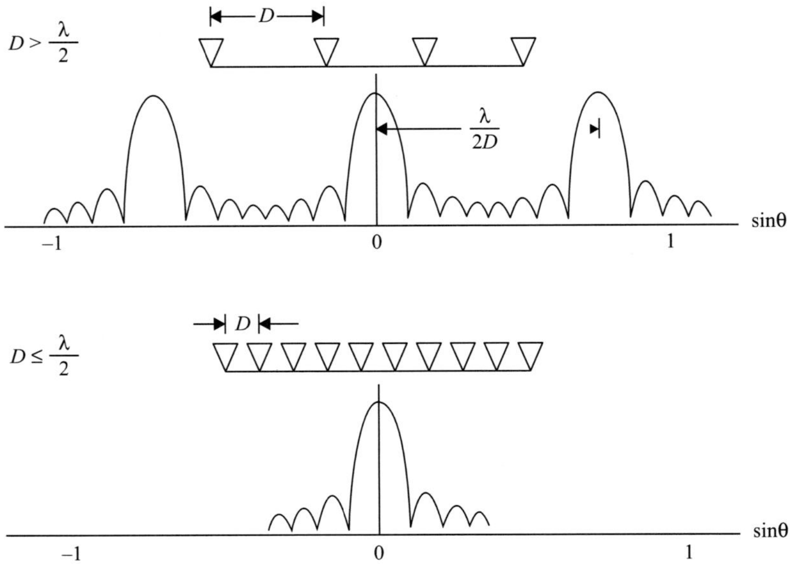

Figure 7.5. Grating Lobes or Aliasing

Notice the additional grating lobes in the antenna pattern when the spacing is greater than $\lambda/2$.

To determine the Nyquist sampling rate, we model the spatial signal sampled at each of the antenna elements with spacing D.

$$s_k = ae^{\frac{2\pi}{\lambda}jD(k-1)\sin(\theta)} \qquad \text{for } k = 1, \ldots, N \tag{7.8}$$

If we were to sample the array continuously instead of discretely at each antenna, the spatial signal can be written as follows:

$$s_k = ae^{2\pi jx\frac{\sin(\theta)}{\lambda}} = ae^{2\pi jxf_s} \tag{7.9}$$

where f_s is the spatial frequency,

$$f_s = \frac{\sin(\theta)}{\lambda} \tag{7.10}$$

Because the Nyquist condition is that the sample rate must be greater than or equal to twice the maximum frequency, we have

$$\frac{1}{D} \geq 2\frac{|\sin(\theta)|}{\lambda} \tag{7.11}$$

Or, to eliminate the possibility of angle ambiguities or grating lobes, the antenna spacing D must satisfy the following Nyquist condition:

$$D \leq \frac{\lambda}{2|\sin(\theta)|} \leq \frac{\lambda}{2} \tag{7.12}$$

For the interferometer, the spacing between antennas D is always much greater than $\lambda/2$ in order to achieve higher angle accuracy. As a result, grating lobes or angle ambiguities in the case of the digital interferometer must be resolved in order to determine the precise angle-of-arrival. The unambiguous angle interval for a conventional interferometer is defined by

$$-\frac{\lambda}{2D} \leq \sin(\theta) \leq \frac{\lambda}{2D} \tag{7.13}$$

The unambiguous angle interval for the orthogonal interferometer is half that for a conventional interferometer.

$$\sin(\theta_{OI}) = \frac{\lambda \Delta\varphi}{4\pi D} = \frac{1}{2}\sin(\theta_{CI}) \tag{7.14}$$

$$-\frac{\lambda}{2D} \leq \sin(\theta_{CI}) \leq \frac{\lambda}{2D} \tag{7.15}$$

$$-\frac{\lambda}{2D} \leq 2\sin(\theta_{OI}) \leq \frac{\lambda}{2D} \tag{7.16}$$

and the unambiguous angle interval for an orthogonal interferometer is defined by

$$-\frac{\lambda}{4D} \leq \sin(\theta_{OI}) \leq \frac{\lambda}{4D} \tag{7.17}$$

7.3.2 *Number of Angle Ambiguities*

The interferometer angle ambiguity actually results from an ambiguity in the differential phase measurement that was described in Section 6.2. The resolved differential phase measurement $\Delta\varphi$ is ambiguous by 2π phase rotations:

$$\begin{aligned} \Delta\varphi &= \frac{2\pi D}{\lambda}\sin(\theta) + 2\pi N \\ &= \beta + 2\pi N \end{aligned} \tag{7.18}$$

where β is the differential phase that is actually measured by the interferometer. Rearranging (7.18), we have

$$\frac{D}{\lambda}\sin(\theta) = \frac{\beta}{2\pi} + N \tag{7.19}$$

Because $0 \leq \beta < 2\pi$, we can assume $\beta = 0$ without loss of generality. Referring to (7.19), we see that the maximum possible value for N occurs at $\theta = 90°$ and that the minimum value for N occurs at $\theta = -90°$. As such, we have

$$N = \frac{D}{\lambda} \sin\left(\frac{\pi}{2}\right) = \frac{D}{\lambda} \tag{7.20}$$

$$-N = \frac{D}{\lambda} \sin\left(-\frac{\pi}{2}\right) = -\frac{D}{\lambda} \tag{7.21}$$

The maximum number of ambiguities M is equal to $2N$, and

$$M = 2N = \frac{2D}{\lambda} \tag{7.22}$$

When the field of view for the interferometer antennas is restricted, the maximum number of ambiguities is reduced. For example when $-30° \leq \theta \leq 30°$, then the number of ambiguities is reduced by a factor of 2, and when $-\lambda/2D \leq \sin(\theta) \leq \lambda/2D$, there is no angle ambiguity.

For an interferometer consisting of multiple receive antenna arrays, the array pattern defines the field of view based on the beamwidth of the array pattern. Let θ_3 denote the beamwidth of the array pattern. The maximum number of effective ambiguities M_e is defined by

$$M_e = \frac{2D}{\lambda} \sin\left(\frac{\theta_3}{2}\right) \tag{7.23}$$

In general, the number of effective ambiguities is determined by the product of the interferometer pattern with ambiguities located at integer multiples of λ/D and the array pattern with beamwdith θ_3. For example, if $D = 100\lambda$, then theoretically (7.22) shows that there are 200 total possible angle ambiguities. However, if the interferometer receive array has a beamwdith $\theta_3 = 0.01$ rad, then only five ambiguities are located within the beamwidth. Thus the array pattern has significantly reduced the number of ambiguities that must be considered in any attempt to resolve the correct ambiquity. Figure 7.6 shows the product of the array factor with the interferometer pattern of five ambiguities located within the 3-dB beamwidth.

7.3.3 Angle Ambiguity Resolution Using Frequency and Spatial Diversity

Equation (7.1) provides insight into how to determine the ambiguity number N. Changing frequency (wavelength) modifies (7.1) to change the relationship among feasible ambiguity numbers. Because the angle θ is independent of frequency, a unique ambiguity number for each frequency gives a unique solution for θ.

$$\sin(\theta) = \frac{\lambda_1(\Delta\varphi_1 + 2\pi N_1)}{2\pi D} \quad \text{and} \quad \sin(\theta) = \frac{\lambda_2(\Delta\varphi_2 + 2\pi N_2)}{2\pi D} \tag{7.24}$$

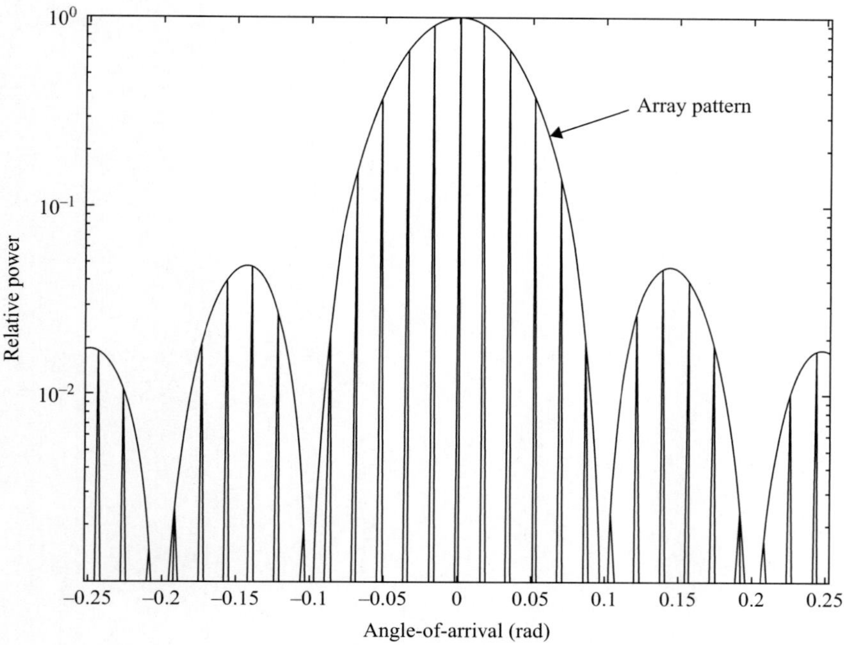

Figure 7.6. Product of the Array Factor with the Interferometer Pattern where $D = 100\lambda$ and $\theta_3 = 0.01$ rad

Also physically changing the separation of the phase centers Δd achieves the same effect for resolving the ambiguity number. However, the performance of both techniques is a function of *SNR*. In Figure 7.7, the Weiss-Weinstein lower bound estimate is computed for interferometers using both frequency and spatial diversity for angle ambiguity resolution. The left plot in Figure 7.7 shows the angle standard deviation versus *SNR* for both frequency and phase center deviation techniques. Notice that a small change in phase center requires less *SNR* than a relatively large change in frequency to resolve the angle ambiguity number. For 15 cm of phase center change, the angle ambiguity is resolved with about 28 dB *SNR* as opposed to 200 MHz of frequency change, which requires about 43 dB *SNR*.

One possible solution to resolve angle ambiguities is to configure one of the arrays to measure angle-of-arrival using monopulse. Monopulse essentially modifies the phase centers of an array to measure angle-of-arrival by computing the quotient of the difference voltage with the sum voltage. The difference voltage is computed by weighting the antenna pattern to produce a null at the array phase center. The right plot in Figure 7.7 shows the standard deviation of angle error versus *SNR* using monopulse angle estimation with one of the three arrays in the interferometer. Note that only about 18 dB *SNR* is required to resolve the angle ambiguity, which is significantly less *SNR* than that required for frequency or phase center agility.

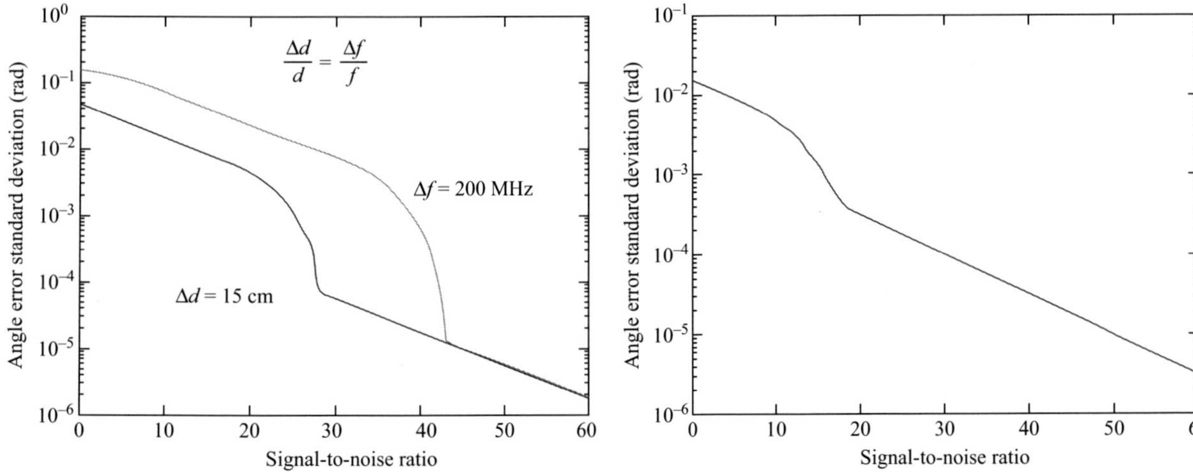

Figure 7.7. Angle Error Standard Deviation Versus SNR for Frequency Agility (light) and Phase Center Deviation (dark) (left); Angle Error Standard Deviation Versus SNR Using Monopulse Angle Estimation (right)

To effectively achieve adequate ambiguity resolution, the array monopulse boresight must be aligned with the interferometer boresight to within about one-quarter of an angle ambiguity. The first ambiguity for the OI design is located in an angle at λ/D from boresight, or about 14.4 mrad. Thus the monopulse antenna must be aligned to within 3.5 mrad with boresight to effectively resolve angle ambiguities. Although a 3.5-mrad alignment requirement is not stressing, it requires an alignment process to be implemented that can assure this level of alignment at all times.

7.3.4 Probability of Correct Ambiguity Resolution

Dybdal [5] derived expressions for the probability of correctly resolving the angle ambiguity from an interferometer using amplitude monopulse measurements from one of the interferometer elements. We will use an analysis similar to that in [5] to derive expressions for the probability of correctly resolving the angle ambiguity from two interferometer architectures shown in Figure 7.8. We first consider using a small interferometer with a baseline d to resolve the angle ambiguity that occurs in a large interferometer with baseline D, as shown in Figure 7.8 (left). Next we consider an architecture similar to [5] where one of the interferometer elements is a phased array that uses monopulse to derive angle measurements, as shown in Figure 7.8 (right).

We now derive an expression for correct ambiguity resolution given that we are using an angle measurement technique with angle error σ. Without loss of generality, we assume that the true angle is 0° and that the probability distribution for the angle θ can be expressed by the following Gaussian distribution:

$$P_\theta = \frac{1}{\sqrt{2\pi\sigma_\theta^2}} e^{-\frac{\theta^2}{2\sigma_\theta^2}} \tag{7.25}$$

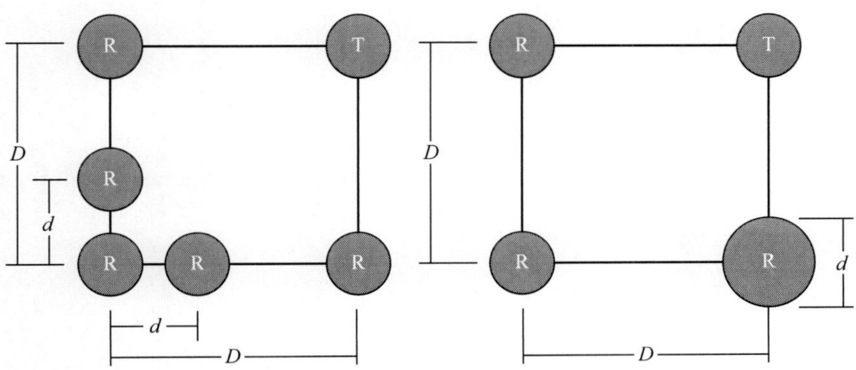

Figure 7.8. Two Architectures Consisting of a Single Transmit Array (T) and Multiple Receive Arrays (R) Using Spatial Diversity for Ambiguity Resolution: Large and Small Interferometer Architecture (left) and Large Interferometer with Monopulse Array (right)

For two-dimensional angle estimation, the probability distribution for the orthogonal angle ϕ is also expressed as a Gaussian distribution.

$$P_\phi = \frac{1}{\sqrt{2\pi\sigma_\phi^2}} e^{-\frac{\phi^2}{2\sigma_\phi^2}} \tag{7.26}$$

Making the following change of variables to polar coordinates,

$$\xi^2 = \theta^2 + \phi^2 \tag{7.27}$$

$$\eta = \tan\left(\frac{\theta}{\phi}\right) \tag{7.28}$$

we have

$$\iint P_\theta P_\phi d\theta d\phi = \iint P_{\xi,\eta} \xi d\xi d\eta \tag{7.29}$$

Referring to Section 7.3.2, we see that, for $-\lambda/2D \leq \sin(\theta), \sin(\phi) \leq \lambda/2D$, there is no angle ambiguity, and because $\sin(\theta) \approx \theta$ and $\sin(\phi) \approx \phi$ over the regions of interest, we define correct resolution as the probability that $\xi \leq \lambda/2D$. Thus, the probability of correct resolution is

$$\begin{aligned} P_{CR} &= P\left(\xi < \frac{\lambda}{2D}\right) \\ &= \int_0^{\lambda/2D} \int_0^{2\pi} P_{\xi,\eta} \xi d\xi d\eta \\ &= 1 - \exp\left(-\left(\frac{1}{2}\left(\frac{\lambda}{2D\sigma}\right)^2\right)\right) \end{aligned} \tag{7.30}$$

where $\sigma_\theta = \sigma_\phi = \sigma$.

7.3.4.1 Large and Small Interferometer Architecture

We determine the probability of correct resolution for the large and small interferometer architectures as shown in Figure 7.8 (left). The angle accuracy for the small interferometer is given by

$$\sigma = \frac{\lambda}{2\pi d \cos(\theta)\sqrt{SNR}} \tag{7.31}$$

where we require the small interferometer to be unambiguous over the field of view (θ_{FOV}) of the large interferometer. Thus we require

$$-\frac{\lambda}{2d} \leq \sin(\theta) \leq \frac{\lambda}{2d} \qquad -\frac{\theta_{FOV}}{2} \leq \theta \leq \frac{\theta_{FOV}}{2} \tag{7.32}$$

or we have

$$\sin\left(\frac{\theta_{FOV}}{2}\right) = \frac{\lambda}{2d} \tag{7.33}$$

Substituting (7.31) into (7.30), we have

$$P_{CR} = 1 - \exp\left(-\left(\frac{1}{2}\left(\frac{\lambda}{2\frac{D\lambda}{2\pi d\cos(\theta)\sqrt{SNR}}}\right)^2\right)\right) \tag{7.34}$$

And the probability of correct resolution for the large and the small interferometer architectures is

$$P_{CR} = 1 - \exp\left(-\left(\frac{1}{2}\left(\frac{d\pi\cos(\theta)\sqrt{SNR}}{D}\right)^2\right)\right) \tag{7.35}$$

We can use (7.35) to determine the small interferometer baseline for a given probability of correct resolution.

$$\frac{d}{D} = \frac{1}{\pi\cos\left(\frac{\theta_{FOV}}{2}\right)\sqrt{SNR}}\sqrt{2\ln(1 - P_{CR})} \tag{7.36}$$

$$\sin\left(\frac{\theta_{FOV}}{2}\right) = \frac{\lambda}{2d} \tag{7.37}$$

Note that (7.36) and (7.37) allow the design of the large and small interferometer architectures for a given probability of correct resolution, the field of view for the large interferometer, and SNR. Substituting (7.37) into (7.36), we have

$$\frac{\lambda}{D} = \frac{2\tan\left(\frac{\theta_{FOV}}{2}\right)}{\pi\sqrt{SNR}}\sqrt{2\ln(1 - P_{CR})} \tag{7.38}$$

Note that (7.38) defines the large interferometer baseline as a function of the field of view, the SNR, and the probability of correct ambiguity resolution. The baseline for the small interferometer is defined by (7.37) as a function of the large interferometer field of view.

7.3.4.2 Large Interferometer with Monopulse Array

We determine the probability of correct resolution for the large and small interferometer architectures as shown in Figure 7.8 (left). The angle accuracy for the large interferometer is given by [6].

$$\sigma = \frac{\theta_3}{k_m\cos(\theta)\sqrt{2SNR_\Delta}} = \frac{0.88\lambda}{k_m d\cos(\theta)\sqrt{2SNR_\Delta}} \tag{7.39}$$

where θ_3 is the 3-dB beamwidth for the monopulse array, k_m is the monopulse slope constant, and SNR_Δ is the SNR in the difference channel of the monopulse array. Substituting (7.39) into (7.30), we have

$$P_{CR} = 1 - \exp\left(-\left(\frac{1}{2}\left(\frac{\lambda}{2\frac{D \cdot 0.88\lambda}{k_m d \cos(\theta)\sqrt{2SNR_\Delta}}}\right)^2\right)\right) \tag{7.40}$$

And the probability of correct resolution for the large and small interferometer architectures is

$$P_{CR} = 1 - \exp\left(-\left(\frac{1}{2}\left(\frac{k_m d \cos(\theta)\sqrt{2SNR_\Delta}}{1.76 \cdot D}\right)^2\right)\right) \tag{7.41}$$

Again, we can use (7.41) to determine the small interferometer baseline for a given probability of correct resolution.

$$\frac{d}{D} = \frac{1.76}{k_m \cos(\theta)\sqrt{SNR_\Delta}}\sqrt{-\ln(1 - P_{CR})} \tag{7.42}$$

Note that (7.42) allows the design of the large interferometer using a monopulse array architecture for a given probability of correct resolution, monopulse slope, and SNR.

7.3.5 Angle Ambiguity Resolution Using Doppler

We now show that Doppler measurements can be used to aid in resolving angle ambiguities for a moving target. Assume that the velocity of a target is constant and that the initial target angle-of-arrival θ_0 is known. For most targets, the constant velocity assumption is valid at least over short time intervals, and the known initial angle assumes that an accurate estimate of angle is made with correct ambiguity resolution. For a two antenna interferometer, we have the following two phase measurements:

$$\begin{aligned} \varphi_1 &= \frac{2\pi}{\lambda} R_1 \\ \varphi_2 &= \frac{2\pi}{\lambda} R_2 \end{aligned} \tag{7.43}$$

where R_i is the range from the ith antenna to the target. Basic interferometry relates phase difference to angle-of-arrival as follows:

$$\Delta\varphi = \frac{2\pi}{\lambda}(R_1 - R_2) = \frac{2\pi}{\lambda} D \sin(\theta) \tag{7.44}$$

From (7.43), we can relate Doppler frequency to the time derivative of phase.

$$f_{\text{Doppler}} = \frac{1}{\pi}\frac{d\varphi}{dt} \tag{7.45}$$

Taking the time derivative of (7.44) and using the result in (7.45),

$$\pi\Delta f_{\text{Doppler}} = \frac{d\Delta\varphi}{dt} = \frac{2\pi}{\lambda} D\cos(\theta)\dot{\theta} \tag{7.46}$$

Now integrating (7.46), we have,

$$\begin{aligned}\sin(\theta(t)) - \sin(\theta_0) &= \int_{s=0}^{t} \frac{d}{ds}\sin(\theta(s))ds \\ &= \int_{s=0}^{t} \frac{\lambda}{2D}\Delta f_{\text{Doppler}}(s)ds\end{aligned} \tag{7.47}$$

Hence,

$$\sin(\theta(t)) = \sin(\theta_0) + \int_{s=0}^{t} \frac{\lambda}{2D}\Delta f_{\text{Doppler}}(s)ds \tag{7.48}$$

We now use (7.48) to resolve the number of integral 2π wraps for a constant velocity target where the initial target angle is resolved and known. First we write

$$\sin(\theta) = \frac{\lambda(\Delta\varphi + 2\pi N)}{2\pi D} \tag{7.49}$$

And for the initial angle, we have

$$\sin(\theta_0) = \frac{\lambda(\Delta\varphi_0 + 2\pi N_0)}{2\pi D} \tag{7.50}$$

where N_0 is known. Substituting (7.49) and (7.50) into (7.48) and solving for N we have

$$N = N_0 + round\left(\frac{1}{2\pi}(\Delta\varphi_0 - \Delta\varphi) + \int_{s=0}^{t} \frac{1}{2}\Delta f_{\text{Doppler}}(s)ds\right) \tag{7.51}$$

where *round* is the function that rounds to the nearest integer. We examine two cases that validate (7.51).

For the case the target is moving radially with respect to the radar at a constant velocity, the angle is not changing as the target is moving and thus $\Delta\varphi = \Delta\varphi_0$, and, because the target is moving with constant velocity, we have $\Delta f_{\text{Doppler}} = 0$. Thus, from (7.51), we see that $N = N_0$, which is exactly the correct result for a radial target.

For the case the target is moving in a circle with respect to the radar at the center, the Doppler frequency is zero, and (7.51) becomes

$$\begin{aligned} N &= N_0 + round\left(\frac{1}{2\pi}(\Delta\varphi_0 - \Delta\varphi) + \int_{s=0}^{t} \frac{\lambda}{2D}\Delta f_{\text{Doppler}}(s)ds\right) \\ &= N_0 + round\left(\frac{1}{2\pi}(\Delta\varphi_0 - \Delta\varphi)\right) \\ &= N_0 + round\left(\frac{D}{\lambda}(\sin(\theta) - \sin(\theta_0))\right) \end{aligned} \tag{7.52}$$

For the case. $\theta_0 = 0$, (7.52) reduces to

$$N = round\left(\frac{D}{\lambda}\sin(\theta)\right) \tag{7.53}$$

which shows the correct result: The integer ambiguity is a function of the scan angle for a target moving in a circular trajectory with the radar located at the center of the circle. Note that, when $D = \lambda/2$, we have $N = 0$ for $-\pi/2 < \theta < \pi/2$, which is the correct result for spatial Nyquist antenna separation.

Equation (7.51) is a fundamental result that uses Doppler frequency differences and phase differences to resolve the integer ambiguity for angle estimation. The Doppler difference measurement accuracy is inversely proportional to the radar dwell time T_{Dwell} and the square root of *SNR*. Assuming that the receiver bandwidth is matched to the waveform bandwidth,

$$\sigma_{\text{Doppler}} = \frac{1}{T_{\text{dwell}}\sqrt{2SNR}} \tag{7.54}$$

If we define the time interval t as

$$t = M \cdot T_{\text{dwell}} \tag{7.55}$$

then we can write

$$\begin{aligned} \int_{s=0}^{t} \Delta f_{\text{Doppler}}(s)ds &= T_{\text{dwell}}\sum_{n=0}^{M}\left(\Delta f_{\text{Doppler}}(nT_{\text{dwell}}) + \varepsilon_n\right) \\ &= T_{\text{dwell}}\sum_{n=0}^{M}\Delta f_{\text{Doppler}}(nT_{\text{dwell}}) + T_{\text{dwell}}\sum_{n=0}^{M}\varepsilon_n \end{aligned} \tag{7.56}$$

Let $\varepsilon = T_{\text{dwell}}\sum_{n=0}^{M}\varepsilon_n$ then

$$\sigma_\varepsilon = \frac{T_{\text{dwell}}\sqrt{M}}{T_{\text{dwell}}\sqrt{SNR}} = \frac{\sqrt{M}}{\sqrt{SNR}} \tag{7.57}$$

Essentially, (7.57) defines the limit for the integrated time t so that (7.51) is an accurate estimate for N. The phase difference $\Delta\varphi$ error has been shown previously to be

$$\sigma_{\Delta\varphi} = \frac{1}{\sqrt{SNR}} \tag{7.58}$$

The Doppler difference error in (7.57) is usually the dominant error source because it is integrated over time. Equation (7.51) is an accurate estimate of N provided that the SNR is sufficiently large and

$$\sigma_{\varepsilon} = \frac{\sqrt{M}}{\sqrt{SNR}} \ll \pi \tag{7.59}$$

Or the phase error requirement is

$$\sigma_{\varphi} = \frac{1}{\sqrt{2SNR}} \ll \frac{\pi}{\sqrt{2M}} \tag{7.60}$$

Comparing (7.57) and (7.60), it can be seen that, for small dwell times ($T \ll 1$), the dominant error source is the error due to Doppler difference. The key is making sure that SNR is sufficiently high that the number of ambiguity slips is a small fraction of the total number of measurements. Equation (7.57) shows that reasonably high SNR is required in order that the ambiguity slips be a small fraction for the total number of measurements. Experience shows that $SNR > 26$ dB is sufficient to meet this criterion, but when $SNR > 36$ dB essentially no ambiguity slips occur.

Figure 7.9 shows the integer changes determined by (7.51) for a moving target starting at a 10-km range and $\theta = 0°$. The target speed is 100 m/s, and the heading is 225° with a radar dwell time of 10 ms. Figure 7.9 shows the errors in estimating N with $SNR = 30$ dB versus the unerred estimate with $SNR = 40$ dB. The figure indicates that (7.51) should be applied over small time intervals t for low SNR. Resetting the initial phase difference and Doppler difference does not allow the error to grow over time when the integration time is small.

Equation (7.51) was derived for 1-D angle estimation but can be easily extended to 2-D angle estimation. For 2-D, two integers must be resolved, N_1 and N_2. In the next section, we derive the angle estimation algorithms for a 2-D interferometer with three antennas. The result in (7.51) can be easily applied to 2-D angle ambiguity resolution where Doppler differences between pairs of antennas can be used to determine N_1 and N_2 in a manner similar to that of (7.51).

7.4 Angle-of-Arrival Determination

Assume that interferometer antennas are configured in an equilateral triangle configuration. Also assume that the three interferometer antennas are located in the x-z plane with z being up and that y is perpendicular to the array plane. Let d define the

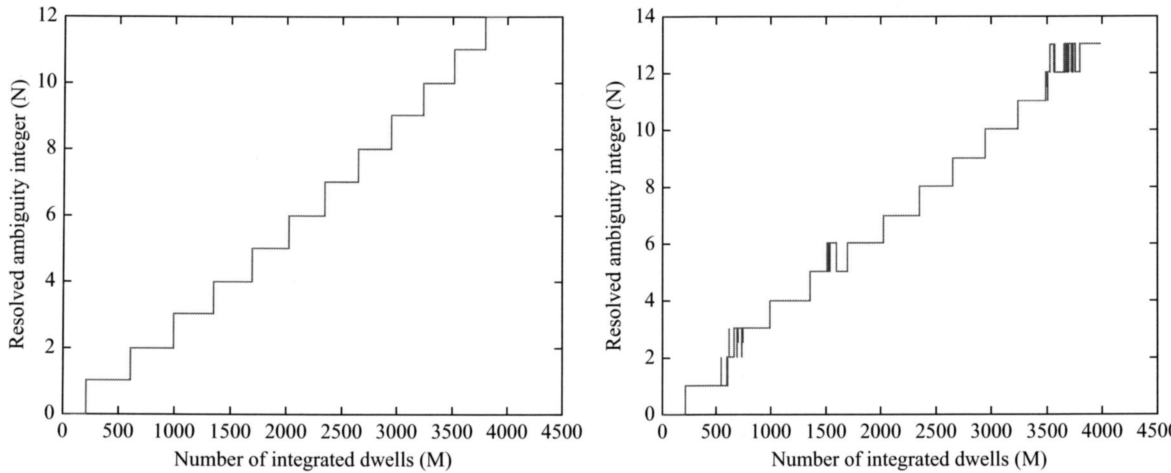

Figure 7.9. Ambiguity Integers Determined by (7.51) for Moving Target: SNR = 40 dB (left) and SNR = 30 dB (right)

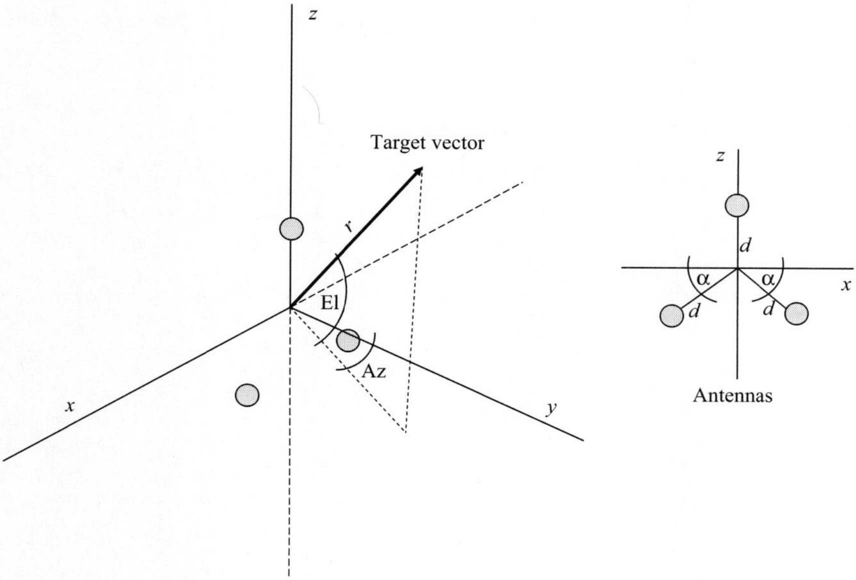

Figure 7.10. Coordinate Frame for Measuring Target Angle

distance from each antenna to the origin, and let α be the rotation angle of the bottom antennas with respect to the x-axis, then the antenna coordinates are as follows:

$$\begin{aligned}\vec{a}_1 &= [-d\cos(\alpha) \quad 0 \quad -d\sin(\alpha)]\\ \vec{a}_2 &= [d\cos(\alpha) \quad 0 \quad -d\sin(\alpha)]\\ \vec{a}_3 &= [0 \quad 0 \quad d]\end{aligned} \tag{7.61}$$

For a more general rotated antenna configuration, see Appendix H.

Let $\vec{r}_t$ be the target position vector located at azimuth angle Az, elevation angle El, and range r.

$$\vec{r}_t = [r\sin(Az)\cos(El) \quad r\cos(Az)\cos(El) \quad r\sin(El)] \tag{7.62}$$

Figure 7.10 illustrates that antenna coordinate frame. The interferometer makes a phase measurement at each antenna:

$$\phi_i = k\|\vec{r}_t - \vec{a}_i\| \tag{7.63}$$

where $k = \frac{2\pi}{\lambda}$.

7.4.1 First-Order Angle Estimation

We first compute the three phase differences associated with the three antennas using a first-order Taylor series approximation for phase as a function of angle in (7.63). The first-order Taylor series approximation of phase difference is equivalent to assuming that the target is in the far field of the interferometer antenna, or, in other words, the phase front across the antennas is sufficiently flat (approximately zero curvature).

$$\begin{aligned}
\Delta\phi_{12} &= \phi_1 - \phi_2 = 2kd\sin(Az)\cos(El)\cos(\alpha) \\
\Delta\phi_{13} &= \phi_1 - \phi_3 = dk\sin(Az)\cos(El)\cos(\alpha) + dk(1+\sin(\alpha))\sin(El) \\
\Delta\phi_{23} &= \phi_2 - \phi_3 = -dk\sin(Az)\cos(El)\cos(\alpha) + dk(1+\sin(\alpha))\sin(El)
\end{aligned} \tag{7.64}$$

The two-dimensional angle-of-arrival (*Az* and *El*) is determined as follows:

$$\begin{aligned}
\sin(El) &= \frac{\Delta\varphi_{13} + \Delta\varphi_{23}}{2dk(1+\sin(\alpha))} \\
\cos(El) &= \sqrt{1-\sin^2(El)} \\
\sin(Az) &= \frac{\Delta\varphi_{12}}{2dk\cos(El)\cos(\alpha)}
\end{aligned} \tag{7.65}$$

Equation (7.65) basically converts phase differences to angle-of-arrival measurements module, resolving the angle ambiguity in both angles.

See Appendix J for implementation in MATLAB® code.

7.4.2 Second-Order Angle Estimation

For some applications, the first-order angle estimation is not sufficiently accurate, possibly due to a variety of reasons, such as the target not being in the far field or being in the far field with a small phase curvature, which imposes an unacceptable error due to angle accuracy requirements. As an example, fire control radar systems that are designed to guide an interceptor to engage a target sometimes require angle accuracy where the first-order estimate is not sufficient. Again, we compute the three phase differences associated with the three antennas, except we use a second-order Taylor series approximation for the phase functions in (7.63).

$$\begin{aligned}
\Delta\phi_{12} &= \phi_1 - \phi_2 \\
&= 2dk\sin(Az)\cos(El)\cos(\alpha) - \frac{2d^2k}{r}\sin(Az)\cos(El)\sin(El)\sin(\alpha)\cos(\alpha) \\
\Delta\phi_{13} &= \phi_1 - \phi_3 \\
&= dk\sin(Az)\cos(El)\cos(\alpha) + dk(1+\sin(\alpha))\sin(El) \\
&\quad - \frac{d^2k}{2r}\sin^2(Az)\cos^2(El)\cos^2(\alpha) - \frac{d^2k}{2r}\sin^2(El)\sin^2(\alpha) \\
&\quad + dk\sin(Az)\cos(El)\sin(El)\sin(\alpha)\cos(\alpha) + \frac{d^2k}{2r}\sin^2(El)
\end{aligned}$$

$$\begin{aligned}\Delta\phi_{23} &= \phi_2 - \phi_3 \\ &= -dk\sin(Az)\cos(El)\cos(\alpha) + dk(1+\sin(\alpha))\sin(El) \\ &\quad - \frac{d^2k}{2r}\sin^2(Az)\cos^2(El)\cos^2(\alpha) - \frac{d^2k}{2r}\sin^2(El)\sin^2(\alpha) \\ &\quad - dk\sin(Az)\cos(El)\sin(El)\sin(\alpha)\cos(\alpha) + \frac{d^2k}{2r}\sin^2(El) \end{aligned} \tag{7.66}$$

Because the preceding equalities are nonlinear in the trigonometric expressions for angle, it is not possible to derive a closed-form expression for angle as in (7.65). For second-order angle estimation, angle is determined by the following iterative algorithm:

- Step 1: Initialize.

$$\sin(El) = \frac{\Delta\phi_{13} + \Delta\phi_{23}}{2dk(1+\sin(\alpha))} \tag{7.67}$$

$$\cos(El) = \sqrt{1 - \sin^2(El)} \tag{7.68}$$

- Step 2: Compute *Az*.

$$\sin(Az) = \frac{\Delta\phi_{12}}{2dk\cos(E)\cos(\alpha) - \dfrac{2d^2k}{r}\cos(El)\sin(El)\sin(\alpha)\cos(\alpha)} \tag{7.69}$$

- Step 3: Compute *El*.

$$\sin(El) = \frac{-B + \sqrt{B^2 - 4AC}}{2A} \tag{7.70}$$

$$A = \frac{d^2k}{r}\cos^2(\alpha)\left(1 + \sin^2(Az)\right) \tag{7.71}$$

$$B = 2dk(1 + \sin(\alpha)) \tag{7.72}$$

$$C = -\frac{d^2k}{r}\sin^2(Az)\cos^2(\alpha) - \Delta\phi_{13} - \Delta\phi_{23} \tag{7.73}$$

$$\cos(El) = \sqrt{1 - \sin^2(El)} \tag{7.74}$$

- Step 4: Return to Step 2, and stop after two iterations.

  See Appendix J for implementation in MATLAB® code.

7.4.3 Interferometer Angle Measurements for Distributed Transmit/Receive Antennas

Assume that interferometer transmit antennas are configured in a triangle configuration and that the origin is chosen to be equidistant from each antenna. Also assume that the three interferometer transmit antennas are located in the *x-y* plane with *y* being up and that z is perpendicular to the array plane. Let d define the distance from any antenna to the origin. Then the transmit antenna coordinates are as follows:

$$
\begin{aligned}
ant_1 &= \begin{bmatrix} d & 0 & 0 \end{bmatrix} \\
ant_2 &= \begin{bmatrix} -\frac{d}{2} & \frac{d\sqrt{3}}{2} & 0 \end{bmatrix} \\
ant_3 &= \begin{bmatrix} -\frac{d}{2} & -\frac{d\sqrt{3}}{2} & 0 \end{bmatrix}
\end{aligned}
\tag{7.75}
$$

The three receive antennas are also configured in a triangle that is a reflection of the transmit antenna triangle. Each receive antenna is the negative of a transmit antenna and is located at the antipodal position with respect to the circle of radius d.

Let $\vec{r}_t$ be the target position vector located at azimuth angle Az, elevation angle El, and range r.

$$\vec{r}_t = \begin{bmatrix} r\sin(Az)\cos(El) & r\sin(El) & r\cos(Az)\cos(El) \end{bmatrix} \tag{7.76}$$

Figure 7.11 illustrates the transmit and receive antenna coordinate frame where the transmit antennas are light gray circles and the receive antennas are dark gray circles. The interferometer makes a phase measurement at each receive antenna from a signal that is transmitted from one of the transmit antennas. We denote φ_{ij} as the phase resulting from transmitting from antenna i and receiving at antenna j. Now define

$$k = \frac{2\pi}{\lambda} \tag{7.77}$$

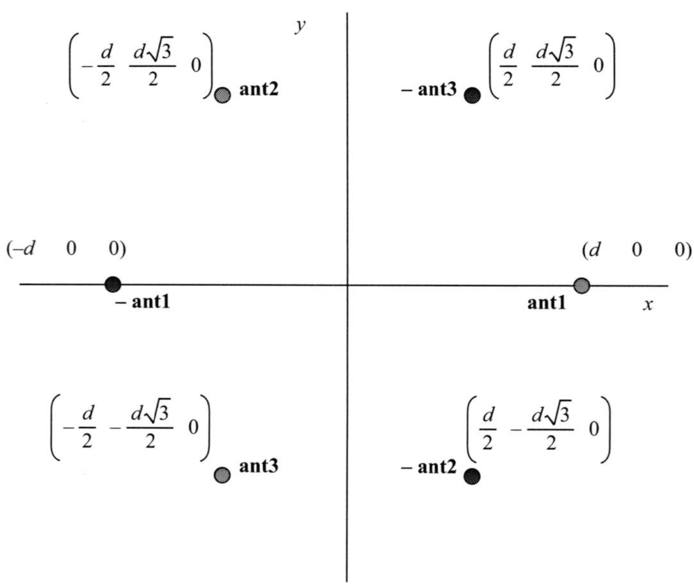

Figure 7.11. Hexagonal Array Structure and Coordinate Frame for Measuring Target Angle

The first three phases result from transmitting and receiving in a clockwise motion around the array.

$$\begin{aligned}\varphi_{13} &= k(\|\vec{r}_t - ant_1\| + \|\vec{r}_t + ant_3\|)\\ \varphi_{21} &= k(\|\vec{r}_t - ant_2\| + \|\vec{r}_t + ant_1\|)\\ \varphi_{32} &= k(\|\vec{r}_t - ant_3\| + \|\vec{r}_t + ant_2\|)\end{aligned} \tag{7.78}$$

The second three phases result from transmitting and receiving in a counter-clockwise motion around the array.

$$\begin{aligned}\varphi_{12} &= k(\|\vec{r}_t - ant_1\| + \|\vec{r}_t + ant_2\|)\\ \varphi_{31} &= k(\|\vec{r}_t - ant_3\| + \|\vec{r}_t + ant_1\|)\\ \varphi_{23} &= k(\|\vec{r}_t - ant_2\| + \|\vec{r}_t + ant_3\|)\end{aligned} \tag{7.79}$$

The last three phases result from transmitting and receiving from antipodal (opposite) antennas.

$$\begin{aligned}\varphi_{11} &= k(\|\vec{r}_t - ant_1\| + \|\vec{r}_t + ant_1\|)\\ \varphi_{22} &= k(\|\vec{r}_t - ant_2\| + \|\vec{r}_t + ant_2\|)\\ \varphi_{33} &= k(\|\vec{r}_t - ant_3\| + \|\vec{r}_t + ant_3\|)\end{aligned} \tag{7.80}$$

7.4.3.1 First-Order Phase Differences

$$\varphi_{32} - \varphi_{23} = 2kd\sqrt{3}\sin(El) \tag{7.81}$$

$$\varphi_{13} - \varphi_{21} = -3kd\sin(Az)\cos(El) \tag{7.82}$$

$$\varphi_{12} - \varphi_{31} = -3kd\sin(Az)\cos(El) \tag{7.83}$$

First-Order Angle

$$\begin{aligned}\sin(El) &= \frac{\varphi_{32} - \varphi_{23}}{2kd\sqrt{3}}\\ \cos(El) &= \sqrt{1 - \sin^2(El)}\\ \sin(Az) &= -\frac{(\varphi_{13} - \varphi_{21}) + (\varphi_{12} - \varphi_{31})}{6kd\cos(El)}\end{aligned} \tag{7.84}$$

See Appendix J for implementation in MATLAB® code.

Angle Accuracy

We now determine the angle accuracy for the orthogonal interferometer and compare the accuracy to the conventional and unconventional interferometers using the hexagonal array structure defined in Figure 7.11. Using the expression for El in (7.84), we have

$$\sigma_{El} = \frac{\sqrt{2}\sigma_\varphi}{2kd\sqrt{3}} = \frac{\sqrt{2}}{2kd\sqrt{3SNR}} = \frac{\lambda}{4\pi d\sqrt{3SNR}} \tag{7.85}$$

and for $El \ll 1$

$$\sigma_{Az} = \frac{2\sigma_\varphi}{6kd} = \frac{1}{3kd\sqrt{2SNR}} = \frac{\lambda}{6\pi d\sqrt{2SNR}} \tag{7.86}$$

Now define the baseline D as follows:

$$D = \|ant1 - ant2\| = \|ant2 - ant3\| = \|ant3 - ant1\| = \sqrt{3}d \tag{7.87}$$

We rewrite the expression for elevation error.

$$\sigma_{El} = \frac{\lambda}{4\pi d\sqrt{3SNR}} = \frac{\lambda}{4\pi D\sqrt{SNR}} \tag{7.88}$$

$$\sigma_{Az} = \frac{\lambda}{6\pi d\sqrt{2SNR}} = \frac{\lambda}{2\sqrt{6}\pi D\sqrt{SNR}} \tag{7.89}$$

For a conventional interferometer with two antennas separated by a distance D, the angle accuracy is

$$\sigma_{El}^{CI} = \frac{\lambda}{2\pi D\sqrt{SNR}} \tag{7.90}$$

$$\sigma_{Az}^{CI} = \frac{\lambda}{\sqrt{6}\pi D\sqrt{SNR}} \tag{7.91}$$

Comparing the orthogonal interferometer with the conventional interferometer accuracy for both elevation and azimuth, the orthogonal interferometer achieves an accuracy improvement of a factor of 2 over the conventional interferometer and the unconventional interferometer using the hexagonal distributed array architecture, as shown in Section 6.3. Also, comparing azimuth and elevation accuracy,

$$\frac{\sigma_{Az}}{\sigma_{El}} = \frac{2}{\sqrt{6}} \tag{7.92}$$

Thus the azimuth accuracy is degraded relative to the elevation accuracy by a factor of $2/\sqrt{6}$.

Unambiguous Angle Interval

We now determine the unambiguous angle interval for both azimuth and elevation for the antenna architecture shown in Figure 7.11. For this architecture, the interferometer baseline is defined as $D = \sqrt{3}d$, and the unambiguous region is defined by

$$\sin(\theta) = \frac{\lambda(\varphi_{23} - \varphi_{32})}{2\pi D} = \frac{2kd\sqrt{3}\sin(El)}{kD} = 2\sin(El) \tag{7.93}$$

$$-\frac{\lambda}{2D} \leq \sin(\theta) \leq \frac{\lambda}{2D} \tag{7.94}$$

$$-\frac{\lambda}{2D} = -\frac{\lambda}{2\sqrt{3}d} \leq 2\sin(El) \leq \frac{\lambda}{2\sqrt{3}d} = \frac{\lambda}{2D} \tag{7.95}$$

and the unambiguous interval for elevation is defined by

$$-\frac{\lambda}{4\sqrt{3}d} \leq \sin(El) \leq \frac{\lambda}{4\sqrt{3}d} \tag{7.96}$$

For azimuth, we use (7.82).

$$\sin(\psi) = \frac{\lambda(\varphi_{21} - \varphi_{13})}{2\pi D} = \frac{3kd\sin(Az)\cos(El)}{kD} = \sqrt{3}\sin(Az)\cos(El) \tag{7.97}$$

$$-\frac{\lambda}{2D} \leq \sin(\psi) \leq \frac{\lambda}{2D} \tag{7.98}$$

$$-\frac{\lambda}{2D} = -\frac{\lambda}{2\sqrt{3}d} \leq \sqrt{3}\sin(Az)\cos(El) \leq \frac{\lambda}{2\sqrt{3}d} = \frac{\lambda}{2D} \tag{7.99}$$

The unambiguous interval for azimuth is defined by

$$-\frac{\lambda}{6d\cos(El)} \leq \sin(Az) \leq \frac{\lambda}{6d\cos(El)} \tag{7.100}$$

Notice that the unambiguous interval for azimuth depends on the elevation and that, for elevation, the unambiguous interval for the orthogonal interferometer using the defined antenna architecture is half that for a conventional interferometer, which agrees with the result for the two-antenna orthogonal interferometer derived in Section 6.3.

7.4.3.2 Second-Order Angle Estimation Using Antipodal Array

It turns out that the symmetry using antipodal antennas cancels the first-order phase differences. However, the antipodal array can be used to compute a second-order angle estimate.

$$\begin{aligned}
\Delta_{12} &= \varphi_{11} - \varphi_{22} = \frac{3kd^2}{4r}\sin^2(Az)\cos^2(El) + \sqrt{3}\sin(Az)\cos(El)\sin(El) + \frac{3kd^2}{4r}\sin^2(El) \\
\Delta_{13} &= \varphi_{11} - \varphi_{33} = \frac{3kd^2}{4r}\sin^2(Az)\cos^2(El) - \sqrt{3}\sin(Az)\cos(El)\sin(El) + \frac{3kd^2}{4r}\sin^2(El) \\
\Delta_{23} &= \varphi_{22} - \varphi_{33} = \frac{\sqrt{3}kd^2}{r}\sin(Az)\cos(El)\sin(El)
\end{aligned} \tag{7.101}$$

Using (7.101) for Δ_{23} to eliminate the middle term in the expression for Δ_{12}, we have fourth-order polynomial in sin(*El*).

$$\frac{3d^2k}{2r}\left(\frac{\sqrt{3}d^2k}{r}\right)^2 \sin^4(El) - (\Delta_{12} + \Delta_{13})\left(\frac{\sqrt{3}d^2k}{r}\right)^2 \sin^2(El) - \frac{3d^2k}{2r}\Delta_{23}^2 = 0 \tag{7.102}$$

Because (7.102) is quadratic in $\sin^2(El)$, we have

$$\sin^2(El) = \frac{-B + \sqrt{B^2 - 4AC}}{2A} \tag{7.103}$$

$$A = \frac{3d^2k}{2r}\left(\frac{\sqrt{3}d^2k}{r}\right)^2 \tag{7.104}$$

$$B = -(\Delta_{12} + \Delta_{13})\left(\frac{\sqrt{3}d^2k}{r}\right)^2 \tag{7.105}$$

$$C = -\frac{3d^2k}{2r}\Delta_{23}^2 \tag{7.106}$$

$$\sin(Az) = \frac{\Delta_{23}}{\frac{\sqrt{3}d^2k}{r}\sin(El)\cos(El)} \tag{7.107}$$

These expressions constitute a closed-form calculation of the second-order angle estimates. However, the square of sin(*El*) is computed, which introduces a sign ambiguity into the estimate of *El*. This ambiguity can be resolved using any of the first-order estimates for *El*. If we denote the clockwise estimate for elevation angle by El_{CW}, then we have

$$\begin{aligned} \sin(El) &= sign(El_{CW})\sqrt{\frac{-B + \sqrt{B^2 - 4AC}}{2A}} \\ \sin(Az) &= \frac{\Delta_{23}}{\frac{\sqrt{3}d^2k}{r}\sin(El)\cos(El)} \end{aligned} \tag{7.108}$$

See Appendix J for implementation in MATLAB® code.

Far Field Versus Near Field

For a target in the near field, the second-order estimate for angle is more precise than any of the first-order methods. However, it has been observed that, when the two first-order estimates are averaged, the precision of the averaged estimates is generally better than the second-order estimate even when the target is in the near field.

It is recommended that only the averaged first-order estimates be used for angle estimation and not the second-order estimates. The second-order estimate can be averaged with the estimates from the two first-order methods to improve accuracy for low *SNR*. However, the caution is that the second-order angle error distribution is not Gaussian and not mean zero due to nonlinear expressions involving phase.

7.5 LFM Stretch Processing

Stretch processing is a processing technique applied to linear frequency modulation (LFM) waveforms that achieves the matched filtered response with a reduced sampling rate, significantly less than the IF bandwidth. The basic principle [7] is derived from the fact that the product (or mixing) of two waveforms results in two new waveforms at half the power of the original waveforms. One resulting waveform is at a frequency equal to the difference in frequencies of the original waveforms, and the other waveform frequency is the sum of the two frequencies. An antialiasing filter is used to isolate the waveform whose frequency is the difference in frequencies.

Figure 7.12 provides an illustration of stretch processing. The mixing operation yields a complex tone whose frequency (often termed the "beat frequency") is proportional to the time delay t_d to the target. The duration of the complex tone is equal to the duration of the transmit pulse, provided the local oscillator is allowed to sweep over a duration equal to the pulse width plus the receive window. Just as in correlation, or two-pass, processing, if the local oscillator does not sweep over this time period, a loss results on signal energy and range resolution.

$$\delta R = \frac{cT}{2BT_{\text{shortened}}} \tag{7.109}$$

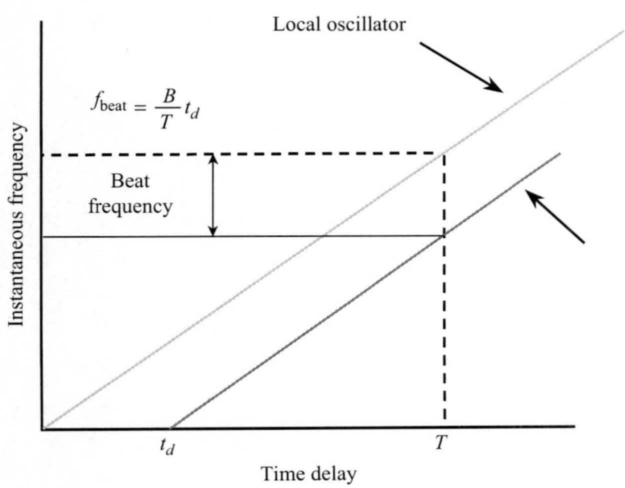

Figure 7.12. Illustration of Stretch Processing

The local oscillator is started at a time delay corresponding to the beginning of the range window.

In stretch processing, the number of digital samples is determined by the IF antialiasing filter that sets the range window. The IF cutoff frequency and the range window are related by the following equation:

$$IF_{\text{cutoff}} = \frac{B}{T}\frac{2\Delta R}{c} \tag{7.110}$$

Figure 7.13 illustrates a typical stretch processing implementation. Notice that only one fast Fourier transform (FFT) is required to implement stretch processing, and, as a result, stretch processing is sometimes referred to as "one-pass processing." The number of digital samples is determined by

$$N = F_s\left[T + \frac{2\,\Delta R_{IF}}{c}\right] \tag{7.111}$$

where $F_s \geq If_{\text{cutoff}}$.

Because two FFTs are required to implement matched filter processing, the total FFT operations are twice the number for matched filter processing than for stretch processing.

7.5.1 Angle-of-Arrival and Stretch Processing

One method of extracting range information from an LFM waveform is to implement stretch processing as opposed to digital match filtering. Stretch processing has the benefit of substantially reducing bandwidth and *ADC* requirements because the

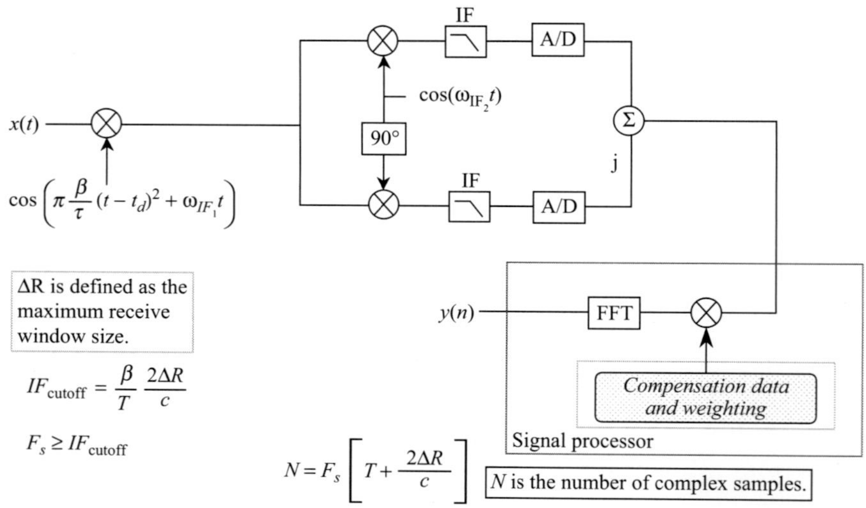

Figure 7.13. Stretch Processing Implementation

received signal is mixed with the return signal. Stretch processing is enabled by a homodyne receiver that mixes two signals to reduce sampling requirements. Because frequency is a linear function of time, phase becomes a quadratic function of time. For a two-antenna interferometer, we define a reference signal in each receiver s_0 and the received target signals s_1 and s_2 as follows:

$$s_0 = \exp\left(2\pi j\left(ft + \frac{1}{2}at^2\right)\right) \tag{7.112}$$

$$s_1 = \exp\left(2\pi j\left(f(t + t_1) + \frac{1}{2}a(t + t_1)^2\right)\right) \tag{7.113}$$

$$s_2 = \exp\left(2\pi j\left(f(t + t_2) + \frac{1}{2}a(t + t_2)^2\right)\right) \tag{7.114}$$

$$t_1 = \frac{R_1}{c} \tag{7.115}$$

$$t_2 = \frac{R_2}{c} \tag{7.116}$$

The stretch processing technique can be described as follows:

$$\begin{aligned} s_1 \otimes s_0' &= \exp\left(2\pi j\left(ft_1 + \frac{1}{2}a\left((t_1 + t)^2 - t^2\right)\right)\right) \\ &= \exp\left(2\pi j\left(ft_1 + \frac{1}{2}a\left(t_1^2 + 2t_1t\right)\right)\right) \end{aligned} \tag{7.117}$$

$$\begin{aligned} s_2 \otimes s_0' &= \exp\left(2\pi j\left(ft_2 + \frac{1}{2}a\left((t_2 + t)^2 - t^2\right)\right)\right) \\ &= \exp\left(2\pi j\left(ft_2 + \frac{1}{2}a\left(t_2^2 + 2t_2t\right)\right)\right) \end{aligned} \tag{7.118}$$

The phase is a function of time and is described by

$$\varphi_1(t) = 2\pi\left(ft_1 + \frac{1}{2}a\left(t_1^2 + 2t_1t\right)\right) \tag{7.119}$$

$$\varphi_2(t) = 2\pi\left(ft_2 + \frac{1}{2}a\left(t_2^2 + 2t_2t\right)\right) \tag{7.120}$$

We will describe two methods of extracting phase.

Method 1: Frequency

$$f_1 = \frac{d\varphi_1}{dt} = 2\pi a t_1 \tag{7.121}$$

$$f_2 = \frac{d\varphi_2}{dt} = 2\pi a t_2 \tag{7.122}$$

$$\begin{aligned}\Delta f &= f_2 - f_1 = 2\pi a(t_2 - t_1)\\ &= 2\pi a\left(\frac{R_2}{c} - \frac{R_1}{c}\right)\\ &= 2\pi a\left(\frac{R_2 - R_1}{c}\right)\\ &= 2\pi a\left(\frac{D\sin(\theta)}{c}\right)\end{aligned} \tag{7.123}$$

$$\sin(\theta) = \frac{c\Delta f}{2\pi D a} \tag{7.124}$$

Method 2: Phase

$$\varphi_1(t_1) = 2\pi\left(f t_1 + \frac{3}{2}a t_1^2\right) \tag{7.125}$$

$$\varphi_2(t_2) = 2\pi\left(f t_2 + \frac{3}{2}a t_2^2\right) \tag{7.126}$$

$$\begin{aligned}\Delta\varphi &= 2\pi\left(f(t_2 - t_1) + \frac{3}{2}a(t_2^2 - t_1^2)\right)\\ &= 2\pi\left(f\frac{D\sin(\theta)}{c} + \frac{3}{2}a\frac{D\sin(\theta)}{c}\frac{2R}{c}\right)\end{aligned} \tag{7.127}$$

Solving for $\sin(\theta)$,

$$\begin{aligned}\sin(\theta) &= \frac{c\Delta\varphi}{2\pi D f}\left(1 + 3a\frac{R}{fc}\right)^{-1}\\ &= \frac{\lambda\Delta\varphi}{2\pi D}\left(1 + 3a\frac{R}{fc}\right)^{-1}\\ &\approx \frac{\lambda\Delta\varphi}{2\pi D}\end{aligned} \tag{7.128}$$

where $R = \frac{R_1 + R_2}{2}$.

Note that, in Method 2, because phase is a function of time, it matters as to when the phase is measured. When sampled at t_1 and t_2, there is a residual phase term that depends on the range to the target. For RF applications where f is large and c is the speed of light, this term is negligible, and the two methods are equivalent. However, for acoustic propagation, the term becomes significant. Equating methods 1 and 2, we have

$$\frac{c\Delta f}{a} = \frac{c\Delta\varphi}{f} \tag{7.129}$$

or

$$\Delta\varphi = \frac{f}{a}\Delta f = \frac{f}{a}2\pi a(t_2 - t_1) = 2\pi f t_2 - 2\pi a f t_1 = \varphi_1 - \varphi_2 \tag{7.130}$$

7.5.2 CW/FMCW Homodyne Processing

As mentioned in Sections 3.2 and 5.1.2, the CW and FMCW waveforms can be implemented in a homodyne receiver due to the fact that they operate at 100 percent duty, making the analog reference available at all times. Figure 7.14 shows a typical homodyne radar design for an interferometer with three receive channels. The design is typical of sports radar applications that require enhanced tracking accuracy. For sports applications, it is not uncommon to track in Doppler as well in order to measure the velocity and spin rate of the ball. For interferometer angle measurement, the receive channels must be calibrated to ensure that phase is balanced. Typically, this requires calibrating the radar using a known external source.

In Figure 7.14, the reference signal is mixed with the incoming signals at each receive antenna to mix down to a baseband signal in each channel. These baseband signals must be balanced for a known target in the far field. One way to achieve this calibration is to inject an RF signal behind the antenna in each receive channel in order to perform an internal receive calibration. Because for sports applications the

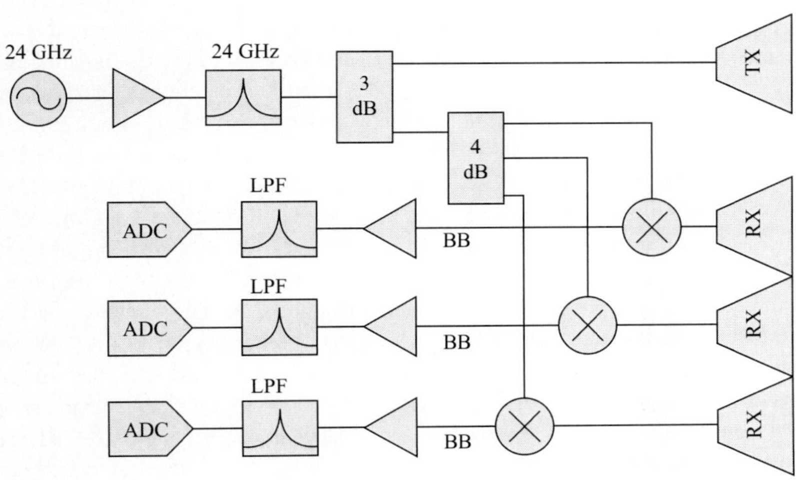

Figure 7.14. 24 GHz Interferometric Homodyne Radar Design

antennas are typically horns or patches, the antennas can be calibrated separately and should maintain calibration over long periods of time. The internal calibration with the external antenna calibration should enable calibrating the entire interferometer system.

7.6 Transmit Interferometry Calibration

One of the issues associated with an orthogonal interferometer (OI) using multiple distributed transmit/receive arrays is that all of the transmitted waveforms from each of the arrays must be time aligned. Errors in time alignment lead directly to phase errors among the channels which translate into angle-of-arrival biases. All phased arrays calibrate and align the arrays on receive, but few to none calibrate and align on transmit. The primary reason is that, for a single phased array, all that really matters is the far-field antenna pattern. In other words, most phased arrays are primarily concerned with how energy is adding up in the far field, primarily to maintain sufficient target sensitivity while operating in various environments that might include clutter or jamming. Therefore, main beam transmit gain is usually the primary concern with root-mean-square sidelobe levels of the transmit antenna pattern usually being a secondary concern.

Phased arrays that use monopulse for angle-of-arrival determination require only that the transmit array provide sufficient energy on target for angle accuracy. Likewise, a conventional interferometer that uses a separate transmit-only array and multiple receive-only arrays requires the transmit array to provide only sufficient energy on target. The target angle-of-arrival is completely determined by the phases from signals on each of the return paths from the target to each of the receive arrays. For a conventional interferometer, the transmit path is common because there is only one transmit array, and angle-of-arrival error is not sensitive to the fact that the transmit timing changes over scan angle. However, the orthogonal interferometer is sensitive to relative timing errors among transmit arrays because it is using the total signal path to determine reduce angle-of-arrival errors.

As opposed to developing an elaborate a priori transmit calibration and alignment procedure that will need to be repeated frequently due to environmental changes, transmit calibration and alignment can actually be accomplished in real time using either the actual target of interest or targets of opportunity. This real-time calibration and alignment procedure is possible because an orthogonal interferometer inherently contains multiple conventional interferometers with single transmit and multiple receive arrays. Figure 7.15 shows an orthogonal interferometer consisting of three transmit/receive apertures. The interferometer configuration on the right shows a conventional interferometer (CI) with one transmit array and two receive arrays, whereas the configuration of the left shows an unconventional interferometer (UI) with two transmit arrays and one receive array. Because the signal phase differences for a given target are the same for both configurations, the angle-of-arrival accuracy is equivalent for both interferometer architectures. Most importantly, the CI phase differences do not contain any relative transmit timing errors, whereas the UI contains transmit errors but no receive errors.

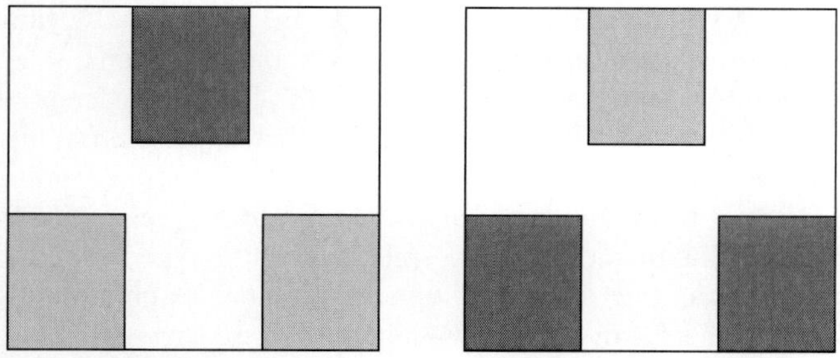

Figure 7.15. Conventional (right) and Unconventional (left) Interferometer Array Architectures Contained Within a Three-Array Orthogonal Interferometer; Light Gray Indicates an Array in Transmit Mode, and Dark Gray Indicates an Array in Receive Mode

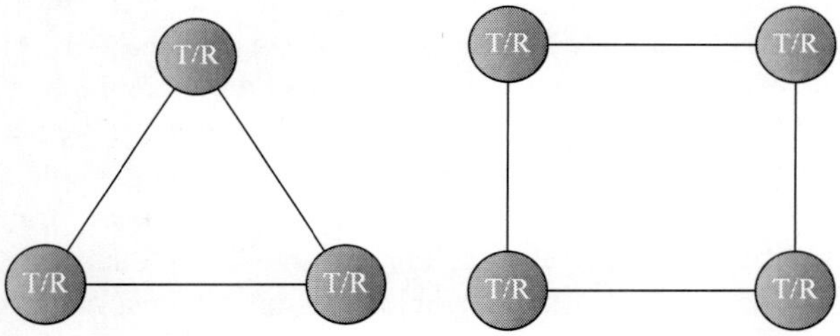

Figure 7.16. Orthogonal Interferometer Architectures with Three Antennas (left) and Four Antennas (right)

The OI experiences errors on both transmit and receive, thereby making it more difficult to calibrate and align the OI channels. Consider an orthogonal interferometer with N ($N = 3$ or 4) antennas, such as the interferometer architectures shown in Figure 7.16. The phases at any antenna can be represented by

$$\varphi_i = 2\pi f\left(\frac{2r_i}{c}\right) + \tau_i + \upsilon_i \qquad i = 1{:}4 \tag{7.131}$$

where f is the operating frequency, r_i is the range from each antenna to the target, τ_i is the transmit error, and υ_i is the receive error. For each antenna i and j, we form the phase differences to compute angle-of-arrival,

$$\Delta\varphi_{ij} = \varphi_i - \varphi_j = 2\pi f\left(\frac{2(r_i - r_j)}{c}\right) + \tau_i - \tau_j + \upsilon_i - \upsilon_j \tag{7.132}$$

Two basic types of bias errors affect angle estimation: internal system bias errors and external system bias errors. Examples of internal system bias errors are phase errors due to cable and receiver channel mismatch, ADC timing errors, transmit timing errors, and antenna amplifier channel mismatch. The dominant source for external system bias error is multipath. Calibration consists of eliminating the internal system bias errors in (7.132) for each phase difference. Receive calibration can be accomplished in nearly real time by signal injection behind each of the antennas prior to the low noise amplifier and by a precise calibration using near- or far-field illumination. Signal injection can be performed during the operation of the radar interferometer, whereas antenna calibration is generally performed once at the factory. Thus we assume that receive calibration has significantly reduced the receive bias errors to a negligible error.

$$\upsilon_i - \upsilon_j = \varepsilon_{ij} \qquad |\varepsilon_{ij}| \ll 1 \tag{7.133}$$

We now show that transmit bias calibration can be accomplished using the inherent interferometer architectures that make up an OI radar. Any OI architecture also consists of multiple CI and UI architectures embedded within the multiple transmit/receive antenna arrays making up the OI architecture. The OI array architecture is designed to transmit and receive N orthogonal waveforms simultaneously from all N antenna arrays, creating N channels at each antenna for a total of N^2 channels. The transmitted waveform from one antenna is received simultaneously from all antennas, creating N number of CI architectures. In addition, all N transmitted waveforms are received simultaneously at each antenna to create N number of UI architectures. The flexibility of the OI architecture to implement multiple architectures provides an inherent capability to calibrate transmit errors.

We assume that the OI consists of four transmit/receive antenna arrays. First consider the four CI systems using a single transmit antenna and the three other remaining antennas. The phases for each of these antennas for each of the four CI radars is given by

$$\gamma_i^k = 2\pi f\left(\frac{r_i + r_k}{c}\right) + \tau_i + \upsilon_k \qquad k = 1:4, \quad i \neq k \tag{7.134}$$

Here we use the lower index to indicate transmit antennas and the upper case index to indicate receive antennas. We can also form four UI systems using three transmit antennas and the remaining receive antenna. The phases for the UI antennas are given by

$$\eta_k^i = 2\pi f\left(\frac{r_i + r_k}{c}\right) + \tau_k + \upsilon_i \qquad i = 1:4, \quad k \neq i \tag{7.135}$$

Now form the phase differences for all of the CI and UI arrays.

$$\Delta\eta_k^{ij} = \eta_k^i - \eta_k^j = 2\pi f\left(\frac{r_i - r_j}{c}\right) + \upsilon_i - \upsilon_j \tag{7.136}$$

$$\Delta\gamma_{ij}^k = \gamma_i^k - \gamma_j^k = 2\pi f\left(\frac{r_i - r_j}{c}\right) + \tau_i - \tau_j \tag{7.137}$$

But now, if we form the double difference, we can eliminate the signal contribution to the phase difference, and we are left with only the error contribution.

$$\Delta\gamma_{ij}^{k} - \Delta\eta_{k}^{ij} = \tau_i - \tau_j + \upsilon_j - \upsilon_i \tag{7.138}$$

Because we assume that the receive bias errors are calibrated, we can see from (7.133) that

$$\tau_i - \tau_j = \Delta\eta_{k}^{i-j} - \Delta\gamma_{i-j}^{k} + \varepsilon_{ij} \tag{7.139}$$

From (7.139) and the preceding analysis, we make the following observations:

1. Transmit and receive phase errors cannot be calibrated simultaneously due to the differential relationship of phase errors in (7.139).
2. Transmit phase errors can be calibrated only to a level equal to the receive phase bias errors correction plus the errors in estimating the phase double difference.
3. Receive errors have to be calibrated independently.
4. Any phase errors common to transmit and receive cancel out in the computation of the calibration coefficient and cannot be mitigated unless calibrated independently.

In (7.139), we have used the double difference of the phase measurement. If the phases are unwrapped, then the computation of the correction terms in (7.139) are straightforward. But if the phases are wrapped, it may not be possible to unwrap the phase properly unless transmit calibration can be applied first. In that case, we may be required to use a smaller baseline interferometer that is unambiguous in phase to determine that calibration correction. Assuming that the correction term is defined by the double phase difference in (7.139), we can write

$$\begin{aligned}\frac{4\pi}{\lambda} D\sin(\theta) &= \measuredangle\Big((e^{\varphi_i} + n_i)(e^{-\varphi_j} + n_j^*)(e^{\gamma_i^k} + n_k)(e^{-\gamma_j^k} + n_k^*)(e^{-\eta_k^i} + n_i^*)(e^{\eta_k^j} + n_j)\Big)\\ &= \measuredangle\Big((e^{\varphi_i} + n_i)(e^{-\varphi_j} + n_j^*)C_{\text{call}}\Big)\end{aligned} \tag{7.140}$$

To compute an accurate estimate of C_{cal}, we must average over M number of measurements due to the additive noise.

$$\begin{aligned}\widehat{C}_{\text{cal}} &= \sum_{p=1}^{M}\Big(\big(e^{\gamma_i^k} + n_k\big)\big(e^{-\gamma_j^k} + \tilde{n}_k^*\big)\big(e^{-\eta_k^i} + n_i^*\big)\big(e^{\eta_k^j} + n_j\big)\Big)_p\\ &\rightarrow e^{\gamma_i^k} e^{-\gamma_j^k} e^{-\eta_k^i} e^{\eta_k^j} + e^{-\eta_k^i} e^{\eta_k^j} E\big(n_k\tilde{n}_k^*\big)\\ &= C_{\text{cal}} + e^{-\eta_k^i} e^{\eta_k^j} E\big(n_k\tilde{n}_k^*\big)\end{aligned} \tag{7.141}$$

where $E(n_k\tilde{n}_k^*)$ is the cross correlation of the noise in receiver k between matched filter i and matched filter j, which is determined by the degree of code orthogonality. Basically, the accuracy in determining the calibration correction is defined by the waveform cross correlation. Thus it is imperative to select low cross-correlation waveforms, as defined in Chapter 5. After calibration correction has been determined, angle is computed using the correction term:

$$\theta = \sin^{-1}\left(\frac{\lambda}{4\pi D}\measuredangle\left(\left(e^{\varphi_i} + n_i\right)\left(e^{-\varphi_j} + n_j^*\right)C_{\text{cal}}\right) + 2\pi m\right) \quad (7.142)$$

Referring to (7.136) and (7.137), we can compute the sum of the phase differences to show the relationships among CI, UI, and OI:

$$\Delta\gamma_{i-j}^{k} + \Delta\eta_{k}^{i-j} = 2\pi f\left(\frac{2(r_i - r_j)}{c}\right) + \tau_i - \tau_j + \upsilon_i - \upsilon_j = \Delta\varphi_{i-j} \quad (7.143)$$

Equation (7.143) provides proof for the following fundamental relationship:

$$\text{OI} = \text{CI} + \text{UI} \quad (7.144)$$

Equation (7.144) verifies that transmit errors are completely determined by UI, receive errors are completely determined by CI, and the combination results in OI, which incurs both transmit and receive errors. Equation (7.143) also shows that, when transmit errors and receive errors are correlated, these errors cancel out due to the differencing of the phase errors.

Now consider the impact of external bias errors, such as phase errors resulting from multipath interference. The error term in (7.132) and (7.143) also applies to external bias errors. If the internal phase errors are calibrated, then the error is completely defined by the external bias error.

$$\text{Bias error} = \tau_i - \tau_j + \upsilon_i - \upsilon_j \quad (7.145)$$

Assume that the transmit and receive external bias errors are totally correlated, making the orthogonal interferometer bias error zero. Thus, we have

$$\tau_i - \tau_j + \upsilon_i - \upsilon_j = 0 \quad (7.146)$$

$$\tau_i - \tau_j = -\left(\upsilon_i - \upsilon_j\right) \quad (7.147)$$

As a result, the CI bias is equal to the negative of the UI bias, and, because OI is the sum of CI and UI, the correlated bias errors for CI and UI exactly cancel out for the OI architecture. This unique result states that OI averages correlated external bias errors resulting from both transmit and receive paths.

Assuming that an OI array has calibrated the internal receive bias errors, the C_{cal} correction becomes the transmit calibration term ($C_{\text{transmit}} = C_{\text{cal}}$) that accounts for relative timing differences among the transmit/receive arrays at any given transmit angle. Because double differencing is required to compute C_{transmit} and angle ambiguities must also be resolved, the transmit calibration process needs sufficient signal-to-noise. Consider the array architecture described in Figure 7.15 with a 1-m array baseline separation and a target located at 10° in azimuth. Then the angle accuracy for both the conventional interferometer and the unconventional interferometer is given by the following:

$$\sigma_{CI} = \frac{\lambda}{2\pi D\sqrt{SNR}} \tag{7.148}$$

For the orthogonal interferometer, the angle accuracy is given by

$$\sigma_{OI} = \frac{\lambda}{4\pi D\sqrt{SNR}} \tag{7.149}$$

And the theoretical ratio of the variances is

$$\text{Ratio} = \frac{\sigma_{OI}}{\sigma_{CI}} = \frac{1}{2} \tag{7.150}$$

Figure 7.17 shows the results of a simulation that computes the ratio of the angle standard deviation versus *SNR* where a one-sigma timing error of 1 ms was added and the real-time transmit calibration was implemented. For 20 dB or greater *SNR*, the standard deviation ratio achieved the theoretical limit, implementing the transmit calibration coefficient to correct for relative timing errors. It should be noted that, for low-*SNR* angles, ambiguity resolution becomes an issue as well as the increased noise due to double differencing. Thus, the transmit calibration coefficient cannot be determined with sufficient precision.

To compute C_{transmit} accurately, the preceding discussion implicitly assumes that the phase error in each array is noise limited as opposed to interference limited. In Chapter 5, we considered orthogonal waveforms that are not truly orthogonal but are only nearly orthogonal, and this residual interference can contribute to angle-of-arrival error. Equation (7.141) shows that, if the waveform cross-correlation interference is sufficiently large to be above the noise floor, then signal-to-interference is the determining factor as opposed to signal-to-noise. The interference levels can be managed by properly selecting waveforms and/or by developing cancellation or nulling techniques in the matched (or possibly mismatched) filter, as suggested in 5.2.4.

7.7 Synthetic Aperture Radar Interferometry

In Section 6.8, we introduced SAR imaging as an application of interferometery. Because SAR interferometry assumes that SAR processing has been accomplished, all the processing issues for SAR are well documented and discussed in other references. In this book, we concentrate only on issues that affect SAR interferometry. Because interferometry is concerned with phase integrity, we focus on SAR interferometry phase issues. For SAR, the interferogram is defined by

$$x_1 x_2^* = a_1 e^{j\varphi_1} a_2 e^{-j\varphi_2} = a_1 a_2 e^{j(\varphi_1 - \varphi_2)} \tag{7.151}$$

The advantage of using the interferogram is that the interferogram phase is likely not to wrap or at least to wrap with minimal 2π integer shifts. Referring to

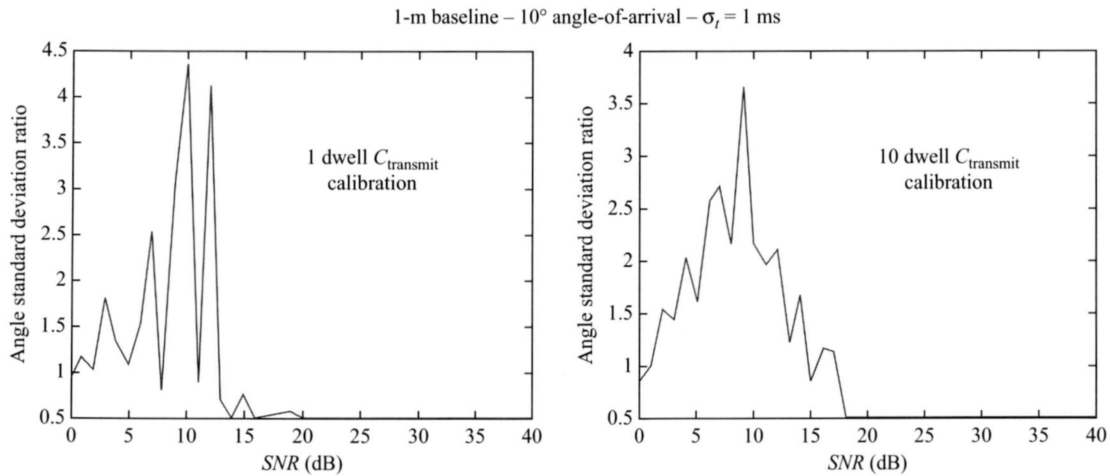

Figure 7.17. Ratio of Standard Deviation Results for Real-Time Transmit Calibration and Alignment Algorithm Versus Signal-to-Noise Ratio (SNR)

Figure 6.6, the phase difference of particular interest is the phase difference associated with the reference point *P*. In processing the height of the point *P*', it assumed that we have a good estimate for the reference phase difference. We will focus on obtaining the reference phase and phase unwrapping.

7.7.1 Reference Phase Determination

In determining the expected phase behavior for a reference body, the phase between the reference body and the orbits needs to be calculated and differenced. Reference bodies can be global ellipsoids generated using WGS84 or ellipsoid estimation techniques to achieve a best local fit. The process is determined in four steps [13].

1. Along orbit 1, determine the ranges to a few uniformly distributed points within the interferogram of interest.
2. Using the points generated in step 1, determine the positions and corresponding ranges along orbit 2 for the time of imaging.
3. For each point in the image, compute the range differences (ΔR), and convert them to phase differences ($\Delta\varphi$).
4. Interpolate the phase differences in the range cell of interest, given the location of the reference body within the range call.

To ensure that a point *P* is located on the reference body, Doppler frequency measurements can be combined with the fact that the point *P* lies on a defined reference body ellipsoid, which allows the precise determination of the location of *P*. Thus, we have the following [13]:

1. The Doppler frequency can be related to the observed range and range rate.

$$f_{\text{Doppler}} = \frac{2|\vec{v}|\sin(\varphi)}{\lambda} = \frac{2\vec{R}\dot{\vec{R}}}{|\vec{R}|} \tag{7.152}$$

2. *P* is required to be located on the reference body and thus must satisfy the ellipsoid defining the reference body.
3. The observed range is defined as the norm of difference between the satellite state vector and the surface point vector *P*.

$$\vec{R} = \text{Orbit location} - P \tag{7.153}$$

Solving the nonlinear system defined by steps 1–3 and using measurements of Doppler and range yield the point *P*.

7.7.2 Phase Unwrapping

The primary observation for a SAR interferometer is the relative phase difference between the point *P* and the two satellite orbit positions. Because the voltage measurement at each orbit location is a complex phasor that rotates around the unit circle, the measured phase is ambiguous by the number of complex phase rotations or by the number of 2π integer rotations. This is the same issue as discussed in the previous section related to phase ambiguities for tracking radar interferometers. The measured wrapped phase is determined by the following [13]:

$$\varphi_{\text{wrap}} = W(\varphi) = [\varphi]_{2\pi} - \pi = \text{mod}\{\varphi + \pi : 2\pi\} - \pi \tag{7.154}$$

where

$$\varphi = \frac{4\pi\Delta R}{\lambda} + \varphi_{\text{noise}} = \varphi_{\text{wrap}} + 2\pi k + \varphi_{\text{noise}} \tag{7.155}$$

W is the phase wrapping operator, $\Delta R = R_1 - R_2$, and $\varphi_{\text{noise}} \in [-\pi, \pi)$ is additive phase noise. The phase unwrap problem can be stated as in for k, given the measurements of φ_{wrap}, R_1, and R_2. For the one-dimensional phase unwrapping discussed in previous sections, an assumption is made that the phase versus time is linear in nature. For the two-dimensional phase unwrapping, the assumption is that the two-dimensional phase difference is zero [14], which can be expressed as

$$\nabla_V \times \nabla_H \varphi = 0 \tag{7.156}$$

where V and H represent two orthogonal spatial directions. Equation (7.156) imposes a smoothness criterion on measured phase, implying that phase is sampled at a sufficient rate to ensure that it does not skip ambiguities from look to look. The gradient $\nabla\varphi$ of the true phase is unknown and must be estimated using the gradient of the unwrapped phase. At a specific pixel with coordinates (i, j), a reasonable approximation to the wrapped phase gradient is given by

$$\hat{\nabla}\varphi^{i,j}_{\text{wrap}} = E\left\{\begin{bmatrix} W\left(\varphi^{i+1,j}_{\text{wrap}} - \varphi^{i,j}_{\text{wrap}}\right) \\ W\left(\varphi^{i,j+1}_{\text{wrap}} - \varphi^{i,j}_{\text{wrap}}\right) \end{bmatrix}\right\} \tag{7.157}$$

In general, $\hat{\nabla}\varphi^{i,j}_{\text{wrap}}$ does not equal the true phase gradient $\nabla\varphi^{i,j}$ due to phase noise and phase errors. Thus, integrating the estimated phase gradient to obtain an estimate of phase can lead to large phase errors after integration. We can use the smoothness criterion to clean up phase residue that may occur due to these error sources. We can form the following metric.

$$\begin{aligned} r = {} & W\left\{\left(\varphi^{i+1,j}_{\text{wrap}} - \varphi^{i,j}_{\text{wrap}}\right)\right\} + W\left\{\left(\varphi^{i+1,j+1}_{\text{wrap}} - \varphi^{i+1,j}_{\text{wrap}}\right)\right\} \\ & + W\left\{\left(\varphi^{i,j+1}_{\text{wrap}} - \varphi^{i+1,j+1}_{\text{wrap}}\right)\right\} + W\left\{\left(\varphi^{i,j}_{\text{wrap}} - \varphi^{i,j+1}_{\text{wrap}}\right)\right\} \end{aligned} \tag{7.158}$$

The value of r can be zero, which translates to no residue, $+1$ cycle which means a positive residue, or -1 cycle, meaning a negative residue [15]. The presence of residues is an indication of an inconsistency with the smoothness condition, which may imply that the phase is undersampled or that the *SNR* is too low. The value of using the metric r is that true phase gradients that exceed one-half cycle can be identified and corrected by connecting nearby positive and negative residues. Several phase-unwrapping algorithms have been developed that utilize search methods to connect phase residues [13]. These methods develop a tree structure or a network of phase residues over a grid of points in the interferogram that identify regions with positive and negative residues. As a result residues are unloaded according to their sign in each region. In regions with low *SNR*, patch unwrapping algorithms have been applied when these low *SNR* regions are surrounded by regions with high *SNR*.

Phase unwrapping for SAR interferometry can be much more difficult than phase unwrapping for a tracking radar interferometer due to the number of points in the interferogram grid. However, the one thing that SAR interferometry usually has over tracking interferometry is that the unwrapping does not have to be accomplished in real time. Thus, the SAR interferometry processing can use significant computational resources with complex algorithms to accomplish phase unwrapping.

7.8 Near-Geostationary Interferometry Tracking

Geostationary satellite tracking has become increasingly important due to the number of these satellites that have been deployed for commercial and military use. Because a geostationary satellite must be located directly over the equator at a specified altitude above the earth, the proliferation of these satellites over the last few decades has resulted in a crowded satellite environment, requiring improvements in orbit estimation accuracy for precise satellite location. The interferometer is ideally suited for near-geostationary tracking due to the enhanced angle accuracy achieved through interferometry. We now show how the restricted orbits of near-geostationary satellites can enable an interferometer that measures only angle-of-arrival or relative phase to determine the orbital parameters for these satellites.

In Figure 7.18, we define a coordinate frame for an idealistic satellite geometry and orbit (ellipse) [16]. The coordinate frame is defined by a longitudinal axis *L*, a range axis *R*, and an axis *Z* that is pointing out of the page. A perfectly geostationary satellite is described by the following:

$$r = a \tag{7.159}$$

$$\theta = \Omega t \tag{7.160}$$

$$z = 0 \tag{7.161}$$

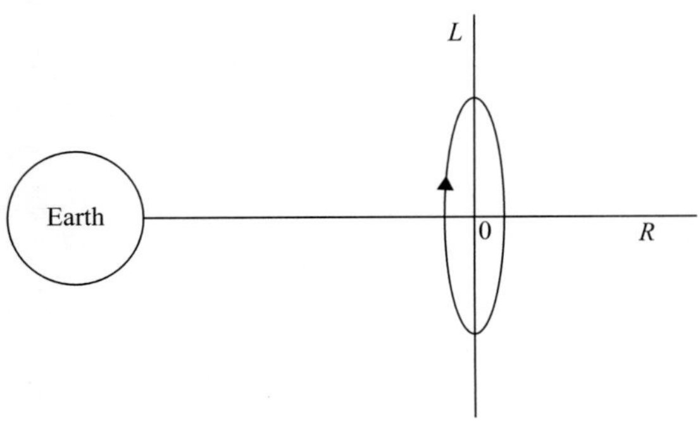

Figure 7.18. *Coordinate Axes for Orbit – L; Longitudinal Axes: R Is the Range Axis Perpendicular to the Earth's Equator, and Z Is the Axis Out of the Page in the Northly Direction*

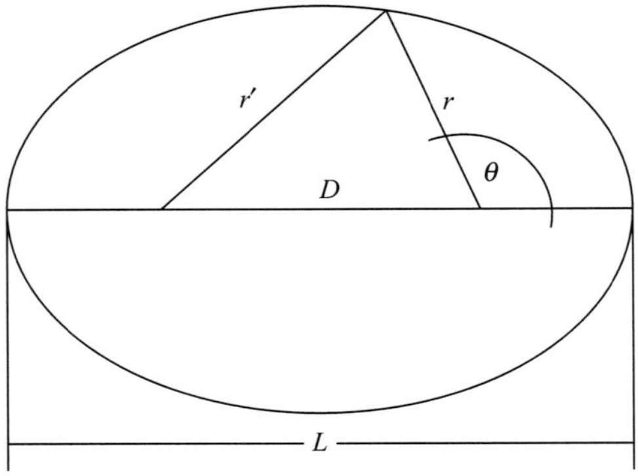

Figure 7.19. Ellipse with Semimajor Axis L/2 with Geometric Parameters

where Ω is the angular rotation rate of the earth (7.292115e-5 rad/s), and a is the distance for the earth to the point 0 in Figure 7.18. For a geostationary satellite, $a = 42{,}164.2$ km.

The ellipse, as shown in Figure 7.19, can be written in the following form:

$$r = \frac{a(1 - e^2)}{1 + e\cos(\theta)} \tag{7.162}$$

where the eccentricity is defined by

$$e = \frac{D}{L} \tag{7.163}$$

and the semimajor axis is defined by

$$a = \frac{L}{2} \tag{7.164}$$

For the near-geostationary orbit, the eccentricity is small, $e \ll 1$, and we can approximate as follows:

$$\begin{aligned} r &\approx \frac{a}{1 + e\cos(\theta)} \cdot \frac{1 - e\cos(\theta)}{1 - e\cos(\theta)} \\ &= \frac{a(1 - e\cos(\theta))}{1 - e^2\cos^2(\theta)} \\ &\approx a(1 - e\cos(\theta)) \end{aligned} \tag{7.165}$$

Thus for near-geostationary orbits, we write

$$r = a(1 - e\cos(\theta)) \tag{7.166}$$

When $\theta = \pi/2$ or $\theta = 3\pi/2$, the radius is equal to a: however, when $\theta = 0$, the radius has decreased by ae, and, when $\theta = \pi$, the radius has increased by ae. As a result, the orbit can be approximated by a circle with a displaced center $d = ae$. Figure 7.20 shows an example of a circular orbit with an off-center origin. Equation (7.166) then becomes

$$r = a - d\cos(\theta) \tag{7.167}$$

The three-dimensional for the motion of a near-geostationary orbit centered at the nominal stationary satellite position are given by

$$R = \Delta a - ae\cos(\Omega t - \alpha) \tag{7.168}$$

$$L = L_0 - \frac{3\Delta a}{2}\Omega t + 2ae\sin(\Omega t - \alpha) \tag{7.169}$$

$$Z = ai\sin(\Omega t - \beta) \tag{7.170}$$

where R, L, and Z are the radial, longitudinal, and north axes, respectively; Ω is the rotation rate of the earth; i is the inclination of the elliptical orbit; L_0 defines the initial longitude position; and α and β define initial elliptical angle. When the geostationary orbit is inclined, the satellite can trace a ground locus in the shape of a figure eight, as shown in Figure 7.21 [16]. The smaller the inclination parameter i, the less likely it will be to observe this figure eight motion. Table 7.1 [16] shows the angular extent of the figure eight motion. What is apparent in Table 7.1 is that significant angle accuracy is required to measure the motion for small inclinations, making this application ideally suited for radar interferometry.

The parameters Δa, e, L_0, i, α, and β are the six parameters that completely define the satellite orbit. The parameters change slowly as a function of time, primarily due to the gravitational forces of the sun and moon. Equation (7.169) has a linear part and a periodic part. Figure 7.22 shows the effect of the linear and periodic parts of (7.169). By observing the longitudinal coordinate of the orbit motion over time, the slope, determined by (7.169), is $3\Delta a/2$, and thus, by estimating the slope, the parameter Δa and the initial longitude L_0 can be determined. By estimating the periodic part, the parameters e and α can be determined. With knowledge of

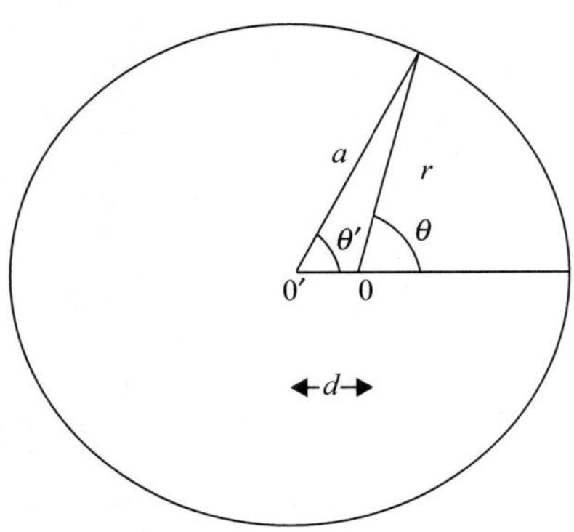

Figure 7.20. Circular Orbit with Off-Center Origin 0

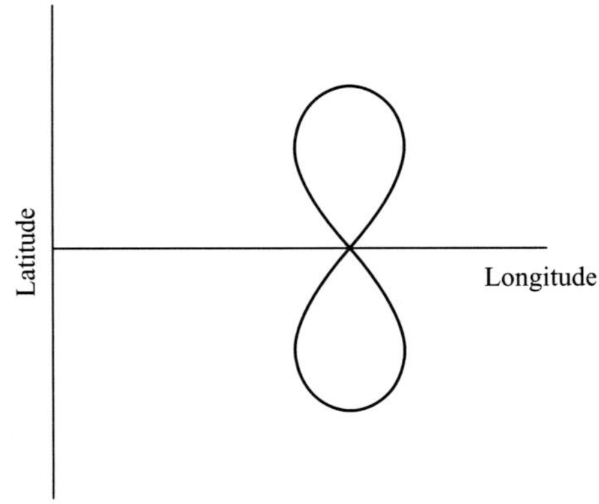

Figure 7.21. Ground Trace of Near-Stationary Orbit in the Shape of a Figure Eight

Table 7.1. Observability of the Figure Eight Near-Geostationary Orbit

Inclination (°)	Angular Width of Figure Eight (°)
0.5	0.002
1	0.009
2	0.035
3	0.14
4	0.56

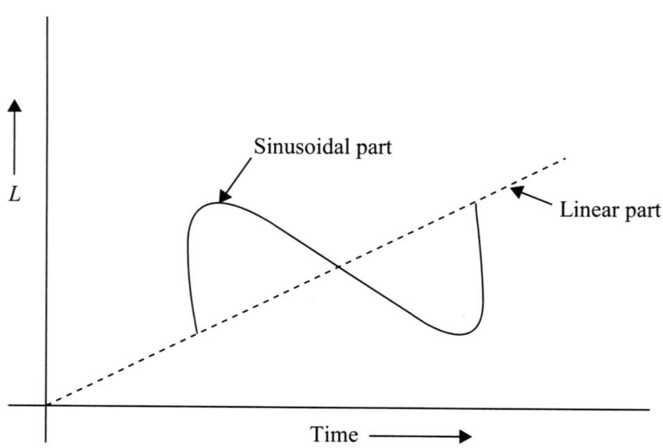

Figure 7.22. Linear and Periodic Part for Longitudinal Motion

the motion along the z-axis, (7.170) shows that the parameters i and β can be derived. In other words, all six orbital parameters for a near-geostationary satellite can be inferred from measurements of satellite motion along two axes.

The period of the sinusoid in (7.169) corresponds to 1 day. Thus it takes 1 day, or one rotation of the earth, to determine the period that allows us to estimate 4 orbital parameters using (7.169). The Z component in (7.170) is similar in that it will also take 1 day to estimate the other two parameters that make up the complete set of orbital parameters. At times, solar pressure is a factor, and there is a need to estimate the ratio of the cross-sectional area A and mass M of the satellite to determine the impact of solar pressure winds on tracking performance. In this case, 2 days of tracking are required to estimate the ratio A/M.

A radar interferometer is ideally suited to determine the satellite motion in two axes [16]. Figure 7.23 shows an idealistic interferometer with baseline AB aligned along the longitudinal axis and baseline AC aligned along the Z-axis. This type of interferometer is typically referred to as an "azimuth-elevation interferometer." Consider a satellite S located in the plane defined by the points A, B, and the origin 0 at an angle ψ with respect to the normal to the interferometer.

We now show that the longitudinal motion denoted by x can be estimated using the interferometer baseline AB [1]. First, the interferometer phase measurements can be used to estimate the angle ψ.

$$\sin(\psi) = \frac{\lambda}{2\pi b_{AB}}(\varphi_B - \varphi_A) = \frac{\lambda}{2\pi b_{AB}}\partial\varphi_{AB} \tag{7.171}$$

But also,

$$\sin(\psi) = \frac{x}{r} \tag{7.172}$$

and

$$\frac{x}{r} = \frac{\lambda}{2\pi b_{AB}}\partial\varphi_{AB} \tag{7.173}$$

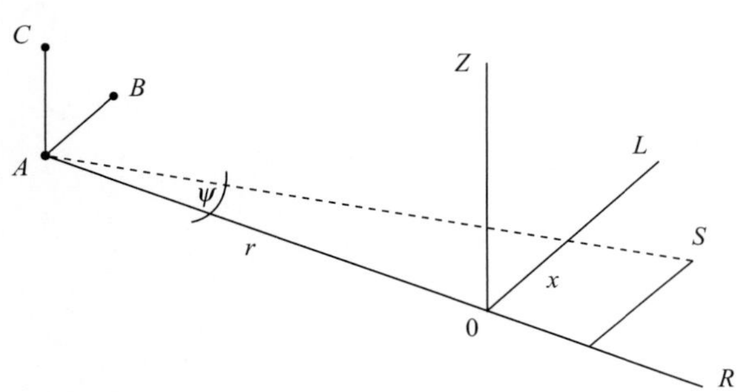

Figure 7.23. Idealistic Interferometer with Satellite Located in the R-L Plane

Thus we see that the interferometer differential phase measurement determines the longitudinal motion.

$$x = \frac{\lambda r}{2\pi b_{AB}} \partial\varphi_{AB} \tag{7.174}$$

Similarly, the interferometer baseline AC can determine the motion along the Z-axis. Figure 7.24 shows the satellite S located in the plane defined by the points A, C, and the origin 0 at an angle γ with respect to the normal to the interferometer. The motion y is determined by the following:

$$y = \frac{\lambda r}{2\pi b_{AC}} \partial\varphi_{AC} \tag{7.175}$$

The idealistic interferometer defined in Figures 7.23 and 7.24 was chosen to show that the interferometer determines satellite motion on two axes and that this motion can, in fact, completely determine the six orbital parameters for a near-geostationary satellite orbit. A realistic interferometer is not necessarily aligned along the L- and Z-axes. However, in [16] it is shown that a linear combination of the phase differences can be used to define motion along any direction. As a result, the techniques derived in this section for the idealistic interferometer can be generalized to any interferometer orientation.

The horizontal interferometer with baseline AB is useful in its own right because four of the six orbital parameters can be estimated from just this 1-D baseline configuration. The addition of the vertical interferometer AC allows all six parameter to be estimated. However, it has been shown [17] that, with the addition of the range measurements, the six orbital parameters can be determined using just the horizontal interferometer, provided the detection directions x and z lie in directions that couple the satellite motion. In this case, the motion along x and z is a linear combination of (7.168), (7.169), and (7.170), and thus all six parameters can be estimated over time. When x and z lie in either the R-L plane or the R-Z plane, then the orbital motion is not sufficiently coupled to determine the orbital parameters.

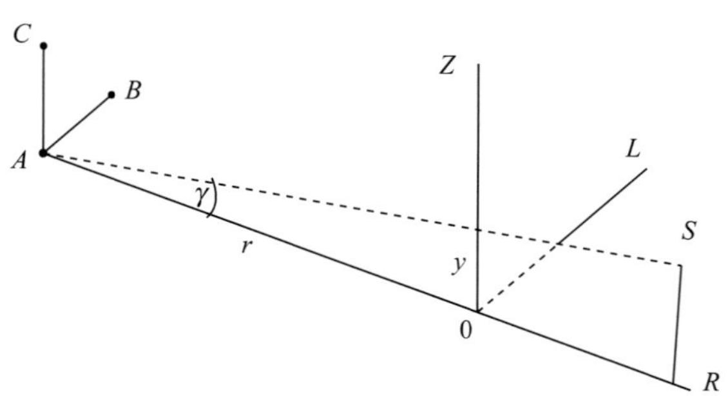

Figure 7.24. Idealistic Interferometer with Satellite Located in the R-L Plane

Because a primary objective for near-geostationary interferometer tracking is to determine the relative locations of two closely spaced satellites, the implementation of differential tracking can provide enhanced tracking accuracy to estimate relative satellite locations. Differential tracking is applicable when two or more satellites can be placed in the same antenna beam for each of the interferometric antennas. Because some system errors are common to both satellite phase measurements, system errors that are common to the differential phase measurements for both azimuth and elevation cancel out in the differential location estimate. The satellite phase measurements need to be resolved in either range or Doppler because the two satellites are not resolvable in angle.

7.9 Adaptive Array Processing

The effect of noise jamming and RF interference is unwanted noise that enters the antenna sidelobes. This noise energy has the effect of increasing the phase noise that results in a degradation of SINR (signal-to-interference-plus-noise ratio). When the interferometer architecture consists of multiple distributed phased arrays, the unwanted noise affects each array and thereby degrades the phase integrity of the interferometer. To mitigate noise jamming and interference, it is necessary to implement sidelobe cancellation. Several types of sidelobe cancelers have been developed that can achieve significant cancellation. The multiple sidelobe canceler [8] was the first canceler developed and can be implemented in most phased arrays. The generalized sidelobe canceler [8] is an implementation of the Weiner-Hopf filter that achieves a minimum mean square error.

For an interferometer, these adaptive cancelers must be implemented on each of the distributed arrays that comprise the interferometer antenna architecture, implying that each antenna array must be divided and digitized into subarrays to allow for the necessary degress of freedom for interference cancellation. For interferometry, the signals must be balanced at each array, which requires that each distributed array implement the exact same canceler with the same antenna weights. It also requires that the SINR be balanced at each array when tracking a target. This increased requirement for calibration requires additional processing and external test targets to maintain the angle measurement accuracy.

7.9.1 The Multiple Sidelobe Canceller

The multiple sidelobe canceller (MSC), the earliest statistically optimum beam former, was developed in the 1950s by Appelbaum [9]. The MSC consists of a main channel and one or more auxiliary channels, as depicted in Figure 7.25. The main channel can be either a single high-gain antenna that uses the output of the auxiliary antenna to cancel the interference in the sidelobe or a data-independent beam former that steers an antenna null or nulls to the angular location of the interference source. The MSC can handle as many jammers or sources as the number of antennas in the auxiliary array. A single-antenna MSC can cancel only one jammer. The auxiliary antenna has a highly directional response, which is pointed in the desired signal

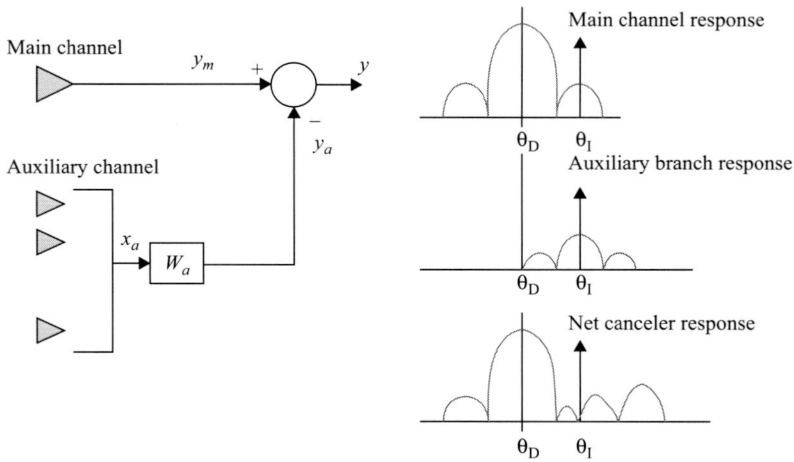

Figure 7.25. Multiple Sidelobe Canceler

direction. Interfering signals enter through both the main channel sidelobes and the auxiliary channels. The MSC [8] chooses the auxiliary channel weights to cancel the main channel interference component, which means that the responses to interferers of the main channel and linear combination of auxiliary channels must balance. The overall system, then, has a response of zero, as illustrated in Figure 7.25. The weights are usually chosen to trade off interference suppression for white noise gain by minimizing the expected value of the total output power.

If the weights are chosen to minimize output power, then cancellation of the desired signal can occur due to the fact that the desired signal also contributes to total output power. One of the problems with the MSC is that, as the desired signal gets stronger, it contributes to a larger fraction of the total output power, and the percentage cancellation of the desired signal increases. The MSC is very effective when the desired signal is weak compared to the interference or when the desired signal is known to be absent during certain time periods. In the latter case, the weights can then be adapted in the absence of the desired signal but applied when the desired signal and interference are present. Again, for the interferometer, the MSC must be implemented in the same manner on each of the distributed antennas.

7.9.2 The Generalized Sidelobe Canceller (GSC)

The Wiener-Hopf solution for the optimal weight vector is given by

$$R_{xx}W_o = r_{xd} \tag{7.176}$$

where W_o is the optimal weight vector, R_{xx} is the correlation matrix of the input process, and r_{xd} is the cross-correlation vector between the input process and the desired response d. Whereas the Weiner-Hopf weight vector is optimal in the mean square error (MSE) sense, a priori knowledge of R_{xx} and r_{xd} is difficult to ascertain, and thus these terms must be estimated using time samples and assumptions of signal and interference stationarity. In addition, errors in these statistics degrade filter performance.

The generalized sidelobe canceler [8] achieves an optimal mean square error solution by implementing the Weiner-Hopf filter using an efficient blocking matrix to separate a desired signal from interference.

Let the normalized target steering vector be defined as

$$s = \frac{v(\vartheta_t, \omega_t, \zeta_t)}{\sqrt{v^H(\vartheta_t, \omega_t, \zeta_t) v(\vartheta_t, \omega_t, \zeta_t)}} \tag{7.177}$$

where $v(\vartheta_t, \omega_t, \zeta_t)$ is the $KJ \times 1$ dimensional space-time steering vector.

Now consider using a unitary $KJ \times KJ$ matrix operator T to transform the data prior to adaptive detection. The unitary nature of this matrix conserves power so that the resulting output *SINR* is seen to be identical to the original direct-form detection processor. The structure of this operator is partitioned as follows:

$$T = \begin{bmatrix} s^H \\ B \end{bmatrix} \tag{7.178}$$

where $\boldsymbol{s}$ is the $KJ \times 1$ 1-D conventional beam former, and B is the full-row rank $N \times KJ$ signal blocking matrix that maps $\boldsymbol{x}$ onto the null-space of $\boldsymbol{s}$ where $N = KJ - 1$ [10]. Hence,

$$Bs = 0 \tag{7.179}$$

so that the matrix B effectively blocks any signal coming from the look direction. Any full-row rank B that satisfies (7.179) and results in an invertible T is a valid signal blocking matrix. The Gram-Schmidt algorithm may be applied, then, to this matrix to generate an orthonormal B and unitary T. An orthonormal B may also be found directly using the singular value decomposition or the QR decomposition [11,12].

The transformation of the radar return x by the operator T in (7.178) yields a vector $\tilde{x}$ with the following form:

$$\tilde{x} = Tx = \begin{bmatrix} s^H x \\ Bx \end{bmatrix} = \begin{bmatrix} d \\ b \end{bmatrix} \tag{7.180}$$

where the scalar-valued beamformed output is denoted by d. Here the N-dimensional vector b is termed the noise subspace and the covariance matrix R_b is expressed by $R_b = E[bb^H] = BRB^H$. The $N \times 1$ cross-correlation vector between the noise subspace data vector and the beam former output are given by $r_{bd} = E[bd^*] = BRs$, where * represents the complex conjugate operator. Define the scalar σ_d^2 as $\sigma_d^2 = s^H Rs$, which represents the variance of the conventional beamformer output.

Next, let T operate on the steering vector s. This operation yields the unit transformed steering vector e_1, given by

$$e_1 = Ts = \begin{bmatrix} 1 & 0 & \cdots & 0 \end{bmatrix}^T \tag{7.181}$$

Then the optimal weight vector in these transformed coordinates is given by

$$w_{gsc} = \frac{R_{\tilde{x}}^{-1} e_1}{e_1^H R_{\tilde{x}} e_1} = \begin{bmatrix} 1 \\ -w \end{bmatrix} \tag{7.182}$$

where $R_{\tilde{x}}$ is the covariance matrix [10]. The partitioning of the matrix operator T leads naturally to the form of the GSC array processor depicted in Figure 7.26. This processor results in a fixed weight of unity for the upper branch and an adaptive weight w of dimension $N = KJ - 1$ in the lower branch. The vector w in (7.182) is provided by the Wiener solution corresponding to the filter depicted in Figure 7.26:

$$w = R_b^{-1} r_{bd} \tag{7.183}$$

The GSC form processor implements the KJ weight vector in (7.182) using the partitioning defined in (7.180). The steady-state performance of the GSC and the direct-form processor are identical, but the adaptive weight vector w in the GSC is of a lower dimension. Hence, the computational requirements for updating this weight vector are reduced, and the GSC form can be considered canonical.

The output of the arrays, using (7.180) and (7.182) may be expressed as follows:

$$y = w_{SINR}^H x = w_{gsc}^H \tilde{x} = (s^H - w^H B)x \tag{7.184}$$

The output of the noise power is found by substitution of (7.183) into (7.184), and the evaluation of the mean-square value of y is

$$P = \frac{1}{s^H R^{-1} s} = \sigma_d^2 - r_{bd}^H R_b^{-1} r_{bd} \tag{7.185}$$

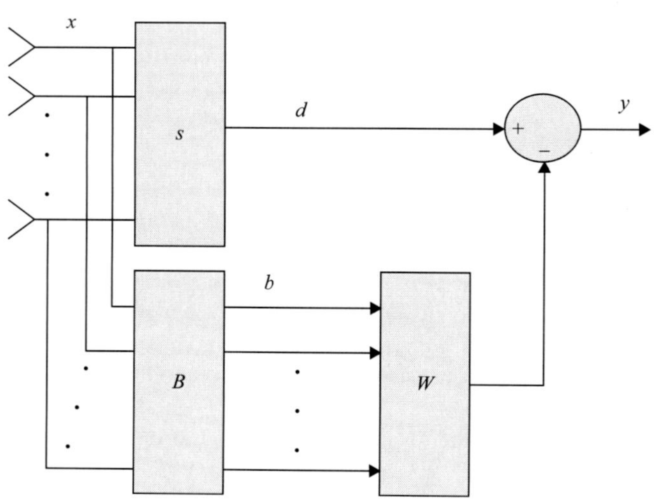

Figure 7.26. The Full-Rank GSC Processor

The GSC requires multiple digital channels, and thus each interferometer antenna array needs to be subdivided into multiple digital subarrays. The subbarray architecture needs to be the same for each distributed antenna in order to maintain phase integrity for interferometer angle estimation. For arrays with a small number of elements, a digital beam forming (DBF) architecture may be preferred to mitigate the grating lobe phenomena that can occur with subarray architectures.

7.9.3 The Orthogonal Space Projection Canceler

The orthogonal space projection (OSP) digital signal processing technique was developed by Propagation Research Associates (patent pending) [18] to mitigate the effects of electronic interference using waveform diversity to enhance spatial cancellation performance. As such, the OSP technique can be naturally implemented in orthogonal interferometers and in conventional interferometers using waveforms from a family of waveforms with low cross-correlation properties. The OSP technique creates both a matched and mismatched channel by projecting the return signal into orthogonal subspaces. The mismatched subspace contains only the interference and receiver noise, and the matched subspace contains the signal plus interference and receiver noise. The output of each channel is combined in the spatial domain or separation space to create a weighting for each of the antenna inputs. A discrete Fourier transform isolates the desired signal from the undesired interference. Figure 7.27 shows the fundamental flow diagram for a general OSP technique. For specific application using beam space, the separation space is defined as angle-of-arrival.

Conventional canceler techniques, such as the generalized sidelobe canceler (GSC), minimum variance distortionless response (MVDR), or eigen-structure techniques such as MUSIC, require the formation and numerical manipulation of a covariance matrix. The OSP essentially creates an orthogonal subspace analogous to a conventional adaptive cancellation blocking matrix that contains only the interference sources and is mismatched to the signal of interest. By creating the matched and mismatched channels using orthogonal projection, the OSP interference cancellation can be achieved in significantly less time than conventional

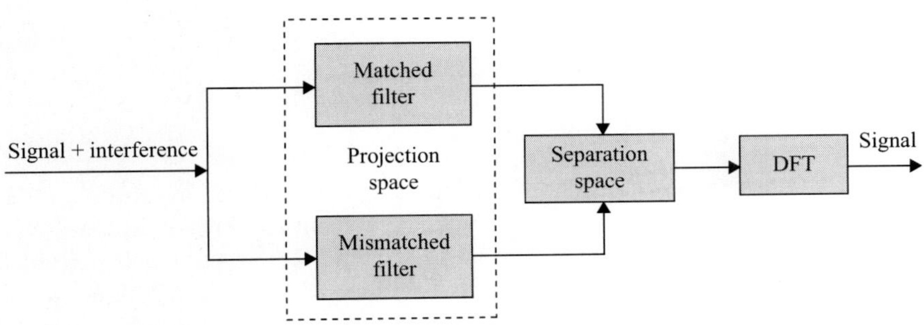

Figure 7.27. Orthogonal Space Projection Processing

adaptive cancellation techniques that require time averaging after compression to form covariance matrices. OSP utilizes fast-time averaging (single-BPSK code), as opposed to conventional adaptive cancellation techniques that utilize slow-time averaging (dwell of multiple codes). The primary advantage to OSP is that it provides effective fast-time interference cancellation, resulting in more robustness to nonstationary effects than to conventional interference cancellation techniques. Thus OSP can be effective in highly dynamic target and interference environments.

The OSP approach addresses the case where interference from one or more sources is present and degrades conventional signal processing performance. Let Π_ρ (e.g., target range) and Π_θ (e.g., target angle-of-arrival) denote parameter spaces, and let Ω_P denote the projection space and Ω_S denote separation space. We define the signal function S and interference function J from the parameter space $\Pi_\rho \times \Pi_\theta$ into the product of projection and separation space.

$$S : \Pi_\rho \times \Pi_\theta \to \Omega_P \times \Omega_S \subset \mathbb{C}^{N \times M} \text{ Signal}$$

$$J : \Pi_\rho \times \Pi_\theta \to \Omega_P \times \Omega_S \subset \mathbb{C}^{N \times M} \text{ Interference}$$

Create a reference signal vector R that is matched to the waveform at each frequency. Assuming the target signal incorporates phase modulation that can be compressed, an orthogonal or nearly orthogonal phase code could be used as a projection operator.

$$R^\perp : \Pi_{\rho_k} \to \Omega_P \subset \mathbb{C}^N$$

We now create the following orthogonal projections using projection space:

$$SJ = R * (S + J)^H : \Pi_\rho \times \Pi_\theta \to \Omega_\rho \times \Omega_S \subset \mathbb{C}^{N \times M} \quad \text{Matched filter}$$

$$SJJ = R^\perp * (S + J)^H : \Pi_\rho \times \Pi_\theta \to \Omega_\rho \times \Omega_S \subset \mathbb{C}^{N \times M} \text{ Mismatched filter}$$

where * indicates convolution. Now create the following modified Capon function in separation space.

$$Y(SJ, SJJ) = \left(SJJ^H SJJ\right)^{-1} SJ^T : \Pi_\rho \times \Pi_\theta \to \Omega_\rho \times \Omega_S \subset \mathbb{C}^{N \times M}$$

We extract the signal parameter estimates from the angle image space

$$RA = DFT(Y) : \Pi_\rho \times \Pi_\theta \to \Omega_\rho \times \Omega_\theta \subset \mathbb{C}^{N \times M}$$

$$\left[\hat{\rho} \quad \hat{\theta}\right] = \max_{\rho,\theta}(|DFT(Y)|)$$

Figure 7.28 shows the results of implementing OSP processing using a modified Capons method to cancel two noise jammers using 255-bit BPSK codes with low cross correlation and angle-of-arrival to separate the target signal of interest from the two jammers using four antennas configured in a 2×2 two-dimensional architecture. The left-hand graph in Figure 7.28 shows the results of array

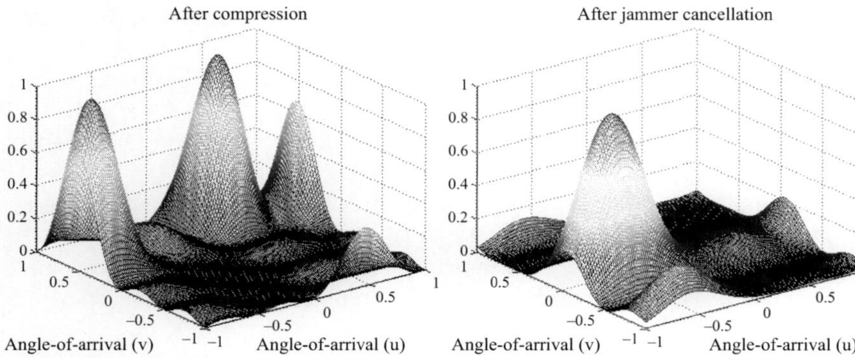

Figure 7.28. The Array Response for a Four-Antenna Array Prior to OPS Cancellation (LEFT) and After OPS Cancellation (RIGHT) for Two Noise Jammers

processing before OSP cancellation, and the right-hand side shows the results after cancellation in the angle-angle separation domain.

References

1. J. L. Eaves and E. K. Reedy, *Principles of Modern Radar*, Van Nostrand Reinhold, New York, 1987.
2. M. I. Skolnik, *Radar Handbook*, 2nd Edition, McGraw Hill, New York, 1990.
3. P. Z. Peebles, *Radar Principles*, John Wiley & Sons, New York, 1998.
4. J. A. Scheer and J. L. Kurtz, *Coherent Radar Performance*, Artech House, Norwood, MA, USA, 1993.
5. R. B. Dybdal, "Monopulse resolution of interferometric ambiguities," *IEEE Trans. AES*, vol. AES-22, no. 2, pp. 177–183, Mar. 1986.
6. D. K. Barton and H. R. Ward, *Handbook of Radar Measurements*, Englewood Cliffs NJ, USA, Prentice-Hall, 1969.
7. J. R. Klauderer, A. C. Price, S. Darlington, and W. L. Albersheim, "The theory and design of chirp radars," *Bell System Tech. J.*, vol. 39, pp. 745–808, July 1996.
8. K. M. Buckley, "Spatial/Spectral Filtering with Linearly Constrained Minimum Variance Beamformers," *IEEE Trans. ASSP*, Vol. 35, March 1987, pp. 240–266.
9. S. P. Applebaum, "Adaptive arrays," Syracuse University Research Corp., Report SURC SPL TR 66-001, Aug. 1966 (reprinted in *IEEE Trans. on AP*, vol. AP-24, pp. 585–598, Sept. 1976).
10. S. Haykin and A. Steinhardt, *Adaptive Radar Detection and Estimation*, John Wiley & Sons, New York, 1968.

11. J. S. Goldstein and I. S. Reed, "Theory of partially adaptive radar," *IEEE Trans. AES*, vol. 33, no. 4, pp. 1309–1325, Oct. 1997.
12. G. Golub and C. F. Van Loam, *Matrix Computations*, Johns Hopkins University Press, Baltimore MD, USA, 1996.
13. R. F. Hanssen, *Radar Interferometry*, Kluwer Academic Publishers, Dordrecht NL, 2001.
14. R. Bambler and P. Hartl, "Synthetic aperture radar interferometry," *Inverse Problems*, vol. 14, pp. R1–R54, 1998.
15. R. M. Goldstein, H. A. Zebker, and C. L. Werner, "Satellite radio interferometer: Two dimensional phase unwrapping," *Radio Science*, vol. 23, no. 4, pp. 713–720, 1988.
16. S. Kawase, *Radio Interferometry and Satellite Tracking*, Artech House, Norwood, MA, USA, 2012.
17. W. W. Smith and P. G. Steffes, "Time delay techniques for satellite interference system," *IEEE Trans AESS*, vol. 25, no. 2, pp. 224–231, 1989.

Chapter 8

Sparsely Populated Antenna Arrays

In this chapter, we discuss sparsely populated arrays. Sparse arrays or unequally spaced arrays were initially developed by Andreason [4], Maffett [5], Lo [8], and others in the early 1960s. Techniques were developed to generate sparse arrays with good antenna pattern characteristics. In [8], a statistical approach was developed that led to generating sparse arrays randomly with an element location probability distribution that provided low sidelobe antenna patterns on the average.

Unz [1] first suggested the use of nonuniform element spacing to control antenna pattern synthesis. King, Packard, and Thomas [2] used unequal array spacings to reduce grating lobes. Harrington [3] presented a method for reducing sidelobe levels of a linear array by using nonuniform element spacing, while retaining uniform excitation, and showed that the sidelobes can be reduced in height to approximately $2/N$ times the main lobe level for an array with N elements. Andreasen [4] computed the various possibilities of unequally spaced linear arrays. Ishimaru [6] expanded on the work of Harrington by using unequal array spacings to control near-in sidelobes at the expense of growing the outer sidelobes. With this method, it is possible to design unequally spaced arrays that produce radiation patterns where the first few sidelobes achieve a predefined level similar to how Taylor weights [7] achieve low sidelobes for a given number of near-in sidelobes. Lo [8] employed a statistical approach that led to generating sparse arrays randomly with element location probability distribution that provided low sidelobe antenna patterns on the average. More recently, Mitra et al. [9–11] used polynomial factorization methods to define a sparse array of transmit elements and a sparse array of receive elements that combine to achieve a two-way uniform excitation pattern. In addition, by proper scaling of the receive elements, a number of elements can be apodized to reduce sidelobe levels. Vaidyanathan et al. [12–15] used interval partitioning with coprime integers to create sparse arrays that possess antenna patterns with low sidelobes. The coprime condition assures that the spatial Nyquist condition is satisfied for coprime arrays to eliminate grating lobes and reduce sidelobes.

In Section 8.2, we use interval partitions to define minimum redundancy partitions and almost minimum redundancy partitions that achieve low sidelobe excitation patterns. These arrays assure that the spatial Nyquist condition is satisfied and also attempt to maximize the number of partition differences for optimized sidelobe control. In Section 8.3, interval partitions are applied to coprime integers to generate coprime arrays in the manner of [12–15]. In Section 8.5, the spatial Nyquist condition and partition difference redundancy are used to develop a numerical sieve to generate arrays with low sidelobe antenna patterns. In Section 8.9, the linear Nyquist condition is generalized to two dimensions to generate 2-D sparse arrays with low sidelobe performance. In Section 8.8, angle-of-arrival estimation methods are developed first for linear sparse arrays that satisfy the spatial Nyquist condition and then generalized to arrays that do not satisfy the spatial Nyquist condition. Finally, the methods of Ishimaru [6] and Mitra [9–11] are developed in Section 8.7 to illustrate two techniques that use formulations of the uniform linear array antenna pattern to create unequally spaced arrays elements.

8.1 Sparse Linear Arrays

We first introduce sparsely populated linear arrays and define terminology. The spatial Nyquist sample interval for linear arrays is one-half wavelength (λ) (see Section 7.3.1). A fully populated linear array with N elements is an array where all N elements are uniformly spaced by one-half wavelength. The length L of a fully populated array is $L = (N - 1)d$.

> *Definition:* We define a *sparsely populated array* of length L as an array with N elements where $(N - 1)d < L$ and $1/d$ define the spatial Nyquist rate for a given unambiguous field of view ($d = \lambda/2$ provides a 180° field of view).

We make a distinction between sparse arrays and thinned arrays. Thinned arrays eliminate around 10 – 20 percent of the elements in an array in order to reduce the cost of the array. Array thinning is particularly applicable for two-dimensional arrays. On the other hand, for sparse arrays, the number of elements that are eliminated from a fully populated array are greater than 20 percent and can be larger than 50 percent. Thus the essence of a sparse array is, in fact, sparseness.

The two-element interferometer is an example of a sparse array when the element separation or baseline is greater than one-half wavelength. Typically, the interferometer baseline is several wavelengths in length. As a result the antenna pattern for an interferometer consists on multiple grating lobes due to the spatial sample rate being less than the Nyquist rate. The long interferometer baseline provides enhanced angle resolution and accuracy at the expense of introducing grating lobes or digital angle ambiguities.

The advantage of a sparse array is that the grating lobes or angle ambiguities can be reduced or eliminated by implementing more than two elements but fewer than the number of elements that would constitute a fully populated array. The sparse array antenna pattern performance can be measured in terms of peak sidelobe, RMS sidelobe, and/or beamwidth. The challenge in sparse array antenna design is to optimize these measures of antenna performance using a minimal number of antenna elements.

8.2 Interval Partitions

The development of sparse spatial and temporal sampling techniques is an active area of research, and interval partitions are directly related to this application.

In this section, we develop a theory of integral length interval partitions that leads to optimized sparse sampling techniques. We start with the mathematical definition of an integer partition and adapt it for our application to sparse arrays.

Definition: A *partition* P of an interval $[0\ L]$ of length L is a sequence of points p_i $I = 0$:M such that

$$0 = p_0 < p_1 < \ldots < p_{M-1} < p_M = L \tag{8.1}$$

Since we are applying the concept of interval partitions to array element spacing, we will adapt the definition as follows:

Definition: A *restricted integer partition* (*RIP*) P of an interval of length L is an interval partition defined by a sequence of integers k_i, $I = 0{:}M < N$, such that

$$0 = k_0 < k_1 < \ldots < k_{M-1} < k_M = N \tag{8.2}$$

and

$$0 = \frac{L}{N}k_0 < \frac{L}{N}k_1 < \ldots < \frac{L}{N}k_{M-1} < \frac{L}{N}k_M = L \tag{8.3}$$

is a partition of L. Typically, L is the length of a linear array where L is a multiple number of half wavelengths ($L = K\lambda/2$), and M is the number of partitions of the array length.

Because the position of the first element is arbitrary, we modify (8.2) for application to sparse arrays. For an array partition, the k_i denote the position number of the antenna elements where $k_0 = 1$ is the first element and k_m is the $(M + 1)$st element. Also for an array, we generally have $K = N$. Therefore, we consider restricted integer partitions of intervals with integer length $N - 1$ defined by the interval $[1\ N]$. The restricted interval partitions P_N of size M with intervals of length $N - 1$ have the form

$$1 = k_0 < k_1 < \ldots < k_{M-1} < k_M = N \tag{8.4}$$

where each $k_i \in [1, 2, \ldots, N]$ for $i = 0, \ldots, M \leq N$ and $k_i \neq k_j$. For a sparse array, the partition defined in (8.4) corresponds to an array with length $(N - 1)\ \lambda/2$ with elements locations defined as follows,

$$\frac{\lambda}{2} = k_0 < k_1\frac{\lambda}{2} < \ldots < k_{M-1}\frac{\lambda}{2} < k_M = N\frac{\lambda}{2} \tag{8.5}$$

Definition: The *partition differences* in a RIP P_N of size M is the set of integers defined by

$$\{k_j - k_i : 0 \leq i < j \leq M\} \tag{8.6}$$

The total number of partition differences defined by a RIP P_N of size M is $\frac{1}{2}M(M+1)$. The total number of distinct partition differences for a RIP P_N is $N - 1$.

$$1, 2, 3, \ldots, N - 1 \tag{8.7}$$

Definition: The number of partition differences in a RIP P_N of an integer N is a *minimum redundancy partition* (*MRP*) if the number of partition differences is equal to the total number of distinct differences for the integer N.

If P_N is a minimum redundancy partition of an integer N of size M, then

$$N - 1 = \frac{1}{2}M(M + 1) \tag{8.8}$$

Example 1: $P_N = \{1, 3, 6, 7\}$ is a minimum redundancy partition of the integer $N = 7$ for $M = 3$ since

$$7 - 6 = 1,\ 3 - 1 = 2,\ 6 - 3 = 3,\ 7 - 3 = 4,\ 6 - 1 = 5,\ 7 - 1 = 6$$

and

$$7 - 1 = 6 = \frac{1}{2} \times 3 \times 4$$

Example 2: $P_N = \{1, 3, 4\}$ is a minimum redundancy partition of the integer $N = 4$ for $M = 2$ since

$$4 - 3 = 1,\ 3 - 1 = 2,\ 4 - 1 = 3$$

and

$$4 - 1 = 3 = \frac{1}{2} \times 2 \times 3$$

However, with the exception of the trivial partition $P_N = \{1, 2\}$, Examples 1 and 2 are the only minimum redundancy partitions.

Proposition 1: Let N be an integer. There are no minimum redundancy partitions P_N unless $N \leq 7$.

Proof: It is trivial to show that $N = 1, 2, 3$, and 7 define minimum redundancy partitions and that $N = 4, 5$, and 6 do not. Let N be an integer, let P be a complete partition of N, and assume that $N > 7$. Because P is complete, it defines all the $N - 1$ differences. In particular, the difference $N - 2$ must be defined, which implies that either 2 or $N - 1$ is contained in P. Without loss of generality, we choose $N - 2$ to be in P.

Because the difference $N - 3$ must be defined, we are forced to choose 3 to be a member of P to avoid redundancy. But now $N - 4$ must be a defined difference. If we choose $N - 3$ to be in P, then we have the redundant difference $N - 1 - (N - 3) = 2$ and $3 - 1 = 2$. So we have defined the differences $1, 2, N - 1, N - 2, N - 3, N - 4$ with the partition $\{1, 3, N - 1, N\}$.

To define the difference $N - 5$, we choose $N - 4$ to be an element of P. But now we face a dilemma. How do we define the difference $N - 6$?

We are forced to choose $N - 8$ or 7 as an element of P. If we choose $N - 8$, then we must have $N - 8 - 3 = N - 6$ or $N - 8 - 1 = N - 6$, both of which are contradictions. If we choose 7, then we must have $N - 1 - 7 = N - 6$ or $N - 2 - 7 = N - 6$ or $N - 4 - 7 = N - 6$. All of these are contradictions. Thus, for N to be a minimum redundancy, we must have $N \leq 7$.

Restricted integer partitions P_N that are *almost minimum redundancy partitions* (*AMRP*) exist where the number of possible differences defined by the restricted integral partition is unique and almost equal to $N - 1$. For a specific integer N, many restricted integral partitions P_N have repeated differences. From this set of overly redundant partitions, those with minimal redundancy are called *minimally overly redundant partitions* (*MORP*).

Example 2: Almost Minimum Redundancy Interval Partitions
Consider the case of partitions P with 5 numbers and $M = 4$ (8.4). The number of partition differences is equal to 10. This suggests we may be able to derive a minimum redundancy partition P_N with $N = 11$. However, it does not take much effort to demonstrate that a minimally redundant partition does not exist. We now determine the minimum integer such that a partition with 4 numbers provides 10 unique differences. It turns out that $N = 13$ provides a restricted integral partition that is almost minimally redundant for a partition size of $M = 4$.

$P = \{1, 3, 6, 12, 13\}$ has 10 unique differences:

$13 - 12 = 1$
$3 - 1 = 2$
$6 - 3 = 3$
$6 - 1 = 5$
$12 - 6 = 6$
$13 - 6 = 7$
$12 - 3 = 9$
$13 - 3 = 10$
$12 - 1 = 11$
$13 - 1 = 12$

As predicted, two differences are missing (4, 8). If we now add one more number to the partition to make a partition of length 5, we can achieve the differences for 4 and 8, but we also introduce redundant differences. Because there are 15 total differences for a partition of length 5 and for $N = 13$, there are 12 total unique differences; we should have 3 redundant differences. If we add the number 9 to the partition P, then we have the following 15 differences that determine a MORP.

$$P = \{1, 3, 6, 9, 12, 13\} \tag{8.9}$$

$13 - 12 = 1$
$3 - 1 = 2$
$6 - 3 = 3$
$12 - 9 = 3$
$9 - 6 = 3$
$13 - 9 = 4$
$6 - 1 = 5$
$12 - 6 = 6$
$9 - 3 = 6$
$13 - 6 = 7$
$9 - 1 = 8$
$12 - 3 = 9$
$13 - 3 = 10$
$12 - 1 = 11$
$13 - 1 = 12$

Notice that we have 2 redundancies for 3 and 1 for 6 for a total of 3 redundancies. Also note that, for the partition P as defined, we have generated a minimally overly redundant partition.

Now consider the case of partitions P with 5 numbers. The number of partition differences is equal to 15, suggesting that we may be able to derive a minimum redundancy partition for the restricted integral partition P_N for $N = 16$; however, again, such a solution does not exist. We now determine the minimum integer such that a partition P_N of size $M = 5$ provides 15 unique differences. Upon inspection, it turns out that $N = 19$ provides an almost minimum redundancy partition for P_{15} with size $M = 5$.

$P = \{1, 3, 6, 10, 18, 19\}$ has 15 unique differences, and P is an almost minimum redundancy partition (AMRP).

8.3 Cyclic Coprime Partitions

The application of coprime integers to sparse array generation has been developed by Vaidyanathan et al. [12–15] and shown to create low sidelobe antenna patterns. We consider restricted integer partitions of intervals with integer length $N - 1$ defined by the interval [1 N]. The restricted integer partitions P_N of size M with intervals of length $N - 1$ have the form

$$1 = k_0 < k_1 < \ldots < k_{M-1} < k_M = N \qquad (8.10)$$

Definition: A *cyclic partition* generated by the integer p is defined by the set of integers $1 + k{*}p \bmod(N)$ $k = 1, 2, \ldots.$

If N and p are coprime, then the partition contains all the integers from 1 to p. If N and p are not coprime, the partition is defined by a subgroup of the integers 1 to p, and the cyclic partitions are uniformly separated. Uniformly separated cyclic partitions generate differences that are multiples of the generating integer p and as such are not able to achieve a set of minimum redundancy differences. However, if we take two cyclic partitions defined by generating integers p_1 and p_2 that are coprimes, then we can generate a partition of an integer with multiple defined differences.

Example 3: Coprime Partitions
Let $N = 16$ and $p_1 = 3$ and $p_2 = 5$. Then the two cyclic partitions generate the following partition for $N = 16$.

$$P_1 = \{1, 4, 6, 7, 10, 11, 13, 16\} \tag{8.11}$$

The 8 points in the partition P_1 generate 28 partition differences. Because there are only 15 unique differences, the partitions differences defined by P_1 are highly redundant but include all of the possible differences with the exceptions of 8, 11, and 13. The following list contains all of the partition differences defined by P_1:

1, 1, 2, 2, 3, 3, 3, 3, 3, 4, 4, 5, 5, 5, 6, 6, 6, 6, 7, 7, 9, 9, 9, 10, 10, 12, 12, 15

If we choose $p_1 = 3$ and $p_2 = 7$, we have the following partition:

$$P_2 = \{1, 4, 7, 8, 10, 13, 15, 16\} \tag{8.12}$$

The 28 partition differences that are defined by the 8 points in the partition are listed here. Note that only two distinct differences are not covered (10 and 13).

1, 1, 2, 2, 3, 3, 3, 3, 3, 4, 5, 5, 6, 6, 6, 6, 7, 7, 8, 8, 9, 9, 9, 11, 12, 12, 14, 15

As a result, it may be concluded that the partition P_2 is preferred over the partition P_1 because P_2 defines all but 2 differences; whereas P_1 defines all but 3.

8.3.1 Application to Spatial Sampling

A restricted integer partition of an interval with integer length can be applied to sparse sampling in either space or time. A minimum redundancy partition allows for the sampling of all frequencies. Redundant differences defined by cyclic coprime partitions add robustness to sampling and can be beneficial in controlling spatial and temporal frequency aliasing.

Below are antenna patterns for various partitions. Figure 8.1 shows the pattern for the minimum redundancy partition $P = \{1, 3, 6, 7\}$. Notice the consistent sidelobe level at about -5 dB below the peak of the mainlobe. The absence of grating lobes indicates that Nyquist sampling is achieved.

Figure 8.2 shows the patterns for the partitions defined in Example 2. The figure on the left in Figure 8.2 is the pattern for the almost minimum redundancy partition $P = \{1, 3, 6, 12, 13\}$. For this partition, the sidelobes are uniform, similar to the minimum redundancy partition in Example 1. The figure on the right side of Figure 8.2 is the pattern when the number 9 is added to the almost minimum

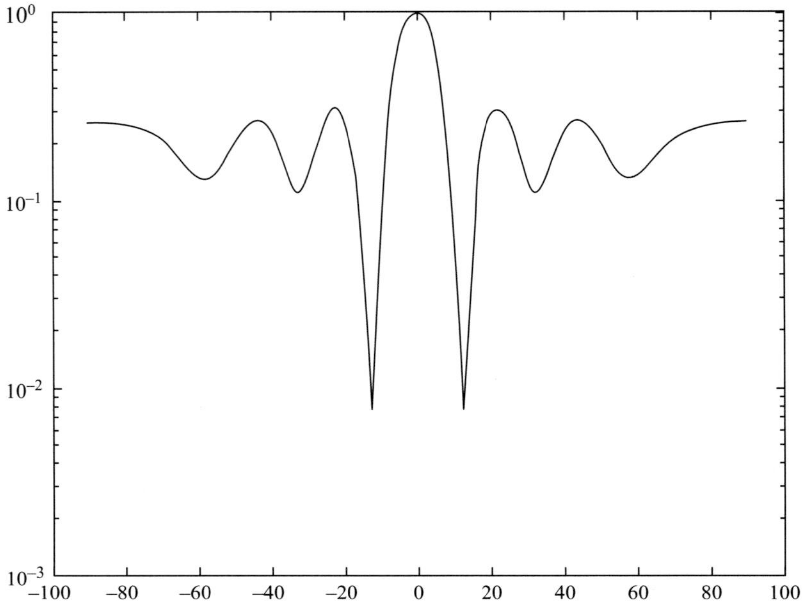

Figure 8.1. Antenna Pattern for the Minimum Redundancy Partition P = {1, 3, 6, 7}

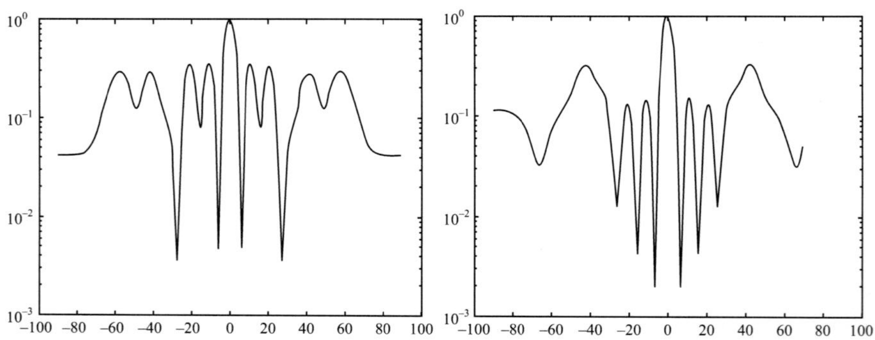

Figure 8.2. Antenna Pattern for the Almost Minimum Redundancy Partitions in Example 2 – P = {1, 3, 6, 12, 13} (left); P = {1, 3, 6, 9, 12, 13} (right)

redundancy partition. Note that the interior sidelobes are significantly reduced as a result of adding the one number to the partition.

Figure 8.3 shows the patterns for two cyclic coprime arrays: one generated by the primes 3 and 5 and the other generated by the primes 3 and 7. Note the irregular sidelobe structure for both arrays. These two coprime partitions do not have particularly low sidelobe structure, especially for the exterior lobes, indicating too much cyclic structure for the partition. To break up the cyclic structure, we

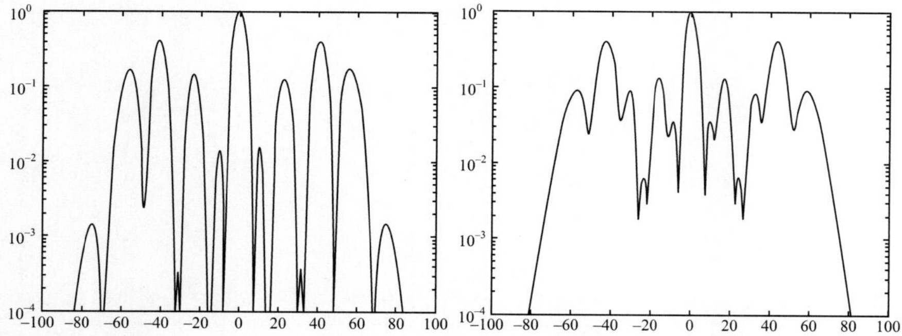

Figure 8.3. Antenna Pattern for the Cyclic Coprime Partitions in Example 3: Pattern Generated Using 3 and 5 Primes, $P_1 = \{1, 4, 6, 7, 10, 11, 13, 16\}$ (left); Pattern Generated Using 3 and 7 Primes, $P_2 = \{1, 4, 7, 8, 10, 13, 15, 16\}$ (right)

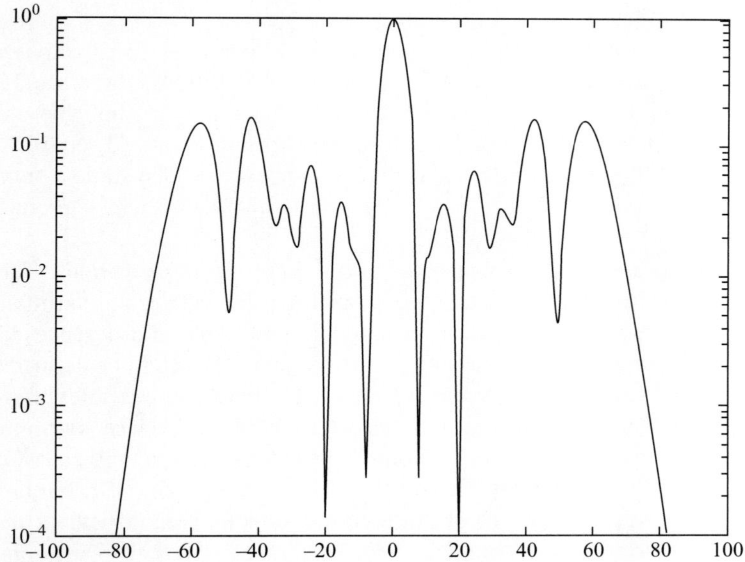

Figure 8.4. Antenna Pattern for the Cyclic Coprime Partitions in Example 3; Pattern Generated Using 3, 5, and 7 Primes, $P_1 = \{1, 4, 6, 7, 8, 10, 11, 13, 15, 16\}$

introduce a third prime. Figure 8.4 shows the pattern for the cyclic coprime array generated by 3, 5, and 7, which essentially adds to the numbers to the partitions defined in Figure 8.3. The coprime array generated by the 3 primes ($P = \{1, 4, 6, 7, 8\ 10, 11, 13, 15, 16\}$) has lower sidelobe structure for the interior sidelobes due to

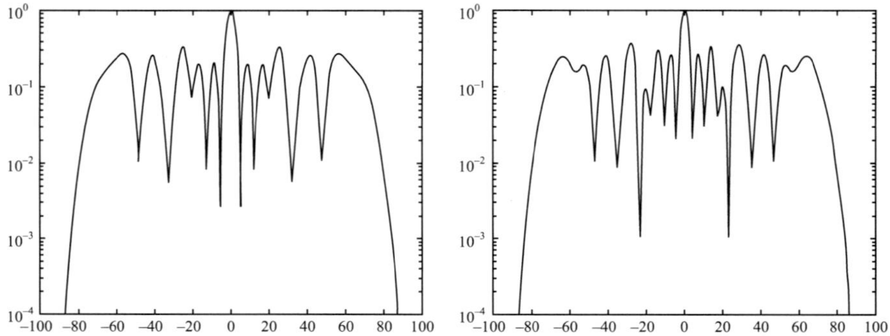

Figure 8.5. Antenna Pattern for $M=5$ Defined by {1, 3, 6, 10, 15, 16} (left); {1, 3, 6, 10, 18, 19} (right)

the redundancy in the difference partitions. In fact, these interior first sidelobes are lower than the first sidelobes for a uniform linear array.

Finally, because the cyclic partitions were all examples of partitions of the number 16, we observe that, for $M=5$ number of partitions, there are 15 total differences. The smallest number with 15 partitions is $N=16$, but this does not yield a minimum redundancy partition. But the partition $P=\{1, 3, 6, 10, 15, 16\}$ produces 13 of the 15 differences with 2 redundancies. The almost minimum redundancy partition for $M=5$ is $P=\{1, 3, 6, 10, 18, 19\}$, which produces 15 unique difference out of a total of 18.

Figure 8.5 shows the antenna patterns for each of these partitions. The peak sidelobe for the first partition is slightly lower than for the second partition, possibly due to the fact that 13 out of 15 is a higher percentage of differences than 15 out of 18. This suggests that a measure for array sparseness might be defined by the ratio of the unique partition differences to the total number of possible differences. However, difference redundancy must also play a factor because the minimum redundancy partition $P=\{1, 3, 6, 7\}$ would have a ratio equal to unity, as would the uniform linear array defined by the partition $P=\{1, 2, 3, 4, 5, 6, 7\}$, but the latter partition has a much lower sidelobe structure due to its redundancy. In fact, the uniform linear array has 21 redundant differences as opposed to the minimum redundancy partitionm which has none.

8.4 Nested Cyclic Partitions

Another example of sparse array partitions is the nested cyclic partition, in which two or more cyclic partitions are nested to create a larger sparse array [16]. The definition of a cyclic partition of the integer N are the integers defined by $1 + k \cdot p_1$ mod(N) for all positive integers k, and if p_1 and p_2 are coprime, then the combined cyclic partitions is called a "cyclic coprime array." We can expand the definition of cyclic partition to define the concept of a nested cyclic partition.

> *Definition:* A nested cyclic partition of the integer N is defined as the set of integers defined by the following sequence of cyclic coprime partitions.
>
> $$P = (1 + k \cdot p_1 \bmod(N-1)) \cup (1 + k \cdot p_2 \bmod(p_1)) \cup (1 + k \cdot p_3 \bmod(p_2)) \cup \ldots \cup (1 + k \cdot p_K \bmod(p_{K-1})) \tag{8.13}$$
>
> where k is a positive integer, and $p_1, p_2, \ldots, p_K$ are all divisors of $N-1$ but not coprime. In fact, nesting occurs when p_{i+1} is a divisor of p_i for $i = 1{:}K-1$. Consider the following three nested coprime partitions of the integer $N = 17$.
>
> *Example 1:* $N = 17$, $[p1, p2] = [4, 1]$, then the nested coprime partition is $P = [1, 2, 3, 4, 5, 9, 13, 17]$.
>
> *Example 2:* $N = 17$, $[p1, p2] = [4, 2]$, then the nested coprime partition is $P = [1, 3, 5, 9, 13, 17]$.
>
> *Example 3:* $N = 17$, $[p1, p2, p3] = [8, 4, 2]$, then the nested coprime partition is $P = [1, 3, 5, 9, 17]$.

Figure 8.6 shows an example of a modified nested cyclic array of GPS antennas that is designed to track occulting GPS satellites. Because the satellites are moving mostly in the elevation plane and atmospheric refraction is mostly in the elevation plane, the linear array is sufficient to compute the excess angle-of-arrival. Because multipath is of concern, the antenna pattern is designed to have low elevation sidelobes at the horizon where multipath is likely to occur. A linear sparse array is preferred to reduce the cost and weight of the antenna fixture while maintaining adequate sidelobe multipath interference rejection and angle resolution.

8.5 Numerical Sieve Methods for Optimized Sparse Array Generation

In Section 8.3, it was suggested that a possible measure for array sparseness for a partition P should include the number of unique partition differences and a measure of minimum redundancy [17]. We now define the following two metrics of a partition of a number N based on the observations. The first metric is designed to measure the relative antenna pattern sidelobes. Essentially, sidelobe levels are affected by the number of partition differences, and the redundancy of partition differences in a partition. Thus we define

$$\Lambda_1 = \frac{1}{\sqrt{2}}\sqrt{\left(\frac{\text{Unique Differences}}{\text{Total Differences}}\right)^2 + \left(\frac{\text{Redundancies} + 1}{\text{Possible Redundancies} + 1}\right)^2} \tag{8.14}$$

Figure 8.6. Linear Sparse GPS Array Using a Modified Nested Coprime Architecture Developed by Propagation Research Associates, Inc. for Atmospheric Refraction Characterization

The second metric is a measure of the minimum redundancy. Thus we want to penalize redundancy of partition differences but encourage the maximum number of differences. As a result, we define

$$\Lambda_2 = \sqrt{\left(\frac{\text{Unique Differences}}{\text{Total Differences}}\right)^2 - \left(\frac{\text{Redundancies}}{\text{Possible Redundancies}}\right)^2} \tag{8.15}$$

The Λ metric is a measure of the performance of a sparse array defined by the partition where $\Lambda \in [0\ \ 1]$. For Λ near 1, the sparse array antenna pattern should approximate a uniform linear array, and for Λ near 0, the antenna pattern should approximate an interferometer with two antenna elements. For the examples in

Figure 8.5, we have a score of $\Lambda = 0.6132$ for the partition $P4 = \{1, 3, 6, 10, 15, 16\}$, whereas for the partition $P5 = \{1, 3, 6, 10, 18, 19\}$, we have $\Lambda = 0.5893$. From Figure 8.5, it can be seen that the peak sidelobe for $P4$ is slightly lower than the peak sidelobe for the nearly complete difference partition, which justifies the slightly lower Λ score. For any uniform linear array, $\Lambda = 1$. The minimum redundancy partition $P = \{1, 3, 6, 7\}$ has a score of $\Lambda = 0.7071$. For an interferometer that is N half-wavelengths in length, the partition is simply $P = \{1, N\}$, and the Λ score is

$$\Lambda_1 = \frac{1}{\sqrt{2}}\sqrt{\left(\frac{1}{N-1}\right)^2 + \left(\frac{1}{\frac{1}{2}(N-2)(N-1)+1}\right)^2} \tag{8.16}$$

$$\Lambda_2 = \frac{1}{N-1} \tag{8.17}$$

For a large baseline interferometer, the Λ score is small and approaches 0 for large N. Also note that, for the special case $N = 2$, the score is $\Lambda = 1$ because this is a degenerate uniform linear array.

> *Problem Statement:* For a given integer partition P containing M numbers the challenge is to find the number N such that the partition P of N has the highest Λ score.

The solution to this problem starts by finding the least integer N that contains M differences by solving the following:

$$N - 1 = \frac{1}{2}M(M+1) \tag{8.18}$$

The next step calls for increasing N to $N+1, N+2, \ldots$ and finding all redundantly nearly complete difference partitions for each iteration. The one with the highest Λ score is the solution. The number of iterations required depends on the value of M, but eventually the Λ values of partitions of $N+K$ decrease in value for large K simply because the denominators in the Λ metric are large relative to the numerators. Let $N+K$ be the integer such that the partition P with M numbers is a nearly complete difference partition.

> *Proposition 2:* The solution to the Problem Statement lies between N and $N+K$.

> *Proof:* Given to the definition of N the solution cannot be a number less than N. Given the definition of $N+K$ the solution cannot be greater than $N+K$ due to the increase in the denominator in the Λ metric relative to numerator values which remain fixed for integers greater than $N+K$.

Table 8.1 shows values for both Λ metrics, as well as the sum of the two metrics for the partitions shown in the preceding examples. Note that the cyclic coprime

Table 8.1. Λ *Metric Values for Various Types of Partitions*

Number	Partition Type	Λ1 Value	Λ2 Value	Λ1 + Λ2
1	Uniform linear array	1	0	1
2	$P = \{1, 16\}$ *Interferometer*	0.0476	0.0660	0.1136
3	$P = \{1, 3, 6, 7\}$ (MRP)	0.7085	1	1.7085
4	$P = \{1, 3, 6, 12, 13\}$ (AMRP)	0.5894	0.8333	1.4227
5	$P = \{1, 3, 6, 9, 12, 13\}$ (MORP)	0.7084	0.9990	1.7074
6	$P = \{1, 3, 6, 10, 15, 16\}$ (MORP)	0.6132	0.8665	1.4797
7	$P = \{1, 3, 6, 10, 18, 19\}$ (AMRP)	0.5500	0.7778	1.3278
8	$P = \{1, 4, 6, 7, 10, 11, 13, 16\}$ (Cyclic coprime generated by 3, 5)	0.5323	0.7156	1.2479
9	$P = \{1, 4, 7, 8, 10, 13, 15, 16\}$ (Cyclic coprime generated by 3, 7)	0.6221	0.8550	1.4771
10	$P = \{1, 4, 6, 7, 8, 10, 11, 13, 15, 16\}$ (Cyclic coprime generated by 3, 5, 7)	0.6936	0.8863	1.5799

arrays (numbers 8 and 9 in the table) perform differently with respect to the metrics. The coprime array generated by the primes 3, 5, and 7 (number 10 in the table) has the highest metric value for the coprime arrays, but this is due to the large number of redundant differences and values in the partition. For smaller size partitions, numbers 3 and 5 in the table have the highest value. The partition defined in number 3 is a complete difference partition (CDP) with no redundancies, and the partition defined in number 5 is a redundantly nearly complete difference partition (RNCDP).

Note the Λ1 value for partition number 4 in the table. It appears low relative to the antenna pattern in Figure 8.2, which is similar to the antenna pattern in Figure 8.1 for partition number 3 in the table. However, the sidelobes for the AMRP (number 4) pattern are uniformly higher than that for the MRP partition (number 3) justifying the lower score for the former pattern.

Two techniques were derived to generate partitions of numbers that are minimum redundancy or minimally overly redundant difference partitions.

8.5.1 Summary of Numerical Sieve Method

Method 1: Consider the partitions with M numbers. The number of difference defined by the partition is $\frac{1}{2}M(M+1)$. The smallest number N with that number of differences is defined by the equation $N - 1 = \frac{1}{2}M(M+1)$. However, we know that there is only one complete difference partition (Example 1), so we find the smallest number $K > N$ such that the partition of K with M numbers generates $N - 1$ unique differences. This partition is a nearly complete difference partition of K.

Method 2: For a given integer N, generate coprime cyclic partitions. The combined coprime partitions generate redundantly nearly complete difference partitions.

Two metrics for evaluating the effectiveness of partitions were proposed as follows:

$$\Lambda_1 = \frac{1}{\sqrt{2}}\sqrt{\left(\frac{\text{Unique differences}}{\text{Total differences}}\right)^2 + \left(\frac{\text{Redundancies}}{\text{Possible redundancies}}\right)^2} \tag{8.19}$$

$$\Lambda_2 = \sqrt{\left(\frac{\text{Unique Differences}}{\text{Total Differences}}\right)^2 - \left(\frac{\text{Redundancies}}{\text{Possible Redundancies}}\right)^2} \tag{8.20}$$

The metric $\Lambda 1$ is a measure of sidelobe performance, whereas $\Lambda 2$ is a measure of the minimal redundancy of the partition. The sum $\Lambda = \Lambda 1 + \Lambda 2$ of the two metrics is also presented as a reasonable measure of sparse arrays. For a given partition P containing M numbers, the challenge is to find the number N such that the partition P of N has the highest Λ score. It is shown that Λ achieves a maximum over a defined interval $[N\, N + K]$, where $N + K$ is the lowest integer that defines a nearly complete difference partition for the partition P and N satisfies

$$N - 1 = \frac{1}{2}M(M + 1) \tag{8.21}$$

Partitions that maximize the Λ metric tend to produce antenna patterns with lower sidelobe values, and the Λ metric can be used to generate sparse arrays with acceptable antenna patterns.

8.6 Sparse Array Antenna Performance

When the number N is large, it becomes numerically prohibitive to search the possible array patterns to determine an optimal array configuration. When optimized, the sum of the two metric values $\Lambda 1$ and $\Lambda 2$ create a feasible set of one-dimensional sparse array architectures. The $\Lambda 1$ and $\Lambda 2$ metrics essentially act as a sieve to select the array configurations with good redundancy and array pattern properties. Using other more conventional metrics such as peak sidelobe or root-mean-square sidelobe performance, an optimized sparse array can be selected with fewer computations than a complete search using the conventional metrics. The advantages of the $\Lambda 1$ and $\Lambda 2$ metrics are (1) the metrics are numerically efficient to compute, and (2) from empirical observation, the metrics define a feasibility set of arrays that contains, among others, coprime, nested cyclic, and reduced redundancy arrays when they exist for a given maximum arrays size of N elements. It should be noted that, for some values of N, there do not exist any coprime arrays, but the $\Lambda 1 + \Lambda 2$ sieve always defines a feasible set of array configurations.

Consider the set of arrays with 10 possible locations. For 10 spaces, there are $256 = 2^8$ array possibilities (8 inner locations either have an antenna or do not, whereas the two outer locations must always have antennas), some of which are equivalent with respect to a reflection about a center location and produce similar antenna patterns ([1, 2, 10] and [1, 9, 10] are reflective equivalent). This set of arrays has a nested pattern, a coprime pattern, and a reduced redundancy pattern available to compare, as shown in Table 8.2.

Only two of the possible unique arrays (total arrays minus reflective equivalent arrays) achieve the maximum $\Lambda 1 + \Lambda 2$ metric for $N = 10$; they are the two reduced redundancy arrays. If the feasible set was defined by the two largest values instead of just the largest value, then the feasible set would contain four arrays configurations, now including the two nested coprime arrays. Figure 8.7 shows the antenna patterns for sparse array contained in the feasible set. The antenna patterns are all different, but if we were to select one solely based on peak sidelobe levels, then the nested coprime array (dark gray) slightly edges out the reduced minimum

Table 8.2. Nested Patterns, Coprime Patterns, and Reduced Redundancy Patterns for an Array with 10 Possible λ/2 Spacing Locations

Nested Array Patterns	**Coprime Array Patterns**	**Reduced Redundancy Array Patterns**
$[p_1, p_2] = [1, 2]$ gives [1, 2, 4, 6, 8, 10]	$[p_1, p_2] = [3, 4]$	[1, 2, 3, 7, 10]
$[p_1, p_2] = [4, 5]$ gives [1, 2, 3, 4, 5, 10]	gives [1, 4, 5, 7, 9, 10]	[1, 2, 5, 8, 10]

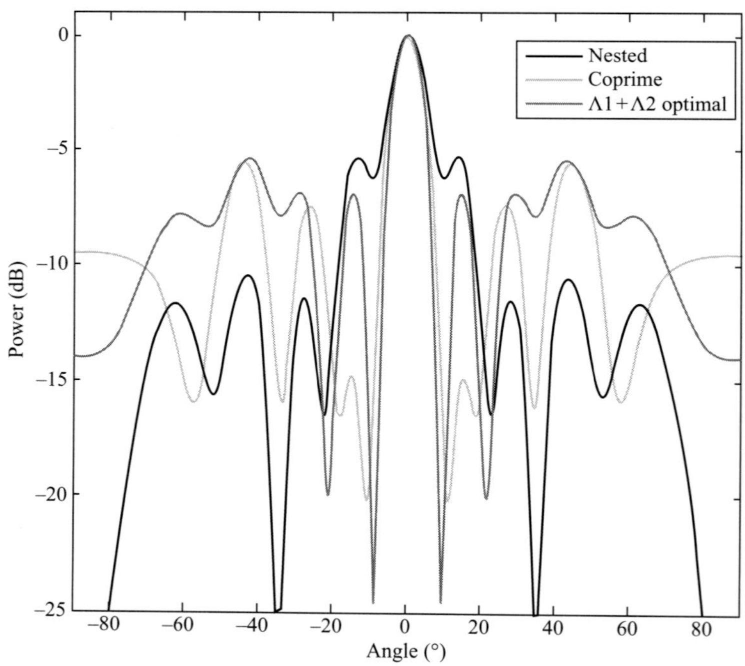

Figure 8.7. Coprime Arrays (light gray) Pattern: Nested Cyclic (dark gray) and Optimal $\Lambda 1 + \Lambda 2$ Patterns (medium gray) for $N = 10$

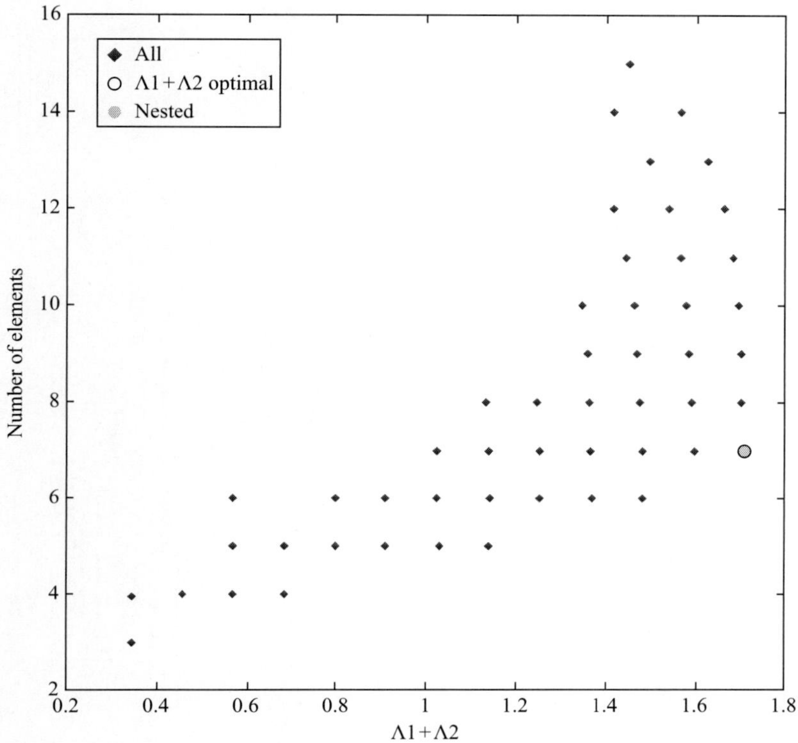

Figure 8.8. *Number of Elements Versus Λ1 + Λ2 Metric for All Arrays in 16 Spaces (small diamonds), Λ1 + Λ2 Optimal Array (dark circle), Nested Cyclic (medium disk)*

redundancy (medium gray) and the coprime array (light gray) at the cost of requiring an extra antenna element.

Now consider the case when $N = 16$. There are no coprime arrays or minimum redundancy arrays, but nested arrays exist. Using the maximum Λ1 + Λ2 metric to create the feasible set, there are only 80 sparse array configurations out of a possible 8,192 total configurations.

Figure 8.8 shows all the possible values for Λ1 + Λ2 versus the number of elements in the sparse array. As can be seen the 80-array configuration, all live, including the two nested coprime arrays, in the dark gray circle to the extreme right on the graph. Also note that these arrays consist of seven elements, providing a reduced redundancy with good pattern performance. Figure 8.9 shows Λ1 + Λ2 versus the peak sidelobe values for all 8,192 possible sparse arrays. Note the maximum Λ1 + Λ2 arrays again live on the extreme lower right of the graph (medium gray dots), indicating that these array exhibit good peak sidelobe performance. The nested coprime arrays also occur on the extreme right but have higher peak sidelobe performance than other sparse arrays in the feasible set.

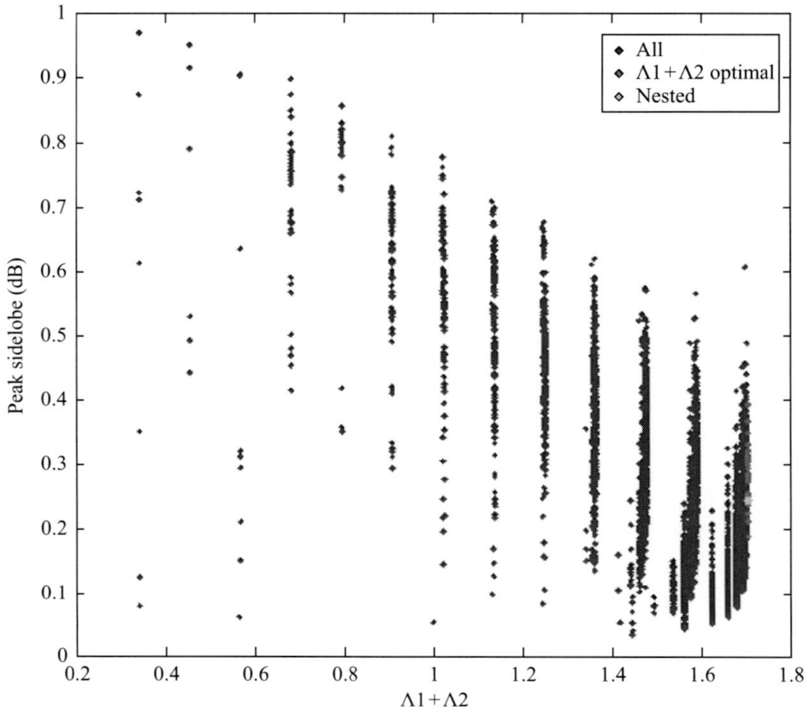

Figure 8.9. Peak Side Lobe Power Versus Λ1 + Λ2 Metric for All Arrays in 16 Spaces (dark), Λ1 + Λ2 Optimal Array (medium), Nested (light)

Also, there are dark gray diamonds with lower peak sidelobe performance with slightly smaller values of Λ1 + Λ2, indicating that we may want to expand the feasible set if peak sidelobe performance is the primary focus. Figure 8.10 is an example pattern for one member in the feasible set defined by the maximum of Λ1 + Λ2. The non-coprime array pattern achieves level first and second peak sidelobes and has better far out sidelobe performance.

8.7 Antenna Pattern Methods

Other approaches for designing sparse array architectures directly manipulate the radiation or antenna pattern to create a desired optimal sidelobe performance [1–6,9–11]. These techniques develop sparse or unequally array element spacings with antenna patterns that attempt to emulate fully populated uniform linear arrays. In the first method in this section, the objective is to develop arrays where the elements have been repositioned to achieve control of sidelobe amplitude levels. In Chapter 4, we introduced Taylor amplitude weights that are used to reduce sidelobe levels in uniform linear arrays where sidelobe levels are reduced at the expense of

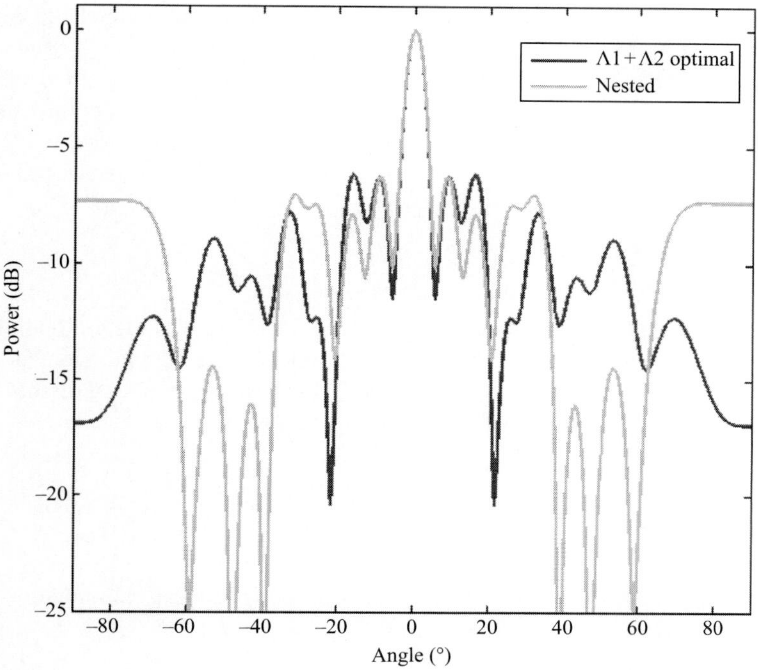

Figure 8.10. Example Patterns: Λ1 + Λ2 Optimal Array (dark), Nested Cyclic (light)

growing the beamwidth of the pattern. The method of unequally spaced antenna elements reduces the first few sidelobe levels in a way similar to the method in defining Taylor weights at the expense of growing the far-out sidelobes.

The second method in this section uses polynomial factorization of the two-way effective antenna pattern to create sparse transmit and receive arrays. By choosing the two-way effective pattern as the pattern for a uniform linear array, the two-way antenna pattern always results in a sinc-like response with a 13.2-dB first sidelobe. Nonunity factorizations are developed that provide amplitude weights for sidelobe amplitude reduction.

8.7.1 Unequally Spaced Arrays

Ishimaru [6] developed a process that creates unequal array element spacings based on maximum sidelobe voltage level control for a given number of sidelobes. The process defined in [6] is focused on manipulation of the elements to control near-in sidelobe levels similar to the construction by Taylor [7]. This manipulation results in a contraction of the array size. In previous sections, we focused on techniques that used sparse array designs for maximum array size in order to achieve enhanced angle resolution and accuracy, but Ishmaru is focused on unequal antenna element spacings created by moving elements in a fully populated uniform linear array. The methods developed by Ishimaru therefore do not generate sparse arrays as defined in this chapter. However, the concept of using array element spacing to control sidelobe levels fits into the general theme of optimizing the antenna pattern response.

The far-field antenna pattern for an array with N elements is given by

$$E(\theta) = \sum_{n=1}^{N} I_n e^{jks_n \sin(\theta)} \tag{8.22}$$

where I_n is the voltage amplitude at the nth array element, N is the total number of radiators, k is the wave number, s_n is the relative location of the nth element along the linear array, and θ is the angle-of-arrival. Ishimaru [6] applies Poisson's sum formula to write an equivalent formulation of (8.22).

$$E(\theta) = \sum_{n=1}^{N} f(n) = \sum_{m=-\infty}^{\infty} \int_{-\infty}^{\infty} f(v) e^{2\pi\, jmv} dv \tag{8.23}$$

By applying Poisson's formula, Ishimaru is able to define a continuous source position function

$$s = s(v) \tag{8.24}$$

that satisfies the following constraints:

$$s_n = s(n) \quad \text{and} \quad n = v(s_n) \tag{8.25}$$

This source function is ultimately the function that defines the array element spacings that achieve sidelobe amplitude level control. Using a change of variable, (8.23) becomes

$$E(\theta) = \sum_{m=-\infty}^{\infty} \int_{s_0}^{s_N} f(s) \frac{dv}{ds} e^{2\pi\, jmv(s)} ds \tag{8.26}$$

$$E(\theta) = \sum_{m=-\infty}^{\infty} E_m(\theta) \tag{8.27}$$

$$E_m(\theta) = \int_{s_0}^{s_N} A(s) \frac{dv}{ds} e^{-j(\psi(s) - 2\pi mv(s))} e^{jks \sin(\theta)} ds \tag{8.28}$$

where

$$I_n = I(s_n) = A_n e^{-j\psi_n} \tag{8.29}$$

Using the normalized change of variables,

$$u = ka\sin(\theta) \tag{8.30}$$

$$2a = s_N - s_0 \tag{8.31}$$

Define the normalized source position function

$$x = x(y) \quad -1 < x < 1 \tag{8.32}$$

and define the normalized source number function

$$y = y(x) \quad -1 < y < 1 \tag{8.33}$$

The position of the nth element is given by

$$s_n = ax(y_n) \tag{8.34}$$

Equation (8.28) becomes

$$E(u) = \sum_{m=-\infty}^{\infty} (-1)^{m(N-1)} E_m(u) \tag{8.35}$$

$$E_m(\theta) = \frac{1}{2}\int_{-1}^{1} A(x)\frac{dy}{dx}e^{-j(\psi(s)-2\pi my)}e^{jk\theta x}dx \tag{8.36}$$

For $-M < n < M$, the length L of the array is given by

$$L = a(x(y_M) - x(y_{-M})) < 2a \tag{8.37}$$

The problem is to reduce the sidelobe levels for a linear array by changing the spacing among the elements of the array. Using the method in the development of Taylor weights [7], we let Q determine the number of sidelobes that will be made equal:

$$\frac{dy}{dx} = f(x) \tag{8.38}$$

$$f(x) = \sum_{q=-Q}^{Q} A_q e^{-jq\pi x} \tag{8.39}$$

$$A_q = \frac{1}{2}\int_{-1}^{1} f(x)e^{jq\pi x}dx \tag{8.40}$$

$$y(x) = x + 2\sum_{q=1}^{Q} A_q \frac{\sin(q\pi x)}{q\pi} \tag{8.41}$$

The position of the nth element is determined by (8.34), and thus (8.41) must be inverted numerically to determine x as a function of y.

Because we are interested in $u << 1$, for unity amplitude $A(x) = 1$ and $\psi(x) = 0$, we have

$$E(u) \approx E_0(u) \tag{8.42}$$

$$E_0(u) = \frac{1}{2}\int_{-1}^{1} \frac{dy}{dx} e^{jux} dx \tag{8.43}$$

which is the radiation pattern for a continuous array with amplitude illumination f defined in (8.39).

If $Q = 4$, then the first four sidelobes on either side of the mainlobe are equal in amplitude for a specified amplitude level, just as seen in Taylor weights for uniform linear arrays. However, the outer sidelobes tend to be higher in amplitude because, when u is relatively large, the approximation in (8.42) is no longer valid. In fact, we have

$$E(u) \approx E_{-1}(u) \quad u > \frac{N\pi}{2} \tag{8.44}$$

and $E(N\pi) = E_{-1}(N\pi)$ is a really good approximation. In [6], Ishimaru provides several examples of array with nonequal element spacing to control sidelobe levels and generalizes the approach to planar arrays.

8.7.2 Polynomial Factorization Method

Polynomial factorization methods were developed by S. Mitra [9–11] and address the two-way antenna pattern response as opposed to the one-way pattern that is addressed on the other sparse array methods. Equation (8.22) is a one-way pattern representation of the far-field radiation pattern and, for a uniform linear array with unity voltage amplitude, can be expressed as

$$E(\theta) = \sum_{n=0}^{N-1} e^{jkdn\sin(\theta)} = \sum_{n=0}^{N-1} x^n \tag{8.45}$$

where θ is the angle-of-arrival, d is the uniform element spacing, and N is the total number of antenna elements in the array. Note that (8.45) is an Nth-degree polynomial in the complex variable x. The two-way pattern results in combining a transmit pattern (P_T) and a receive pattern (P_R) where the overall effective aperture P_{eff} is given by [4]:

$$P_{eff}(x) = P_T(x)P_R(x) \tag{8.46}$$

where

$$P_T(x) = \sum_{n=0}^{T-1} w(n)x^n \tag{8.47}$$

$$P_R(x) = \sum_{n=0}^{R-1} w(n)x^n \tag{8.48}$$

where $w(n)$ represents the transmit and receive aperture function.

8.7.2.1 Unity Aperture Illumination Function

The problem at hand is to determine factorizations of (8.46) that conform to (8.47) and (8.48), where we assume that P_{eff} has a desired form [4–7]. A commonly employed array design is when $w(n)=1$ for all n, resulting in a uniform effective aperture function. In this case, P_{eff} has unity coefficients and the form shown in (8.45). The method is based on the factorization of the Nth-degree polynomial

$$P_N(x) = \sum_{n=0}^{N} x^n \tag{8.49}$$

The number of coefficients in P_N is $L = N + 1$.

The fundamental result that allows factorizations of (8.49) can be stated in the following proposition, which is proved in [4]:

Proposition: If L is expressible as a product of $K + 1$ irreducible integers L_k with $0 \le k \le K$, then $P_N(x)$ can be expressed in the form

$$P_N(x) = \prod_{k=0}^{K} P_{M_k}(x^{S_k}) \tag{8.50}$$

where

$$S_k = \prod_{l=0}^{k} L_{l-1} \tag{8.51}$$

$$L_{-1} = 1 \tag{8.52}$$

with $M_k = L_k - 1, 0 \le k \le K$.

The factorization of a specified L into a product of irreducible integers can be carried out using Euclid's algorithm that finds the greatest common divisor of two positive integers.

Example 1: Let $L = 18$ then $P_{17}(x) = \sum_{n=0}^{17} x^n$ and $L = 18 = 2 \times 3 \times 3$. There are three possible factorizations containing the smallest number of coefficients [4].

Factorization 1: $L_0 = 2,\ L_1 = 3,\ L_2 = 3$

$$P_{17}(x) = (1 + x)(1 + x^2 + x^4)(1 + x^6 + x^{12}) \tag{8.53}$$

There are three possible factorizations.

Array Design 1:

$$P_T(x) = 1 + x^2 + x^4 + x^6 + x^8 + x^{10} + x^{12} + x^{14} + x^{16} \tag{8.54}$$

$$P_R(x) = 1 + x \tag{8.55}$$

The total number of elements is $L_0 + L_1 L_2 = 11$.

Array Design 2:

$$P_T(x) = 1 + x + x^2 + x^3 + x^4 + x^5 \tag{8.56}$$

$$P_R(x) = 1 + x^6 + x^{12} \tag{8.57}$$

The total number of elements is $L_0L_1 + L_2 = 9$.

Array Design 3:

$$P_T(x) = 1 + x + x^6 + x^7 + x^{12} + x^{13} \tag{8.58}$$

$$P_R(x) = 1 + x^2 + x^4 \tag{8.59}$$

The total number of elements is $L_0L_2 + L_1 = 9$.

Factorization 2: $L_0 = 3,\ L_1 = 2,\ L_2 = 3$

$$P_{17}(x) = \left(1 + x + x^2\right)\left(1 + x^3\right)\left(1 + x^6 + x^{12}\right) \tag{8.60}$$

There are three possible factorizations.

Array Design 4:

$$P_T(x) = 1 + x + x^2 + x^3 + x^4 + x^5 \tag{8.61}$$

$$P_R(x) = 1 + x^6 + x^{12} \tag{8.62}$$

The total number of elements is $L_0L_1 + L_2 = 9$.

Array Design 5:

$$P_T(x) = 1 + x + x^2 + x^6 + x^7 + x^8 + x^{12} + x^{13} + x^{14} \tag{8.63}$$

$$P_R(x) = 1 + x^3 \tag{8.64}$$

The total number of elements is $L_0 + L_1L_2 = 11$.

Array Design 6:

$$P_T(x) = 1 + x^3 + x^6 + x^9 + x^{12} + x^{15} \tag{8.65}$$

$$P_R(x) = 1 + x + x^2 \tag{8.66}$$

The total number of elements is $L_0L_1 + L_2 = 9$.

Factorization 2: $L_0 = 3,\ L_1 = 2,\ L_2 = 3$

$$P_{17}(x) = \left(1 + x + x^2\right)\left(1 + x^3 + x^6\right)\left(1 + x^9\right) \tag{8.67}$$

There are three possible factorizations.

Array Design 7:

$$P_T(x) = 1 + x + x^2 + x^3 + x^4 + x^5 + x^6 + x^7 + x^8 \tag{8.68}$$

$$P_R(x) = 1 + x^9 \tag{8.69}$$

The total number of elements is $L_0 + L_1L_2 = 11$.

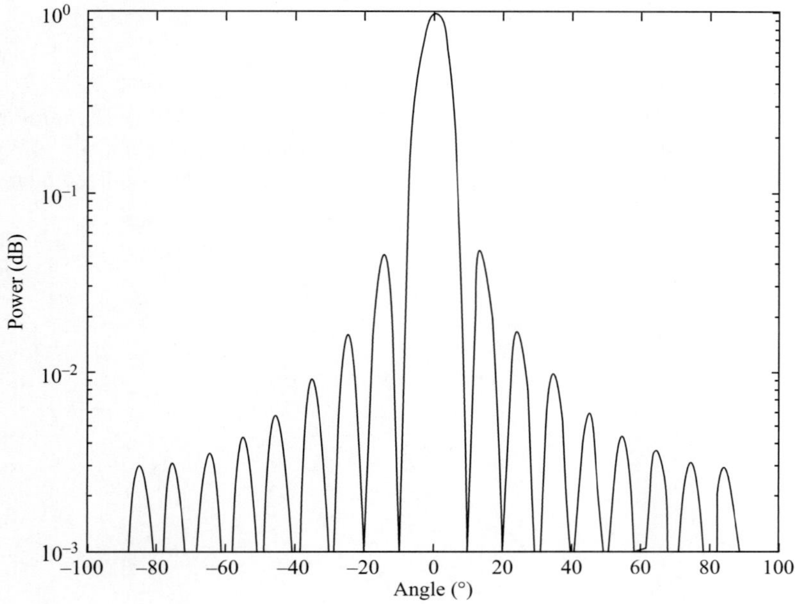

Figure 8.11. Antenna Pattern for Each of the Array Designs in Example 1

Array Design 8:

$$P_T(x) = 1 + x + x^2 + x^9 + x^{10} + x^{11} \tag{8.70}$$

$$P_R(x) = 1 + x^3 + x^6 \tag{8.71}$$

The total number of elements is $L_0L_1 + L_2 = 9$.

Array Design 9:

$$P_T(x) = 1 + x^3 + x^6 + x^9 + x^{12} + x^{15} \tag{8.72}$$

$$P_R(x) = 1 + x + x^2 \tag{8.73}$$

The total number of elements is $L_0L_1 + L_2 = 9$.

The two-way radiation pattern responses for the array designs 1–9 are the same as the response of an 18-element uniformly spaced array, which is shown in Figure 8.11.

8.7.2.2 Linearly Tapered Effective Aperture Function

In [10–11], linearly tapered aperture functions are shown to be generated by scaling one of the factorizations of the $P_{eff}(x)$.

$$P_{eff}(x) = P_T(x)P_R(x) \tag{8.74}$$

where

$$P_1(x) = \frac{1}{R}\sum_{n=0}^{R-1} x^n \tag{8.75}$$

$$P_2(x) = \sum_{n=0}^{S-1} x^n \tag{8.76}$$

The number of elements in the effective aperture is $N = R + S - 1$. The number of apodized elements is $2(R - 1)$ with $R - 1$ apodized elements at the beginning of P_{eff} and $R - 1$ apodized elements at the end. The values of the apodized elements are as follows:

$$\frac{1}{R}, \frac{2}{R}, \ldots, \frac{R-1}{R} \tag{8.77}$$

The parameter S must satisfy $S > R - 1$.

Example: We choose $R = 2$ and $S = 18$ and

$$P_1(x) = \frac{1}{2}(1 + x) \tag{8.78}$$

$$P_2(x) = \sum_{n=0}^{17} x^n \tag{8.79}$$

Using the Factorization 3 in the preceding example, we have

$$P_{17}(x) = \left(1 + x + x^2\right)\left(1 + x^3 + x^6\right)\left(1 + x^9\right) \tag{8.80}$$

Hence, P_{eff} has the form

$$\begin{aligned} P_{eff}(x) &= \frac{1}{2}(1 + x)\left(1 + x + x^2\right)\left(1 + x^3 + x^6\right)\left(1 + x^9\right) \\ &= 0.5 + x + x^2 + x^3 + \cdots + x^{16} + x^{17} + 0.5x^{18} \end{aligned} \tag{8.81}$$

One possible design for transmit and receive arrays is given by

$$P_T(x) = (1 + x)\left(1 + x^3 + x^6\right)\left(1 + x^9\right) \tag{8.82}$$

$$P_R(x) = \frac{1}{2}\left(1 + x + x^2\right) \tag{8.83}$$

In Figure 8.12, the apodization of the two outside antenna elements in Example 2 provides lower antenna sidelobes than the pattern in Figure 8.11.

8.8 Sparse Array Angle-of-Arrival

The objective of sparse array antennas is the development of lower-cost antennas without significantly degrading performance. The advantage for sparse arrays is that large array apertures can be realized with a fewer number of elements, resulting in smaller beamwidths with enhanced angle accuracy. However, the type of sparse array can dictate the method used to derive angle estimates. In general, it is difficult to use monopulse angle estimation for most sparse arrays, but other angle estimation techniques can be applied. For example, sparse array architectures can be generated that use multiple interferometers for improved angle accuracy;

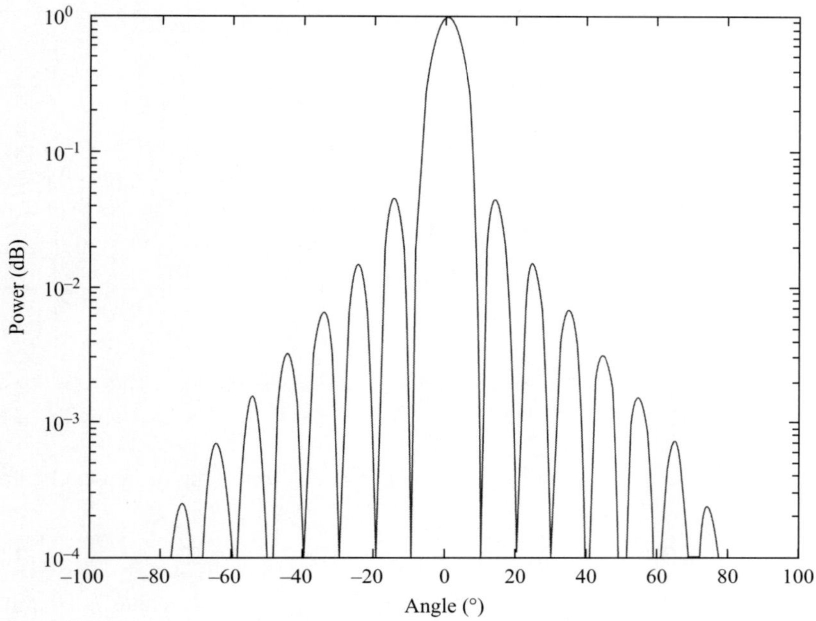

Figure 8.12. Antenna Pattern for Each of the Array Designs in Example 2

enhanced resolution techniques that use the array covariance can be implemented for any sparse array architectures.

8.8.1 Sparse Array Monopulse

For many sparse arrays, the angle-of-arrival can be determined by using a maximum likelihood algorithm. The maximum likelihood algorithm to determine the angle-of-arrival for a single target is the discrete Fourier transform (DFT). A maximum likelihood solution is achieved by determining the angle that achieves the maximum value for the absolute value of the DFT of the array voltage values. Due to interference and other error sources, the peak of the DFT can be distorted resulting in angle error, and, because the peak is located on the main beam of the DFT response, the search algorithm to achieve high accuracy can be computationally burdensome due to the flatness of the search surface.

Using monopulse to determine angle-of-arrival is problematic due to the sparseness of the array. The array differencing for a linear array requires that the array be divided into two subarrays forming a differencing operation. For sparse arrays, it is not always possible to define two distinct subarrays that can provide an array difference with good monopulse slope characteristics. In general, finding a null is preferred over finding a peak due to the flatness of the DFT mainlobe, but the degradation in monopulse slope due to the array sparseness leads to reduced accuracy. In some cases, where the sparse array exhibits some symmetry,

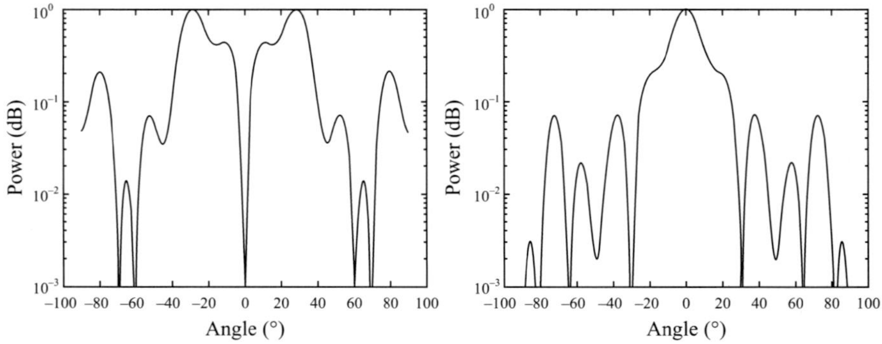

Figure 8.13. Difference and Sum Antenna Patterns for Nested Cyclic Array P = {1, 6, 7, 8, 9, 10, 11, 16}

monopulse techniques may be used. For example, the following nested cyclic array is symmetric about its center element, the number eight (8).

$$P = \{1, 6, 7, 8, 9, 10, 11, 16\} \tag{8.84}$$

The sum and difference pattern for this nested cyclic array, using a Hamming weighting to reduce antenna sidelobes, is shown in Figure 8.13.

The basic equation for monopulse angle accuracy was derived in Chapter 4 as

$$\sigma_\theta = \frac{\theta_{BW}}{k\sqrt{2SNR}} \tag{8.85}$$

For a sparse array monopulse estimate, the monopulse slope k_m (Figure 8.14), and the *SNR* are degraded due to the fewer number of elements. The beamwidth can also be affected due to the increased sidelobes.

As shown in Figures 8.13 and 8.14 for the nested cyclic array as defined, the monopulse sum beam has broadened, and the monopulse slope is 0.28, which is significantly smaller than the monopulse slope for a uniform linear array.

8.8.2 Sparse Array Interferometry

We now show how interferomteric method can be applied to two-dimensional sparse arrays that include both angle estimation and ambiguity resolution. We first develop the approach using three- and four-element sparse arrays and then extend it to arrays with an arbitrary number of elements. Angle ambiguity resolution is first derived using specific sparse arrays where at least two elements are separated by a half wavelength and then extended to general sparse arrays.

8.8.2.1 Interferometry Applied to Three- and Four-Element Sparse Arrays

Consider the case of a three-element array with spacing defined in units of half wavelengths ($\lambda/2$) as

$$P = \{1,\ 2,\ N+1\} \tag{8.86}$$

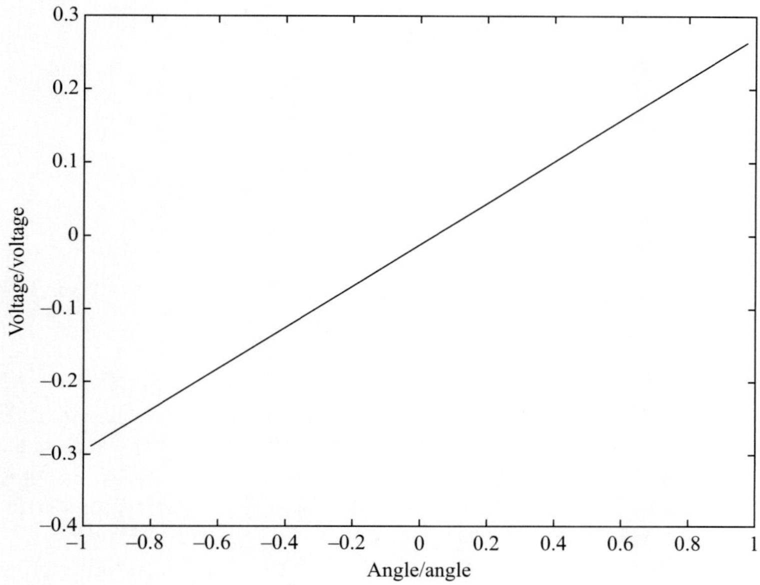

Figure 8.14. Monopulse Slope for Sparse Array Monopulse Estimation

where N is a relatively large number. Note that P is essentially two nested interferometers: one that satisfies spatial Nyquist and one with baseline equal to N half wavelengths that does not.

$$P = \{1, 2\} \quad \text{and} \quad P = \{1, N+1\} \tag{8.87}$$

As such we should be able to use the principles of interferometry to determine angle-of-arrival. Let φ_i denote the phases at each of the three elements. Then, due to phase wrapping (see Section 7.3), we have

$$\varphi_1^w = [\varphi_1 + \pi]_{2\pi} - \pi \tag{8.88}$$

$$\varphi_2^w = [\varphi_1 + \pi - \pi \sin(\theta)]_{2\pi} - \pi \tag{8.89}$$

$$\varphi_3^w = [\varphi_1 + \pi - N\pi \sin(\theta)]_{2\pi} - \pi \tag{8.90}$$

We can compute the two interferometer phase differences:

$$\Delta\varphi_{12} = \varphi_1^w - \varphi_2^w = [\varphi_1 + \pi]_{2\pi} - [\varphi_1 + \pi - \pi \sin(\theta)]_{2\pi} \tag{8.91}$$

$$\Delta\varphi_{13} = \varphi_1^w - \varphi_3^w = [\varphi_1 + \pi]_{2\pi} - [\varphi_1 + \pi - N\pi \sin(\theta)]_{2\pi} \tag{8.92}$$

But since the first interferometer satisfies a spatial Nyquist condition, we have

$$\Delta\varphi_{12} = \varphi_1^w - \varphi_2^w = [\varphi_1 + \pi]_{2\pi} - [\varphi_1 + \pi - \pi \sin(\theta)]_{2\pi} = \pi \sin(\theta) \tag{8.93}$$

and

$$N\Delta\varphi_{12} = N\pi\sin(\theta) = 2K\pi + \varphi_{residual} \tag{8.94}$$

$$\begin{aligned}\Delta\varphi_{13} &= [\varphi_1 + \pi]_{2\pi} - [\varphi_1 + \pi - N\pi\sin(\theta)]_{2\pi}\\ &= [\varphi_1 + \pi]_{2\pi} - [\varphi_1 + \pi - 2\pi K - \varphi_{residual}]_{2\pi}\\ &= [\varphi_1 + \pi]_{2\pi} - [\varphi_1 + \pi - \varphi_{residual}]_{2\pi}\\ &= \varphi_{residual}\end{aligned} \tag{8.95}$$

$$N\Delta\varphi_{12} = N\pi\sin(\theta) = 2K\pi + \Delta\varphi_{13} \tag{8.96}$$

$$\sin(\theta) = \frac{\Delta\varphi_{13} + 2K\pi}{\pi N} \tag{8.97}$$

Equation (8.97) agrees exactly with (7.1) when the interferometer baseline is $D = N\lambda/2$. Thus solving (8.94) for K and substituting into (8.97), we have used the smaller unambiguous interferometer to unwrap the phase of the larger ambiguous interferometer. In Chapter 7 we showed that spatial diversity provided a more robust way of resolving angle ambiguities.

Now consider the following sparse array with three embedded interferometers.

$$P = \{1, 2, N_1 + 1, N_2 + 1\} \tag{8.98}$$

where $2 << N_1 << N_2$.

$$\begin{aligned}\varphi_1^w &= [\varphi_1 + \pi]_{2\pi} - \pi\\ \varphi_2^w &= [\varphi_1 + \pi - \pi\sin(\theta)]_{2\pi} - \pi\\ \varphi_3^w &= [\varphi_1 + \pi - N_1\pi\sin(\theta)]_{2\pi} - \pi\\ \varphi_4^w &= [\varphi_1 + \pi - N_2\pi\sin(\theta)]_{2\pi} - \pi\end{aligned} \tag{8.99}$$

$$\begin{aligned}\Delta\varphi_{12} &= \varphi_1^w - \varphi_2^w = [\varphi_1 + \pi]_{2\pi} - [\varphi_1 + \pi - \pi\sin(\theta)]_{2\pi}\\ \Delta\varphi_{13} &= \varphi_1^w - \varphi_3^w = [\varphi_1 + \pi]_{2\pi} - [\varphi_1 + \pi - N_1\pi\sin(\theta)]_{2\pi}\\ \Delta\varphi_{14} &= \varphi_1^w - \varphi_4^w = [\varphi_1 + \pi]_{2\pi} - [\varphi_1 + \pi - N_2\pi\sin(\theta)]_{2\pi}\end{aligned} \tag{8.100}$$

$$\begin{aligned}N_1\Delta\varphi_{12} &= N_1\pi\sin(\theta) = 2K_{N_1}\pi + \varphi_{residual}^{N_1} = 2K_{N_1}\pi + \Delta\varphi_{13}\\ N_2\Delta\varphi_{12} &= N_2\pi\sin(\theta) = 2K_{N_2}\pi + \varphi_{residual}^{N_2} = 2K_{N_2}\pi + \Delta\varphi_{14}\end{aligned} \tag{8.101}$$

$$\begin{aligned}\sin(\theta) &= \frac{\Delta\varphi_{14} - \Delta\varphi_{13} + 2K_{N_1}\pi - 2K_{N_2}\pi}{(N_1 - N_2)\pi}\\ &= \frac{\Delta\varphi_{43} + 2(K_{N_1} - K_{N_2})\pi}{(N_1 - N_2)\pi}\end{aligned} \tag{8.102}$$

and for the difference equation

$$\sigma_\theta^D = \frac{1}{\pi(N_2 - N_1)\sqrt{SNR}} \tag{8.103}$$

and also using the sum

$$\sin(\theta) = \frac{\Delta\varphi_{14} + \Delta\varphi_{13} + 2K_{N_1}\pi + 2K_{N_2}\pi}{(N_1 + N_2)\pi} = \frac{\Delta\varphi_{14} + \Delta\varphi_{13} + 2\pi(K_{N_1} + K_{N_2})}{(N_1 + N_2)\pi} \tag{8.104}$$

The estimate of θ using (8.104) can be more accurate than the estimate using (8.102) due to the larger denominator. One consideration is that (8.104) contains the phase φ_1^w twice as follows:

$$\sin(\theta) = \frac{\varphi_4^w - \varphi_1^w + \varphi_3^w - \varphi_1^w + (K_{N_1} + K_{N_2})2\pi}{(N_1 + N_2)\pi} = \frac{\varphi_4^w + \varphi_3^w - 2\varphi_1^w + (K_{N_1} + K_{N_2})2\pi}{(N_1 + N_2)\pi} \tag{8.105}$$

and

$$\sigma_\theta^S = \frac{\sqrt{3}}{\pi(N_1 + N_2)\sqrt{SNR}} \tag{8.106}$$

Note that $\sigma_\theta^S < \sigma_\theta^D$ when $N_1 > \left(2 - \sqrt{3}\right)N_2$.

Now consider the more general four-element case

$$P = \{1, 2, N_1 + 1, N_2 + 1, N_3 + 1\} \tag{8.107}$$

Then (8.100) becomes

$$\begin{aligned} N_1\pi\sin(\theta) &= 2K_{N_1}\pi + \varphi_{residual}^{N_1} = 2K_{N_1}\pi + \Delta\varphi_{12} \\ N_2\pi\sin(\theta) &= 2K_{N_2}\pi + \varphi_{residual}^{N_2} = 2K_{N_2}\pi + \Delta\varphi_{13} \\ N_3\pi\sin(\theta) &= 2K_{N_3}\pi + \varphi_{residual}^{N_3} = 2K_{N_3}\pi + \Delta\varphi_{14} \end{aligned} \tag{8.108}$$

And for a given K_{N_1} we can compute

$$\begin{aligned} N_2(2K_{N_1}\pi + \Delta\varphi_{12}) &= N_1N_2\pi\sin(\theta) = N_1\left(2K_{N_2}\pi + \varphi_{residual}^{N_2}\right) = 2N_1K_{N_2}\pi + N_1\Delta\varphi_{13} \\ N_3(2K_{N_1}\pi + \Delta\varphi_{12}) &= N_1N_3\pi\sin(\theta) = N_1\left(2K_{N_3}\pi + \varphi_{residual}^{N_3}\right) = 2N_1K_{N_3}\pi + N_1\Delta\varphi_{14} \end{aligned} \tag{8.109}$$

or solving for K_{N_2} and K_{N_3} we have

$$\begin{aligned} K_{N_2} &= \text{round}\left(\frac{N_2K_{N_1}}{N_1} - \frac{N_1\Delta\varphi_{13} - N_2\Delta\varphi_{12}}{2\pi N_1}\right) \\ K_{N_3} &= \text{round}\left(\frac{N_3K_{N_1}}{N_1} - \frac{N_1\Delta\varphi_{14} - N_3\Delta\varphi_{12}}{2\pi N_1}\right) \end{aligned} \tag{8.110}$$

From above we form

$$\begin{aligned} s_2 &= \frac{\Delta\varphi_{13} + 2\pi K_{N_2}}{N_2\pi} \\ s_3 &= \frac{\Delta\varphi_{14} + 2\pi K_{N_3}}{N_3\pi} \end{aligned} \tag{8.111}$$

But theoretically

$$s_2 = \sin(\theta) = s_3 \tag{8.112}$$

Using (8.112) we can form the error function

$$error = |s_2 - s_3| \tag{8.113}$$

The algorithm to unwrap the phase ambiguity for the general case is as follows:

1. Start with $K_{N_1} = 0$ and $L_{N_1} = 0$.
2. Using (8.110) define K_{N_k} and L_{N_k} for all $k = 1, 2$.
3. Define s_k using K_{N_k} and t_k using L_{N_k} in (8.110).
4. Compute *error* in (8.113) using both s_k and t_k.
5. If $error < \varepsilon$ compute angle estimate using the equation below.
6. If $error > \varepsilon$ $K_{N_1} = K_{N_1} + 1$ and $L_{N_1} = L_{N_1} - 1$.

The angle estimate is computed using either a sum or difference algorithm as defined above. The sum algorithm is defined by

$$\sin(\theta) = \frac{\Delta\varphi_{12} + \Delta\varphi_{13} + \Delta\varphi_{14} + 2\pi(K_{N_1} + K_{N_2} + K_{N_3})}{(N_1 + N_2 + N_3)\pi} \tag{8.114}$$

8.8.2.2 Interferometry Applied to General Sparse Arrays

All of the above algorithms can be extended to the general $M + 1$ element case

$$P = \{1, N_1 + 1, N_2 + 1, \ldots, N_M + 1\} \tag{8.115}$$

However, we will begin by showing how this extension can be developed using the following restricted case.

$$P = \{1, 2, N_1 + 1, N_2 + 1, \ldots, N_M + 1\} \tag{8.116}$$

We first unwrap the phases by solving the following system of equations for the integers $K_1, K_2, \ldots, K_M$:

$$\begin{aligned} N_1\Delta\varphi_{1N_1} &= 2K_1\pi + \Delta\varphi_{1N_1} \\ N_2\Delta\varphi_{1N_2} &= 2K_2\pi + \Delta\varphi_{1N_2} \\ &\vdots \\ N_M\Delta\varphi_{1N_M} &= 2K_M\pi + \Delta\varphi_{1N_M} \end{aligned} \tag{8.117}$$

In order to guarantee a solution to (8.117), it is necessary that N_i not divide $N_i + 1$ for $i = 1{:}\ M - 1$. Then we solve for the angle by solving the system of equations.

$$\begin{aligned} N_1\pi \sin(\theta) &= 2K_1\pi + \Delta\varphi_{1N_1} \\ N_2\pi \sin(\theta) &= 2K_2\pi + \Delta\varphi_{1N_2} \\ &\vdots \\ N_M\pi \sin(\theta) &= 2K_M\pi + \Delta\varphi_{1N_M} \end{aligned} \tag{8.118}$$

Without loss of generality, we assume that M is even and that solving the system of equations using differences results in the following expression for angle-of-arrival:

$$\begin{aligned} \sin(\theta) &= \frac{\sum_{i=M/2+1}^{M} \Delta\varphi_{1N_i} - \sum_{i=1}^{M/2} \Delta\varphi_{1N_i} + 2\pi \sum_{i=M/2+1}^{M} K_i - 2\pi \sum_{i=1}^{M/2} K_i}{\pi \sum_{i=M/2+1}^{M} N_i - \pi \sum_{i=M/2+1}^{M} N_i} \\ &= \frac{\sum_{i=1}^{M/2} \Delta\varphi_{N_{M/2+i}N_i} + 2\pi \sum_{i=1}^{M/2} (K_{M/2+i} - K_i)}{\pi \sum_{i=1}^{M/2} (N_{M/2+i} - N_i)} \end{aligned} \tag{8.119}$$

The angle accuracy using an embedded interferometer sparse array can be determined to be

$$\sigma_\theta^D = \frac{\sqrt{M}\sigma_\varphi}{\pi \left| \sum_{i=1}^{M/2} (N_{M/2+i} - N_i) \right|} = \frac{\sqrt{M}}{\pi\sqrt{2SNR} \left| \sum_{i=1}^{M/2} (N_{M/2+i} - N_i) \right|} \tag{8.120}$$

Using the summation in (8.118), we have

$$\sin(\theta) = \frac{\Delta\varphi_{11} + \Delta\varphi_{12} + \cdots + \Delta\varphi_{1M} + 2\pi(K_1 + K_2 + \cdots + K_M)}{N_1 + N_2 + \cdots + N_M} \tag{8.121}$$

and

$$\sigma_\theta^S = \frac{\sqrt{M^2 + M}}{\pi\sqrt{2SNR} \sum_{i=1}^{M} N_i} \tag{8.122}$$

As an example, consider the sparse array defined by

$$P = \{1, 2, 11, 21, 99, 101\} \tag{8.123}$$

The angle error using the difference solution (8.120) for the sparse array is smaller than the error for the interferometer defined by the extreme elements $P = \{1, 101\}$, but the angle error using the summation solution is larger.

$$\sigma_\theta^D = \frac{2}{\pi \cos(\theta)\sqrt{2SNR}(99+101-21-11)} = \frac{1}{84\cos(\theta)\pi\sqrt{2SNR}} < \frac{1}{100\pi\cos(\theta)\sqrt{SNR}}$$
$$\sigma_\theta^S = \frac{\sqrt{20}}{\pi \cos(\theta)\sqrt{2SNR}(99+101+21+11)} = \frac{\sqrt{20}}{232\cos(\theta)\pi\sqrt{2SNR}} > \frac{1}{100\pi\cos(\theta)\sqrt{SNR}} \tag{8.124}$$

Again (8.119) and (8.121) do not involve the phase at the second antenna position φ_2^w and can be applied to the general case (8.117) which is only used to resolve the angle ambiguities in (8.117); however, angle ambiguity resolution must be approached differently than the case for $N_1 = 1$. For the general case we have

$$\begin{aligned} N_1\Delta\varphi_{12} &= 2K_1\pi + \Delta\varphi_{1N_1} \\ N_2\Delta\varphi_{12} &= 2K_2\pi + \Delta\varphi_{1N_2} \\ &\vdots \\ N_M\Delta\varphi_{12} &= 2K_M\pi + \Delta\varphi_{1N_M} \end{aligned} \tag{8.125}$$

And mimicking (8.109) we have

$$\begin{aligned} N_1\Delta\varphi_{12} &= 2K_1\pi + \Delta\varphi_{1N_1} \\ N_1N_2\Delta\varphi_{12} &= N_1\left(2K_2\pi + \Delta\varphi_{1N_2}\right) = N_2\left(2K_1\pi + \Delta\varphi_{1N_1}\right) \\ &\vdots \\ N_1N_M\Delta\varphi_{12} &= N_1\left(2K_M\pi + \Delta\varphi_{1N_M}\right) = N_M\left(2K_1\pi + \Delta\varphi_{1N_1}\right) \end{aligned} \tag{8.126}$$

And solving for K_{N_k} for $k > 1$

$$\begin{aligned} K_{N_2} &= \text{round}\left(\frac{N_2K_{N_1}}{N_1} - \frac{N_1\Delta\varphi_{13} - N_2\Delta\varphi_{12}}{2\pi N_1}\right) \\ &\vdots \\ K_{N_M} &= \text{round}\left(\frac{N_MK_{N_1}}{N_1} - \frac{N_1\Delta\varphi_{1M} - N_M\Delta\varphi_{12}}{2\pi N_1}\right) \end{aligned} \tag{8.127}$$

Again we can define

$$s_k = \frac{\Delta\varphi_{1k+1} + 2\pi K_{N_k}}{N_k\pi} \tag{8.128}$$

And since $s_k = \sin(\theta)$ for all $k > 1$, we define an error function as before

$$error = diff \cdot [s_k]' \tag{8.129}$$

where

$$diff = \overbrace{\begin{bmatrix} 1 & -1 & 1 & -1 & \ldots & -1 \end{bmatrix}}^{M} \tag{8.130}$$

The algorithm to unwrap the phase ambiguities is similar to the one defined above.

1. Start with $K_{N_1} = 0$ and $L_{N_1} = 0$.
2. Using (8.127) define K_{N_k} and L_{N_k} for all $k > 1$.
3. Define s_k using K_{N_k} and t_k using L_{N_k} in (8.127).
4. Compute *error* in (8.129) using both s_k and t_k.
5. If *error* $< \varepsilon$ compute angle estimate using the equation below.
6. If *error* $> \varepsilon$ $K_{N_1} = K_{N_1} + 1$ and $L_{N_1} = L_{N_1} - 1$.

Thus, we have derived the result for the general sparse array for ambiguity resolution and angle estimation using interferometric techniques.

The accuracy for angle estimation using differencing or summation derived in (8.120) and (8.122) can be smaller than the accuracy for the interferometer defined by $P = \{1, N_M\}$. Thus, a sparse array using interferometric methods to derive angle can be more accurate than an interferometer using the two outside elements in the sparse array which will depend upon the total number of elements and the selection of the element separations. As an example choose $P = \{1, 14, 18, 20\}$, then the standard deviation of the angle error for a 30° angle-of-arrival with 20 dB SNR is 5.81 milli-radians and the accuracy for the interferometer $P = \{1, 20\}$ is 6.12 milli-radians. The interferometric sparse array methods essentially trade angle accuracy and angle ambiguity resolution and thus sparse array angle ambiguity resolution can be applied to any two-element interferometer contained in the sparse array for potentially enhanced angle accuracy.

8.8.2.3 Sparse Array Interferometry Using Monopulse

Another way of achieving spatial diversity is by using nested cyclic arrays where we have a small array embedded in an interferometer. Again, consider the following nested cyclic array:

$$P = \{1, 6, 7, 8, 9, 10, 11, 16\} \tag{8.131}$$

We can implement monopulse on the smaller embedded array $\{6, 7, 8, 9, 10, 11\}$ to estimate the angle-of-arrival for the interferometer $\{1, 16\}$. Because

$$\sin(\theta_{monopulse}) = \frac{\Delta\varphi_{1,16} + 2K\pi}{15\pi} \tag{8.132}$$

we can use the monopulse angle to estimate the number of phase wraps K by finding the integer K that minimizes.

$$\min_K \left\{ \left| \sin(\theta_{monopulse}) - \frac{\Delta\varphi_{1,16} + 2K\pi}{15\pi} \right| \right\} \tag{8.133}$$

Once K is determined, we can reapply (8.132) to determine the angle-of-arrival.

$$\sin(\theta_{interferometer}) = \frac{\Delta\varphi_{1,16} + 2K\pi}{15\pi} \tag{8.134}$$

Since the phase measurements are erred due to noise, (8.114) determines an accurate estimate of *K*, provided that the phase error is small enough not to add or subtract a phase ambiguity. We have shown that the phase error of differential phase is equal to the inverse of the square root of *SNR*. When the phase difference $\Delta\varphi_{1,16}$ is close to $\pm 2\pi$ higher, *SNR* is required to assure that an ambiguity has not been introduced. Thus, it may be necessary to determine the unambiguous angle over several measurements, especially if the target is moving.

8.8.3 Sparse Array Angle Estimation Using the Array Covariance

Angle estimation techniques that use an array covariance can be implemented with sparse arrays to determine angle-of-arrival. For example, the MUSIC algorithm uses the noise space eigen-space defined by the array covariance to determine angle-of-arrival. First consider the minimum redundancy array $P = \{1, 3, 6, 7\}$. Figure 8.15 shows the MUSIC response for two targets located at angles $-20°$ and $30°$.

Next, consider the sparse array defined by the partition $P = \{1, 3, 6, 12, 13\}$. Figure 8.16 shows the MUSIC response for two targets located at angles $-30°$ and $40°$.

Figures 8.15 and 8.16 show that using an eigen-space technique such as MUSIC produces angle estimates for sparse arrays. However, because the array covariance requires multiple pulses or looks to compute the time required to determine an angle

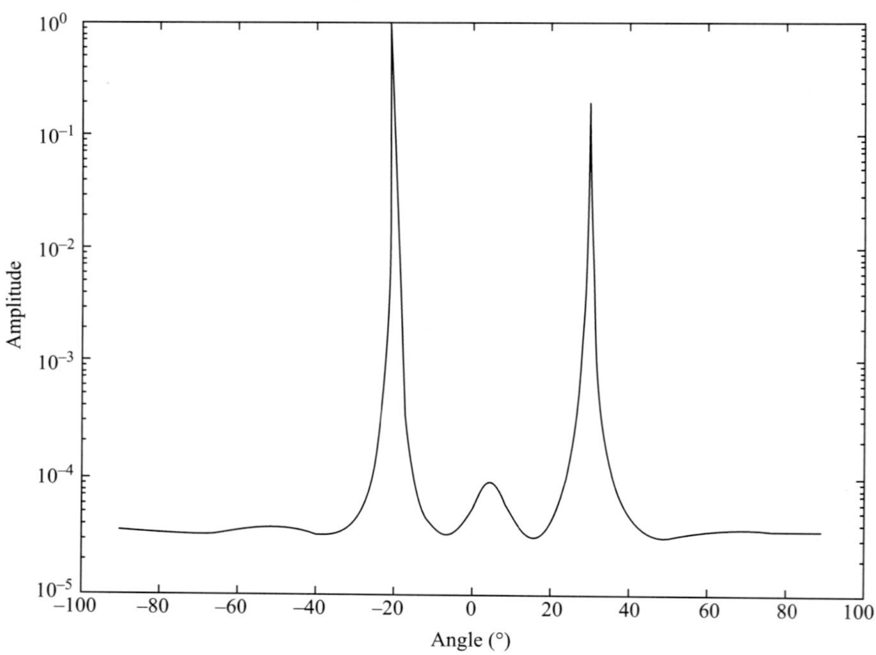

Figure 8.15. MUSIC Response for P = {1, 3, 6, 7} with Two Targets Located at −20° and 30°

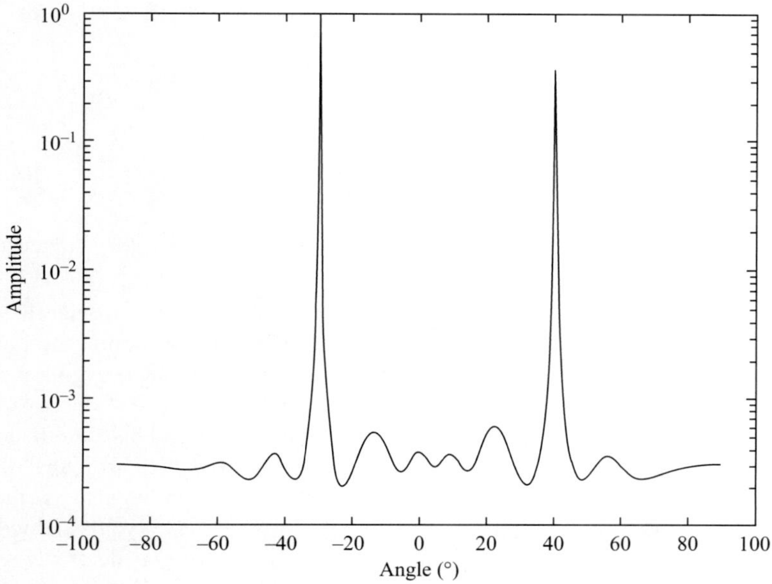

Figure 8.16. MUSIC Response for $P = \{1, 3, 6, 12, 13\}$ with Two Targets Located at −30° and 40°

estimate, using these techniques can be an issue. Sparse array angle estimation using an array covariance technique therefore depends on the application.

8.9 Two-Dimensional Sparse Arrays

Two-dimensional sparse array generating processes are more difficult to define than their one-dimensional counterparts due to the infinite number of possible locations for elements in a two-dimensional plane. To place structure on the two-dimensional problem, we start by generalizing the concept of spatial Nyquist sampling. For the 1-D array, we define the fundamental Nyquist interval as an interval of length d where $2\sin^{-1}(2d/\lambda)$ defines the array field of view and λ is the signal wavelength. If the desired field of view is 180°, then $d = \lambda/2$. The one-dimensional interval is the only choice for the definition of Nyquist sampling; however, two dimensions offer several choices depending on the geometry of the array structure: lattice structures that are triangular, rectangular, hexagonal, or any other basic geometric polygon. Figure 8.17 shows three basic lattice structures that define a Nyquist sampling for a 2-D array. Each structure determines the array shape for the sparse arrays generated using these Nyquist lattices.

> *Definition:* A *2-D Nyquist Lattice* is an array of elements located in the vertices of a regular polygon and at the center (barycenter) of the polygon where the distance from the center element to the elements located on the vertices is equal to d.

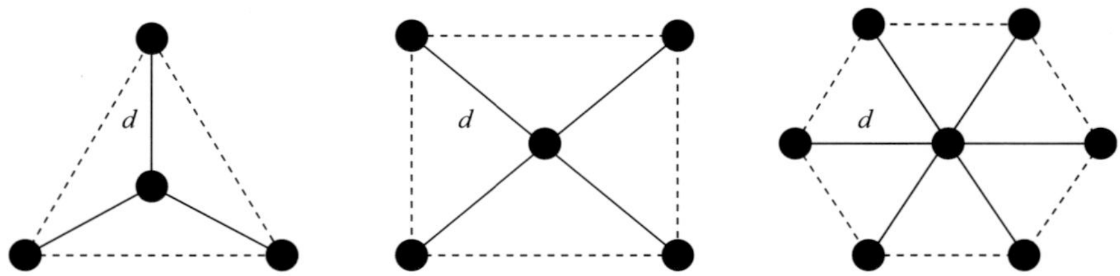

Figure 8.17. Fundamental Nyquist Lattices for 2-D Sparse Array Generation

We include the center element in order to define an array center for convenience. It is also possible to exclude the center element and define the Nyquist lattice with only the elements located on the vertices of the polygon that are separated by the distance *d*.

Because the triangular lattice structure is the most basic and leads to the lowest number of antenna elements, we use a triangular Nyquist lattice structure to generate 2-D sparse arrays. Instead of the 1-D Nyquist interval, we can also use the 1-D interval partitions to generate 2-D spare arrays using the 2-D Nyquist lattice. For example, we can start with the minimum redundancy 1-D array generated from the partition $\{1, 3, 6, 7\}$ or its symmetric partition $\{1, 2, 5, 7\}$. The first interval $\{1, 2\}$ is the basic Nyquist interval, and for the 2-D case, we use the basic Nyquist lattice. For the interval $\{1, 5\}$, we use the triangular lattice that is four times the basic Nyquist lattice (4*d*), and for the interval $\{1, 7\}$ we use a triangular lattice that is 6 times as large as the Nyquist lattice. This process leads to the construction of a minimum redundancy 2-D sparse array.

Figure 8.18 shows the structure for the 10-element 2-D minimum redundancy array. The area of the largest triangle is $27\sqrt{3}d^2$. The area of the smallest triangle is

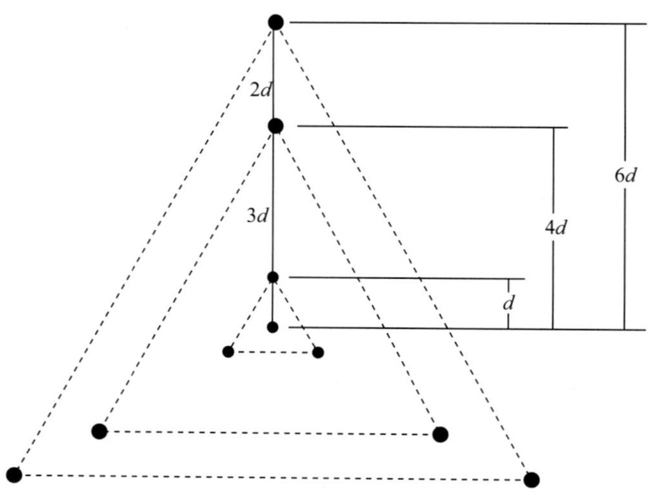

Figure 8.18. Two-Dimensional Minimum Redundancy Array Using a Triangular Nyquist Lattice

$3\sqrt{3}d^2/4$, and thus a possible $37 = 36 + 1$ total elements could be contained in the largest square, but the minimum redundancy array uses only 10 of the possible 37 elements. Also, if the rectangular or hexagonal Nyquist lattices were used to generate a minimum redundancy array, we would see an overall rectangular or hexagonal shape in the sparse array. Figure 8.19 shows a rectangular minimum redundancy 2-D array with 13 elements. The area of the largest rectangle is $72d^2$, and the area of the smallest rectangle is $2d^2$. Thus, the large array can contain a maximum of $37 = 36 + 1$ elements, but the rectangular Nyquist lattice uses 13 elements out of a possible 37.

The antenna pattern for the 2-D minimum redundancy array can be computed using the 2-D DFT. In Figure 8.20, we compute the 2-D pattern for the rectangular and triangular Nyquist lattice minimum redundancy 2-D arrays.

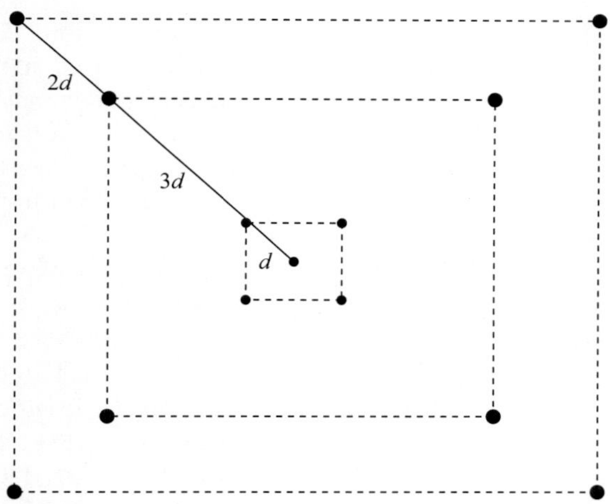

Figure 8.19. 2-D Rectangular Nyquist Lattice Minimum Redundancy Array

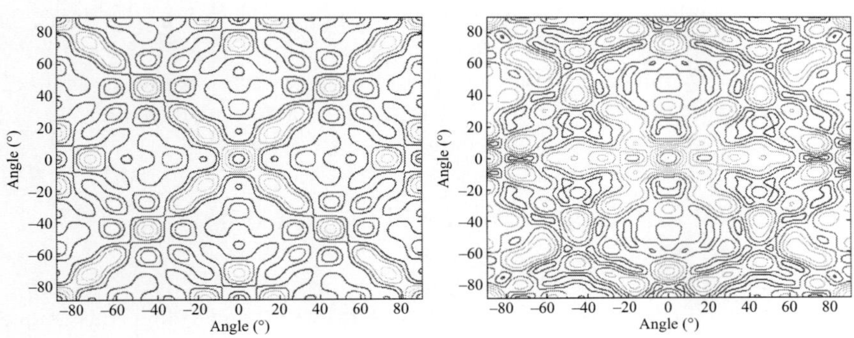

Figure 8.20. Antenna Pattern for Rectangular (left) and Triangular (right) Nyquist Lattice Minimum Redundancy 2-D Arrays

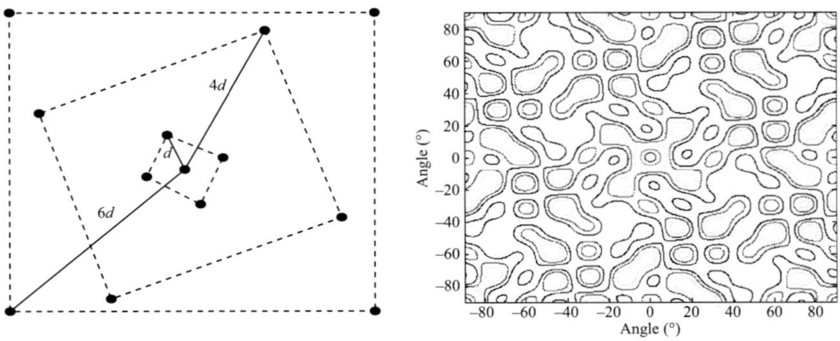

Figure 8.21. Rotated Rectangular Lattices for 2-D Minimum Redundancy Array (left) and Antenna Array Response (right)

The advantage to using the Nyquist lattice approach for 2-D sparse array generation is that the theory of 1-D partition intervals can be applied to the 2D sparse array design. Thus, 1-D coprime and nested cyclic arrays can be generalized to 2-D sparse arrays. Also, because there is a clear definition of Nyquist, the sieve method can be used to generate more optimal sparse arrays in a numerically efficient manner. The Nyquist lattice technique generates a symmetric sparse array that has symmetric antenna pattern responses.

However, it is possible to generate nonsymmetric sparse arrays using the Nyquist lattice approach by rotating each interior lattice by a random angle. Figure 8.21 shows an example using the rectangular Nyquist lattice where the two interior lattices are rotated by a random angle. It should be noted that rotations of the lattices preserve the distances from the vertices of each lattice to the array center, thus preserving interval partitions. Nonsymmetric sparse arrays break the symmetry of the antenna pattern sidelobes to mitigate directional sidelobe effects such as principal plane lobes. Figure 8.21 shows the lack of symmetry in the antenna pattern for the rotated rectangular lattice for a minimum redundancy array compared to the symmetric rectangular lattice in Figure 8.20.

Another advantage for polygon Nyquist lattices is the ability to use techniques developed for angle-of-arrival estimation for a 1-D sparse array. In particular, the nested sparse array interferometry technique discussed in Section 8.6 can be used to resolve angle ambiguities because the Nyquist lattice is unambiguous in angle-of-arrival. The phase unwrapping for 2-D is more complicated than the 1-D and depends on the type of polygon used in the Nyquist lattice.

8.10 Multiple-Input and Multiple-Output (MIMO) Sparse Arrays

When a radar whose antenna consists of multiple antenna elements transmits unique, nearly orthogonal waveforms from each antenna element and processes some or all of the transmitted waveforms received at each antenna element, the radar is referred to a multiple-input and multiple-output (MIMO) radar [18]. For MIMO radar applications, it is usually convenient to implement a sparse array architecture due to the cost and complexity of transmitting and receiving multiple waveforms from each array element. We now apply interferometric principles to MIMO sparse array architectures, similar to the methods and architectures defined in Section 8.8.2.

The basic angle accuracy is derived in a fashion similar to that of angle accuracy for the orthogonal interferometer. For the two-antenna orthogonal interferometer, the angle accuracy was shown in (6.51) to be

$$\sigma_\theta^{OI} = \frac{\lambda}{4\pi D\cos(\theta)}\sigma_{\Delta\phi} = \frac{\lambda}{4\pi D\cos(\theta)\sqrt{SNR}} \tag{8.135}$$

The extra factor of 2 in the denominator, compared to a conventional interferometer, resulted from the two-way phase rotation of the signal (out and back) for the orthogonal interferometer as compared to the conventional interferometer (return only). Using the methodology introduced in Section 8.8.2, we can extend the results for sparse array with M embedded interferometers to the MIMO case. In (8.116), we used the following antenna element location notation to describe a sparse array with M embedded interferometers.

$$P = \{1, 2, N_1 + 1, N_2 + 1, \ldots, N_M + 1\} \tag{8.136}$$

and the angle accuracy for a conventional sparse array was derived to be

$$\sigma_\theta = \frac{\sqrt{M}}{\pi\sqrt{2SNR}\left|\sum_{i=1}^{M/2}(N_{M/2+i} - N_i)\right|} \quad \text{(conventional sparse array)} \tag{8.137}$$

Case 1. Each antenna element transmits a unique waveform but receives and processes only that one unique waveform on receive.

In this case, the antenna is operating like the orthogonal interferometer, with two elements, and the out and back phase rotations introduce an extra factor of 2 in the denominator, as was the case for the orthogonal interferometer.

$$\sigma_\theta = \frac{\sqrt{M}}{2\pi\sqrt{2SNR}\left|\sum_{i=1}^{M/2}(N_{M/2+i} - N_i)\right|} \quad \text{(MIMO sparse array)} \tag{8.138}$$

Case 2. Each of the N antenna elements transmits a unique waveform and receives and processes all N waveforms. For this case, let $\varphi_{i,k}$ denote the phase measured at the kth receiver using the ith transmitter signal. Then $[\varphi_{i,k}]$ is an $(M+2)\times(M+2)$ symmetric matrix that contains all the measured phases in the sparse array derived from all of the transmitted waveforms. In Case 1, we used only the phases on the diagonal of the matrix to derive an expression for angle-of-arrival. For the MIMO array, even though the same receiver is used for multiple phase measurements, the matched filter for each different waveform projects the noise into nearly orthogonal subspaces. Thus, we can assume that the phase measurements are independent and that the cross correlation among waveforms defined by the Welch bound is below the thermal noise. If this is not the case, then the phase measurements are no longer noise limited, and angle accuracy can no longer be described by random noise. Computing the phases we have

$$\begin{aligned}
\varphi_{1,1}^{w} &= [\varphi_1 + \varphi_1 + \pi]_{2\pi} - \pi \\
\varphi_{2,2}^{w} &= [\varphi_1 + \varphi_1 + \pi - 2\pi \sin(\theta)]_{2\pi} - \pi \\
\varphi_{1,2}^{w} &= [\varphi_1 + \varphi_1 + \pi - \pi \sin(\theta)]_{2\pi} - \pi \\
\overset{2,1w}{\varphi} &= [\varphi_1 + \varphi_1 + \pi - \pi \sin(\theta)]_{2\pi} - \pi \\
\varphi_{i+2,k+2}^{w} &= [\varphi_1 + \varphi_1 + \pi - (N_i + N_k)\pi \sin(\theta)]_{2\pi} - \pi \quad i = 1 : M, \quad k = 1 : M
\end{aligned} \tag{8.139}$$

where φ_1 is the one-way phase at the first element in the array.

$$\begin{aligned}
\Delta\varphi_{1,2} &= \varphi_{1,1}^{w} - \varphi_{2,2}^{w} = [\varphi_1 + \varphi_1 + \pi]_{2\pi} - [\varphi_1 + \varphi_1 + \pi - 2\pi \sin(\theta)]_{2\pi} \\
\Delta\varphi_{i+2,k+2} &= \varphi_{1,1}^{w} - \varphi_{i+2,k+2}^{w} = [\varphi_1 + \varphi_1 + \pi]_{2\pi} \\
&\quad - [\varphi_1 + \varphi_1 + \pi - (N_i + N_k)\pi \sin(\theta)]_{2\pi}
\end{aligned} \tag{8.140}$$

$$\begin{aligned}
(N_i + N_k)\Delta\varphi_{1,2} &= 2(N_i + N_k)\pi \sin(\theta) = 2K_{N_{i,k}}\pi + \varphi_{residual}^{N_{i,k}} \\
&= 2K_{N_{i,k}}\pi + \Delta\varphi_{i+2,k+2}
\end{aligned} \tag{8.141}$$

Using the symmetry of the phase matrix $[\varphi_{i,k}]$, we use only the phase measurements associated with the upper triangle of the matrix, including the diagonal. We now define a difference vector in order to define $\sin(\theta)$ using (8.141). The *diff* vector is a vector from a Hadamard matrix (matrix consisting of the same number 1s and -1s with orthogonal rows, [19]) of order $L \times L$, where $L = M(M + 1)/2$. For the *diff* vector, we insist that the number of $+1$s and -1s be equal. There are $L - 1$ number of such vectors and for a given element spacing, *diff* vectors minimize the angle accuracy. Thus the *diff* vector is length M^2 consisting of a number of $+1$s, and -1s.

$$\begin{aligned}
&2\pi \sin(\theta) \sum_{k=1}^{M} \sum_{i=k}^{M} diff(M(k-1)+i)(N_i + N_k) \\
&= 2\pi \sum_{k=1}^{M} \sum_{i=k}^{M} diff(M(k-1)+i)K_{i,k} \\
&+ \sum_{k=1}^{M} \sum_{i=k}^{M} diff(M(k-1)+i)\Delta\varphi_{i+2,k+2}
\end{aligned} \tag{8.142}$$

$$\sin(\theta) = \frac{2\pi \sum_{k=1}^{M} \sum_{i=k}^{M} diff(M(k-1)+i) K_{i,k} + \sum_{k=1}^{M} \sum_{i=k}^{M} diff(M(k-1)+i) \Delta\varphi_{i+2,k+2}}{2\pi \sum_{k=1}^{M} \sum_{i=k}^{M} diff(M(k-1)+i)(N_i+N_k)} \tag{8.143}$$

The *diff* vector is chosen to maximize the term in the denominator, and the expression for angle accuracy becomes

$$\sigma_\theta = \frac{\sqrt{M(M+1)}\sigma_\varphi}{2\pi \cos(\theta) \left| \sum_{k=1}^{M} \sum_{i=k}^{M} diff(M(k-1)+i)(N_i+N_k) \right|} \tag{8.144}$$

$$\sigma_\theta = \frac{\sqrt{M(M+1)}}{2\pi \cos(\theta) \left| \sum_{k=1}^{M} \sum_{i=k}^{M} diff(M(k-1)+i)(N_i+N_k) \right| \sqrt{2SNR}} \tag{8.145}$$

Looking at (8.126), three factors can determine angle accuracy: (1) the number of embedded interferometers (M), (2) the location of the elements (N_i), (3) the *diff* vector.

Example: Consider the sparse array defined by

$$P = \{1, 2, 11, 21, 99, 101\} \tag{8.146}$$

$M = 4$ and $N_1 = 11$, $N_2 = 21$, $N_3 = 99$, $N_4 = 101$

For Case 1, we have

$$\sigma_\theta = \frac{1}{168\pi\sqrt{2}} \qquad \text{(Case 1)} \tag{8.147}$$

which is half the error shown in (8.124) for a conventional sparse array. For Case 2, we choose an optimal *diff* vector and angle accuracy is defined as

$$\sigma_\theta = \frac{\sqrt{5}}{524\sqrt{2}\pi} < \frac{1}{168\pi\sqrt{2}} \qquad \text{(Case 2)} \tag{8.148}$$

Note that Case 2 has better angle accuracy than Case 1 by a factor of approximately $\sqrt{2}$. Also note that Case 1 has angle accuracy two times better than angle accuracy for a conventional sparse array, as defined in (8.124); this result agrees with the noise-limited performance difference between a conventional interferometer and an orthogonal interferometer established in Section 7.6.

For two-dimensional MIMO sparse arrays, there exists multiple interferometers that can more easily resolve phase ambiguities with the proper choice of baselines similar to the discussion in Section 8.8. An interferometer architecture that decouples azimuth and elevation measurements is best suited to implement the techniques developed in Section 8.8.2. We can also define a bootstrap algorithm that first resolves the ambiguities for the embedded conventional interferometers contained in an orthogonal interferometer or MIMO array and then resolves the MIMO ambiguities using the resolved conventional ambiguities. Usually monopulse or smaller interferometer arrays can aid the bootstrap process. The bootstrap technique can be made more efficient by selecting baseline distances that are relatively prime integer multiples of the half wavelength in order to assure that the boundaries of the various ambiguity regions do not coincide. By choosing the interferometer baselines this way, the angle ambiguity regions are decorrelated providing an efficient bootstrap process that can use smaller arrays to resolve larger arrays over a given field of view.

References

1. H. Unz, "Linear arrays with arbitrarily distributed elements," *IRE Transactions on Antennas and Propagation*, vol. AP-8, pp. 222–223, Mar. 1960.
2. D. D. King, R. F. Packard, and R. K. Thomas, "Unequally spaced, broadband antenna arrays," *IRE Transactions on Antennas and Propagation*, vol. AP-8, pp. 380–385, Jul. 1960.
3. R-F. Harrington, "Sidelobe reduction by nonuniform element spacing," *IRE Transactions on Antennas and Propagation*, pp. 187–192, Mar. 1961.
4. M. G. Andreason, "Linear array with variable interelement spacings," *IRE Transactions on Antennas and Propagation*, vol. 10, pp. 137–143, Mar. 1962.
5. A. Maffett, "Array factors with nonuniform spacing parameter," *IRE Transactions on Antennas and Propagation*, vol. 10, no. 2, pp. 131–136, Mar. 1962.
6. A. Ishimaru, "Theory of unequally-spaced arrays," *IEEE Trans. AP*, vol. 10, no. 6, pp. 691–702, Nov. 1962.
7. T. T. Taylor, "Design of line-source antennas for narrow beamwidth and low sidelobes," *IRE TRAN*, vol. AP-3, pp. 16–28, Jan. 1955.
8. Y. T. Lo, "A mathematical theory of antenna arrays with randomly spaced elements," *IEEE Trans. Antennas Propagat.*, vol. AP-12, pp. 257–268, May 1964.
9. S. K. Mitra, M. Tchobanou, and G. Jovanovic-Dolecek, "A simple approach to the design of sparse antenna array," *Proc. 2004 IEEE International Symposium on Circuits & Systems*, Vancouver, BC, Canada, May 2004.
10. S. K. Mitra, G. Jovanovic-Dolecek, and M. K. Tchobanou, "On the design of one-dimensional sparse arrays with apodized end elements," *Proc. 12th European Signal Processing Conference*, pp. 2239–2242, Vienna, Austria, September 2004.

11. S. K. Mitra, M. K. Tchobabou, and M. I. Bryukhanov, "A general method for designing sparse antenna arrays," *Proc. of Euro Conf on Circuit Theory and Design*, vol 2. pp. 263–266, Aug. 28–Sept. 2, 2005.
12. P. P. Vaidyanathan and P. Pal, "Sparse co-prime sensing with multi-dimensional lattice arrays," *IEEE DSP/SPE Workshop*, pp. 425–430, Jan. 2011.
13. P. Pal and P. P. Vaidyanathan, "Two dimensional nested arrays on lattices," *International Conference on Acoustics, Speech, and Signal Processing—ICASSP*, pp. 2548–2551, 2011.
14. P. P. Vaidyanathan and P. Pal, "Sparse sensing with co-prime samplers and array," *IEEE TRANS SP*, vol. 59, no. 2, Feb. 2011.
15. P. P. Vaidyanathan and P. Pal, "Theory of sparse co-prime sensing in multiple dimensions," *IEEE Trans SP*, vol 59, no. 8, Aug. 2011.
16. P. Pal and P.P. Vaidyanathan "Nested arrays: A novel approach to array processing with enhanced degrees of freedom," *IEEE Trans. SP*, vol. 58, no. 8, pp. 4167–4180, Aug. 2010.
17. E. J. Holder and G. M. Hall, "Numerical sieve methods for selecting optimal sparse arrays," Propagation Research Associates Internal Technical Memorandum, May 2012.
18. J. Li and P. Stoica, *MIMO Radar Signal Processing*, John Wiley & Sons, Hoboken, NJ, USA, 2009.
19. K. J. Horadam, *Hadamard Matrices and Their Applications*, Princeton University Press, Princeton, NJ, USA, 2006.

Chapter 9
Interferometer Angle-of-Arrival Error Effects

The angle error equations derived to this point have reflected only the error performance due to thermal noise effects. However, angle-of-arrival performance can be affected by various sources of error other than thermal noise. In this chapter, we focus on errors that affect angle accuracy as opposed to angle precision. Because an interferometer determines angle-of-arrival from the output of phase or voltage from two antennas, the sensitivity of angle measurement to errors can be significant. Typical types of errors that can affect the performance of angle estimation using interferometry are:

- Multipath
- Glint
- ADC timing jitter
- I&Q detector imbalances
- Quantization effects
- Channel transfer function mismatch
- Channel timing errors
- Dispersion effects

The effect of phase noise is a function of how each interferometer method uses phase to determine angle-of-arrival. For the digital interferometer, phase and angle are related as follows:

$$\Delta\phi = \phi_1 - \phi_2 = \frac{2\pi}{\lambda} D \sin(\theta) \tag{9.1}$$

and, when the phase errors are uncorrelated between two phase measurements, we have

$$\sigma_\theta^I = \frac{\lambda}{2\pi D \cos(\theta)} \sigma_{\Delta\phi} = \frac{\lambda}{\sqrt{2}\pi D \cos(\theta)} \sigma_\phi \tag{9.2}$$

The expression for angle error in (9.2) is derived in Chapter 6 under the assumption that the phase noise is identically distributed and independent with mean zero. This assumption is generally valid for thermal noise but is not generally valid for other error sources that can degrade interferometer angle estimation. In this chapter, we consider error sources other than thermal noise and quantify the impact on interferometric angle estimation. In some cases, such as specular multipath, we present a discussion of ways to detect and mitigate this type of multipath using distributed antenna arrays.

9.1 Specular Multipath

Ground bounce specular multipath [1–4] occurs when the transmitted signal reflects from the ground and is received by each interferometer antenna, causing interference with the desired direct path signal. In general, multipath causes both amplitude fluctuations and phase distortion. Barton [4,9] has formulated an expression for the angle error due to multipath as follows:

$$\sigma_\theta^{multipath} = \frac{\rho\theta_3}{\left(8G_{se}(peak)\right)^{1/2}} \tag{9.3}$$

where σ_m is the one-sigma angle error due to multipath, θ_3 is the one-way 3-dB beamwidth, ρ is the reflection coefficient, and $G_{se}(peak)$ is the ratio ratio of the antenna gains for target and multipath signals.

The reflection coefficient is a function of the electrical properties of the reflecting surface as well as the surface roughness. We define the polarized reflection coefficients as [10]:

$$\Gamma_v = \frac{\varepsilon\sin(\psi) - \sqrt{\varepsilon - \cos^2(\psi)}}{\varepsilon\sin(\psi) + \sqrt{\varepsilon - \cos^2(\psi)}} \tag{9.4}$$

$$\Gamma_h = \frac{\sin(\psi) - \sqrt{\varepsilon - \cos^2(\psi)}}{\sin(\psi) + \sqrt{\varepsilon - \cos^2(\psi)}} \tag{9.5}$$

$$\Gamma_c = \frac{\Gamma_v + \Gamma_h}{2} \tag{9.6}$$

where ψ is the grazing angle, and the permittivity ε consists of the relative permittivity ε_r, and the surface conductivity σ as follows:

$$\varepsilon = \varepsilon_r - j60\sigma\lambda \tag{9.7}$$

The surface roughness can be defined using an exponential model [10].

$$\rho_s = e^{-2\left(\frac{2\pi\sigma_h\sin(\psi)}{\lambda}\right)^2} \tag{9.8}$$

And the total multipath reflection coefficient is as follows:

$$\rho = \Gamma_{v,h,c}\rho_s \tag{9.9}$$

The preceding definition for ρ assumes a flat surface but can be modified for a spherical surface by inserting a divergence term D that models the reflecting wave divergence due to the earth's curvature.

$$\rho = \Gamma_{v,h,c} D \rho_s \tag{9.10}$$

where

$$D \frac{1}{\sqrt{1 + \frac{2d_1 d_2}{r_e d \sin(\psi)}}} \tag{9.11}$$

where d, d_1, and d_2 are defined in Figure 9.1 for a curved earth multipath geometry.

Multipath effects can be mitigated by using frequency agility, polarization agility, high-resolution waveforms, clutter fences that absorb multipath signals, and signal processing using complex angle processing [5].

9.1.1 Multipath Mitigation Using the Orthogonal Interferometer

Using the MIMO characteristics of the orthogonal interferometer, multipath effects can be mitigated by selectively processing multiple signals. For certain multipath environments where the reflecting surface is highly specular, the orthogonal interferometer architecture may be able to provide mitigation using spatial degrees of freedom to decorrelate the multipath. For example, if we assume that the multipath is mostly in the elevation plane, an array of vertical elements using orthogonal waveforms can decorrelate the multipath in a structured way that allows the effects to be reduced. An interferometer requires an array architecture with antennas located at different heights above the reflecting surface. For example, the standard interferometer antenna configuration shown in Figure 6.4 can be rotated to decorrelate multipath due to the height of the antennas above the reflecting surface. Figure 9.2 shows a rotated version of the orthogonal interferometer architecture.

Multipath mitigation can be achieved by integrating over multiple signal paths that occur due to both direct paths and reflected paths. Figure 9.3 shows a planar geometry for two arrays separated in the vertical dimension, creating four total paths (two direct and two reflected), three of which are uncorrelated. For three

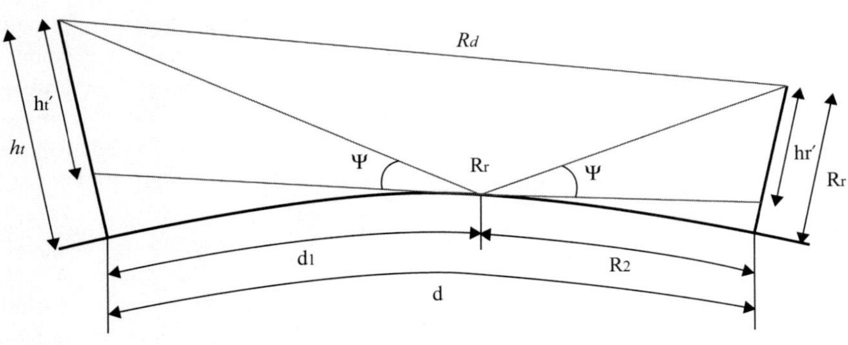

Figure 9.1. Curved Earth Multipath Geometry

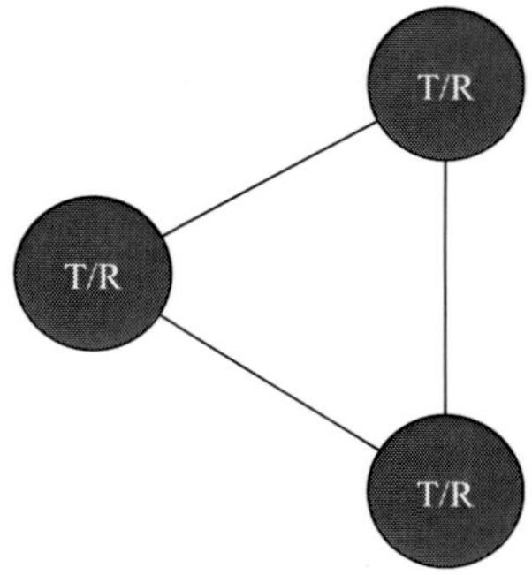

Figure 9.2. Orthogonal Interferometer with Three Transmit/Receives (T/R), Oriented for Multipath Decorrelation

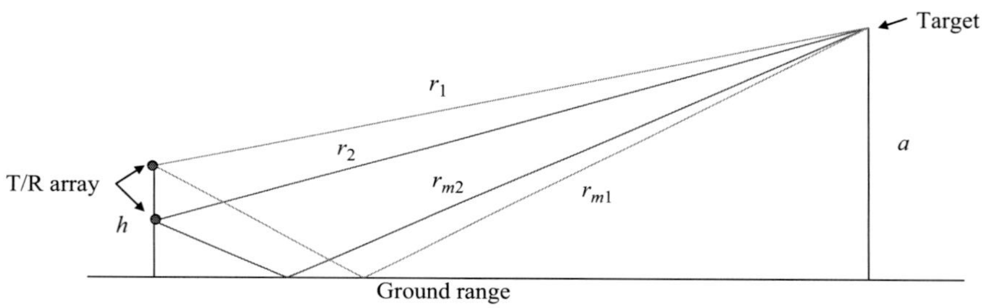

Figure 9.3. Two-Dimensional Multipath Array Geometry with Four Signal Paths

arrays that are each transmitting a unique waveform and each array receiving all three waveforms, nine signal paths are created by an orthogonal interferometer. Assuming a flat-earth reflecting plane and a rotated OI design, six of the nine paths are uncorrelated. Thus, six paths provide a different multipath signal with slightly different phases. By noncoherently integrating all signal paths, radar performance in a multipath environment can be improved. Figure 9.4 shows the improvement in detection performance using a simple flat-earth multipath model for three arrays at different heights and a target at 100-m altitude. Note that the spatial degrees of freedom can take advantage of constructive multipath to increase signal performance an average of 5-dB improvement using an orthogonal interferometer with noncoherent integration processing.

For an interferometer to estimate angle-of-arrival, it turns out that only three "good" paths are required out of the six total independent paths. Actually only two good paths are required to measure a single angle for use in updating a Kalman track filter. The track filter can therefore be fed with one or two angle measurements, depending on the signal-to-noise ratios of the various paths. Figure 9.5 shows that the number of paths with no signal loss due to multipath is always greater than three paths for the flat-earth multipath model and target geometry used

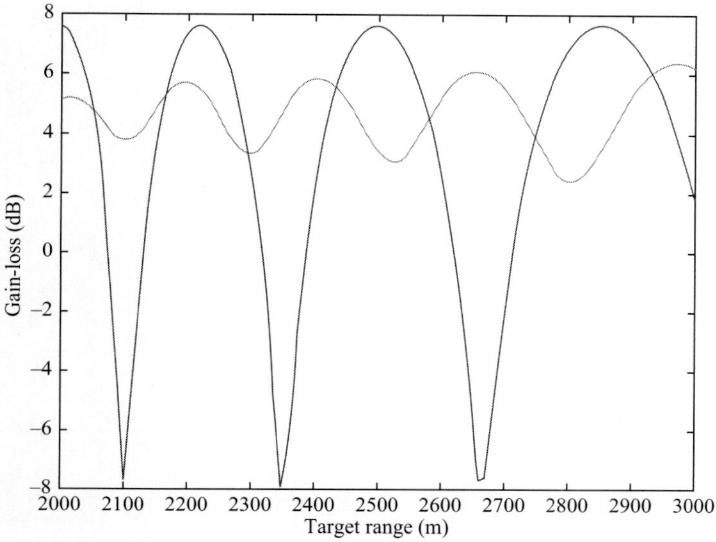

Figure 9.4. OI (light gray) Versus CI (dark gray) Performance in Multipath

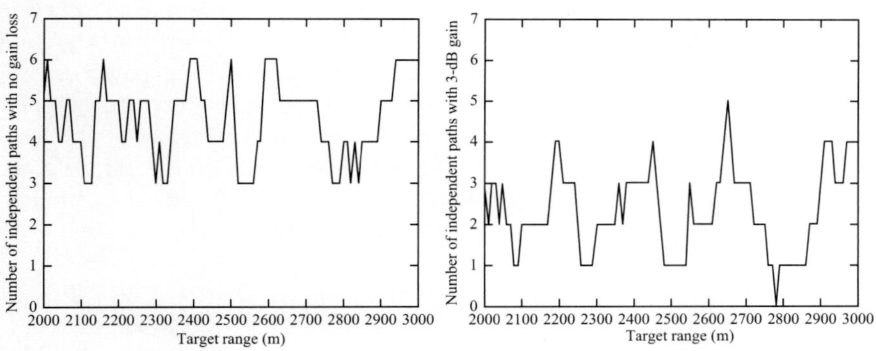

Figure 9.5. Number of Paths with No Signal Gain Loss (left) and with 3-dB Signal Gain (right)

in Figure 9.4. Figure 9.5 also shows the number of paths that have 3-dB gain due to in-phase multipath addition, providing good phase quality for interferometry. If the three transmit arrays are separated from the three receive arrays, then nine total independent path are possible, allowing for improved multipath and angle interferometry performance.

Figure 9.6 shows the difference between the noncoherent sum of three channels and the individual channels in a multipath environment using a flat-earth

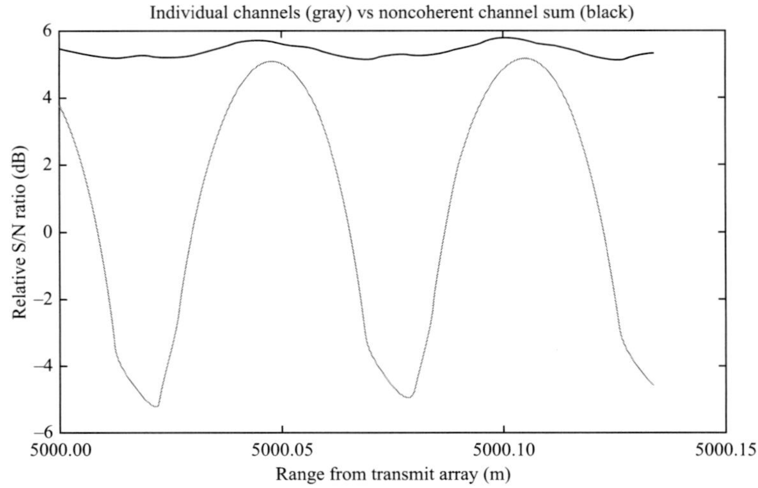

Figure 9.6. Simulated Performance of Noncoherent Multipath Mitigation Compared with Single Antenna Multipath Performance

orthogonal interferometer simulation and orthogonal interferometer processing. Note that, for the individual channels, the receive signal power rises to a peak when the incident reflection and the multipath signal are in phase. When the two return signals add destructively (180° out of phase), the receive signal power is at a minimum. By summing the three receiver signals from the orthogonal interferometer noncoherently, the resulting signal power is maintained at a high level for detection. The three receivers are physically separated and decorrelate the multipath effects. Because all three antennas have been used for signal integration, interferometric angle processing cannot be accomplished.

9.1.2 Multipath Mitigation Using Sparse Arrays

We now consider using distributed arrays to resolve the direct path signal from the specular multipath signal. The approach assumes that the radar and target are elevated above a flat earth and that four receive antennas are required to implement the algorithm, providing the orientation of the reflecting plane is known. If this orientation is not known, then the algorithm requires a minimum of seven receive antennas. One of the receive antennas is required to measure range to the target.

Assume that a radar is located at a height h above a flat earth and that the target is located at altitude a above the same flat earth. If the ground range from the radar to the target is r, the direct path signal sig_d can be expressed as

$$sig_d = A\, e^{\frac{2\pi j \sqrt{r^2+(a-h)^2}}{\lambda}} \tag{9.12}$$

The indirect or specular multipath signal sig_m results from radar energy reflecting from the ground to the radar and can be expressed as

$$sig_m = A\rho\, e^{\frac{2\pi j\sqrt{r^2+(a+h)^2}}{\lambda}} \tag{9.13}$$

where ρ is the ground reflection coefficient.

Multipath interference results when the direct path signal and the multipath combine at the radar to mutually interfere [1–4]. The combined signal can be expressed as follows:

$$\begin{aligned} sig_d &= sig_d + sig_m \\ &= Ae^{\frac{2\pi j\sqrt{r^2+(a-h)^2}}{\lambda}} + A\rho\, e^{\frac{2\pi j\sqrt{r^2+(a+h)^2}}{\lambda}} \end{aligned} \tag{9.14}$$

Using a second-order Taylor series expansion, the phase of the direct and multipath signals can be approximated as follows:

$$\sqrt{1+x^2} \approx 1 + \frac{1}{2}x^2 \tag{9.15}$$

$$\sqrt{r^2+(a-h)^2} \approx r\left(1 + \frac{(a-h)^2}{2r^2}\right) \tag{9.16}$$

$$\sqrt{r^2+(a+h)^2} \approx r\left(1 + \frac{(a+h)^2}{2r^2}\right) \tag{9.17}$$

Using these approximations, we write the combined signal as

$$\begin{aligned} sig &= Ae^{\frac{2\pi\, jr}{\lambda}\left(1+\frac{(a-h)^2}{2r^2}\right)} + A\rho\, e^{\frac{2\pi\, jr}{\lambda}\left(1+\frac{(a+h)^2}{2r^2}\right)} \\ &= e^{\frac{2\pi\, jr}{\lambda}}\left(Ae^{\frac{2\pi\, jr}{\lambda}\left(\frac{(a-h)^2}{2r^2}\right)} + A\rho\, e^{\frac{2\pi\, jr}{\lambda}\left(\frac{(a+h)^2}{2r^2}\right)}\right) \\ &= e^{\frac{2\pi\, jr}{\lambda}} e^{\frac{2\pi\, j}{\lambda}\left(\frac{a^2+h^2}{2r}\right)}\left(Ae^{-\frac{2\pi\, jah}{\lambda r}} + A\rho\, e^{\frac{2\pi\, jah}{\lambda r}}\right) \end{aligned} \tag{9.18}$$

At the output of a square law detector, the signal is expressed as

$$|sig|^2 = A^2 + A^2\rho^2 + 2A^2\rho\cos\left(\frac{4\pi ah}{\lambda r}\right) \tag{9.19}$$

We now consider the following trigonometric relationship:

$$\begin{aligned} y_N &= \sin(N\theta) \\ &\dots \\ y_N &= 2\cos(\theta)y_{N-1} - y_{N-2} \\ y_{N-1} &= 2\cos(\theta)y_{N-2} - y_{N-3} \end{aligned} \tag{9.20}$$

and

$$\cos(\theta) = \frac{y_N - y_{N-1} + y_{N-2} - y_{N-3}}{2(y_{N-1} - y_{N-2})} \tag{9.21}$$

Note that (9.21) can be used to solve for $\cos(\theta)$ in the more general expression

$$y_N = c + b\sin(N\theta) \tag{9.22}$$

because the constant term c cancels in both the numerator and denominator. Thus (9.21) can be applied to (9.19).

We can now use (9.21) and (9.22) to estimate target parameters using specular multipath modulation in a sparse array. Place four antennas stacked directly above each other in height and uniformly spaced at a distance d apart. Let h_1 be the antenna that is lowest in height; then $h_2 = h_1 + d$, $h_3 = h_2 + d$, and $h_4 = h_3 + d$. Furthermore, let

$$p(i) = A^2 + A^2\rho^2 + 2A^2\rho\cos\left(\frac{4\pi a h i}{\lambda r}\right), \qquad \text{for } i = 1, 2, 3, 4 \tag{9.23}$$

Then

$$a_r \equiv \frac{a}{r} = a\cos\left(\frac{p(1) - p(2) + p(3) - p(4)}{2(p(2) - p(3))}\right)\left(\frac{\lambda}{4\pi d}\right) \tag{9.24}$$

Assume that the radar makes a measurement of the direct path range (r_d).

$$r_d = \sqrt{r^2 + (h-a)^2} + r_{\text{error}} \tag{9.25}$$

where r_{error} is uniformly distributed over the interval $[-r_{\text{error}}/2\ r_{\text{error}}/2]$.

Then an estimate of ground range (r_{est}) can be derived by solving (9.24) and (9.25) simultaneously,

$$r_{\text{est}} = \frac{\left(a_r h + \sqrt{r_d^2(1 + a_r^2) - h^2}\right)}{1 + a_r^2} \tag{9.26}$$

and an estimate of altitude (a_{est}) is derived by

$$a_{\text{est}} = a_r r_{\text{est}} \tag{9.27}$$

In a noisy environment, (9.24) tends to be inaccurate and thus averaging may be required to recover the desired accuracy.

9.1.2.1 Estimating Receiver Height for Practical Applications

In the preceding analysis, it is assumed that the height of the radar above the reflecting plane is known. For many applications, this assumption is not practical. Using the expression for frequency,

$$f = \frac{c}{\lambda} \tag{9.28}$$

where c is the speed of light and substituting in (9.19), then

$$|sig|^2 = A^2 + A^2\rho^2 + 2A^2\rho\cos\left(\frac{4\pi ahf}{cr}\right) \tag{9.29}$$

Now changing the signal frequency at a given receiver provides an estimate of the quantity ah/r. By changing the spatial height, each frequency provides an estimate of the quantity af/r as in (9.24). Because frequency is known, an estimate of height h_{est} is obtained as follows:

$$h_{\text{est}} = f\frac{{}^{ah}/_{r}}{{}^{af}/_{r}} \tag{9.30}$$

The elevation angle (el) of the target relative to the radar receiver is computed by

$$el = \tan^{-1}\left(\frac{a_{\text{est}} - h_{\text{est}}}{r_{\text{est}}}\right) \tag{9.31}$$

9.1.2.2 Estimating the Orientation of the Reflecting Plane

In the preceding analysis, the orientation of the reflecting plane was assumed to be at a right angle to the linear array of four elements. In practice, this orientation angle is not 90° and in general is not known. To determine this angle, a second array is required that is not colinear with the first array. The two contiguous colinear arrays could have a shared antenna, as in Figure 9.7. Thus a minimum of seven antennas is required for this configuration.

Let d be the spacing between antenna elements, and let θ be the orientation of the linear array with the line normal to the reflecting plane. The output of the square law detector at each antenna of the linear array is given by

$$|sig|^2 = A^2 + A^2\rho^2 + 2A^2\rho\cos\left(\frac{4\pi a(h - d\cos(\theta))}{\lambda(r - d\sin(\theta))}\right) \tag{9.32}$$

Using the first-order Taylor series expansion

$$\frac{1}{r - \Delta r} \approx \frac{1}{r}\left(1 + \frac{\Delta r}{r}\right) \tag{9.33}$$

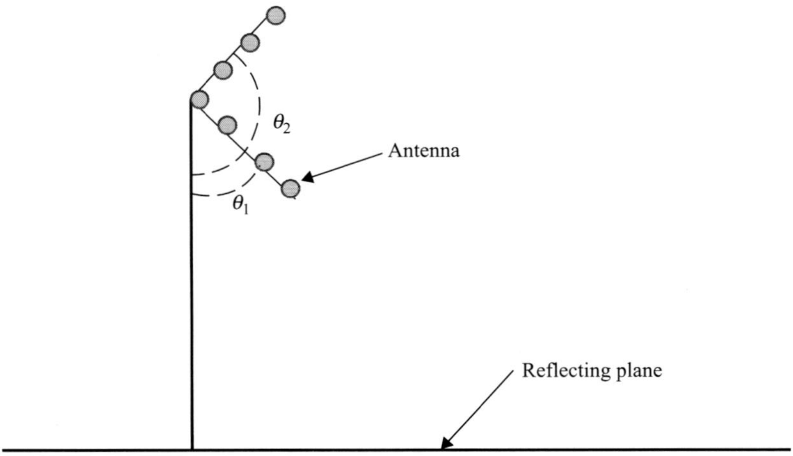

Figure 9.7. Two Contiguous Colinear Antenna Arrays Mounted Using a Shared Antenna

and eliminating terms that are of order $1/r^2$, (9.32) can be written as

$$|sig|^2 = A^2 + A^2\rho^2 + 2A^2\rho\cos\left(\frac{4\pi a}{\lambda r}\left(\frac{h}{r} - \frac{dk\cos(\theta)}{r}\right)\right) \quad (9.34)$$

Regarding (9.34) as a function of antenna position k, the frequencies of the cosine modulation can thus be estimated. Let f_1 and f_2 be the frequencies of the cosine modulation as a function of antenna position for each of the two noncolinear arrays.

$$f_1 = \frac{4\pi a\cos(\theta_1)}{\lambda r} \quad \text{and} \quad f_2 = \frac{4\pi a\cos(\theta_2)}{\lambda r} \quad (9.35)$$

Now determine the orientation of the first array as a function of the measured frequencies f_1, f_2 and the known angular separation of the two arrays $\Delta\theta = \theta_2 - \theta_1$.

$$\frac{f_2}{f_1} = \frac{\cos(\theta_2)}{\cos(\theta_1)} \quad (9.36)$$

$$\begin{aligned}\frac{f_2}{f_1}\cos(\theta_1) &= \cos(\theta_2)\\ &= \cos(\theta_1 + \Delta\theta)\\ &= \cos(\theta_1)\cos(\Delta\theta) - \sin(\theta_1)\sin(\Delta\theta)\end{aligned} \quad (9.37)$$

and

$$\tan(\theta_1) = \frac{\cos(\Delta\theta) - \dfrac{f_2}{f_1}}{\sin(\Delta\theta)} \quad (9.38)$$

and for each array, the altitude above the reflecting plane is determined by,

$$a_{\text{est}} = \frac{a_r r_{\text{est}}}{\cos(\theta)} \quad (9.39)$$

where for each array a_r and r_{est} are as computed, and θ is the angle between each linear array and the line normal to the reflecting plane.

In a simulation, ground range was varied from 20 km down to 10 km and the following parameters were used:

$a = 10$

$\lambda = 0.03$ m

$d = 1$ m

$\rho = 0.9$

$SNR = 6$ dB @ 20 km, one pulse

$M = 4$

$r_{error} = 0.2$ m

Assuming a 10° orientation reflecting plane and the same parameters, Figure 9.8 shows the resulting estimate of altitude using the estimate of orientation angle.

For certain multipath environments where the surface is highly specular, the interferometer architecture may be able to quantify the degree of multipath that is present and may even provide mitigation. Assuming that the multipath is mostly in the elevation plane, an interferometer with separated antennas can decorrelate the multipath in a structured way that allows the effects to be quantified.

9.1.3 Quantification of Multipath Using Interferometry

Assuming that a flat-earth reflecting plane, a single-bounce multipath can be modeled by integrating over multiple signal paths that occur due to both direct

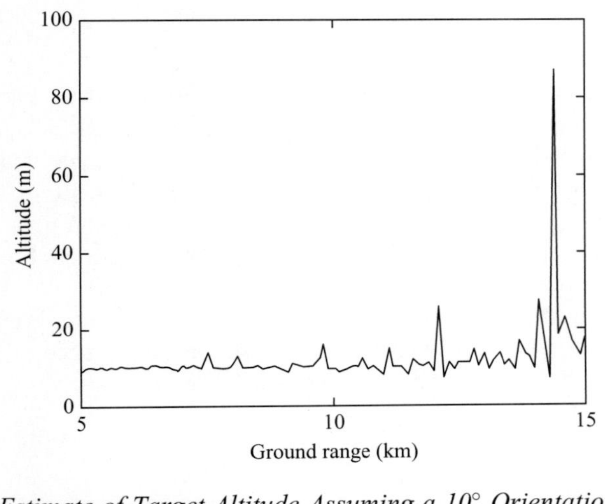

Figure 9.8. Estimate of Target Altitude Assuming a 10° Orientation of the Reflecting Plane

paths and reflected paths. Referring to Figure 9.3 and (9.18), the power of the signal is expressed as

$$P = |V|^2 = A^2 + A^2\rho^2 + 2A^2\rho\cos\left(\frac{4\pi ah}{\lambda r}\right) \tag{9.40}$$

Equation (9.40) shows that the power of the returned signal depends on the geometry of the radar and target environment. In the absence of multipath ($\rho = 0$), the power is constant regardless of the radar-target environment. As a result, (9.40) shows that, for specular multipath, the variation in signal power as a function of time provides an indicator of the presence of multipath.

9.1.3.1 Multipath Identification and Quantification

Because multipath is decorrelated over antenna position, target position, and operating frequency, we can utilize these parameters to identify the presence of multipath. Assume that the antennas are calibrated so that the gain for each of the three receive antennas is the same (or at least known) and the receiver characteristics are the same (or at least known). Then, in a nonmultipath environment, the power output of each receiver would be nearly the same, assuming that the antennas are calibrated. Of course, there will be some variation in outputs due to element and channel mismatch, but we assume that these variations are relatively small. When multipath is introduced, the power values are significantly different because the antennas are in different relative locations. We should see a larger difference in elevation than azimuth due to the ground bounce multipath.

Let h_i indicate the height of the ith antenna, and let $r(t)$ indicate the ground range to the target. Then the power at the ith antenna is

$$P = |V|^2 = A^2 + A^2\rho^2 + 2A^2\rho\cos\left(\frac{4\pi ah_i}{\lambda r_{(t)}}\right) \tag{9.41}$$

9.1.3.2 Power Difference Metrics

Consider the conventional interferometer architecture shown in Figure 9.9. For a moving target, form the power difference between two antennas separated in elevation and the power difference between the two bottom antennas that are separated in azimuth. In a perfect world, these differences should be zero. As the target moves along its trajectory, we accumulate these differences in time. The statistics and power spectrum of these differences tell us something about how multipath is affecting us throughout the trajectory.

Assume that antenna 1 is the highest antenna in the interferometer antenna architecture, consisting of three antennas located at the vertices of an equilateral triangle, and form the differences for the three radar antennas.

$$D_1(t_k) = (P_1(t_k) - P_3(t_k)) = 2A^2\rho\left(\cos\left(\frac{4\pi ah_1}{\lambda r(t_k)}\right) - \cos\left(\frac{4\pi ah_3}{\lambda r(t_k)}\right)\right) \tag{9.42}$$

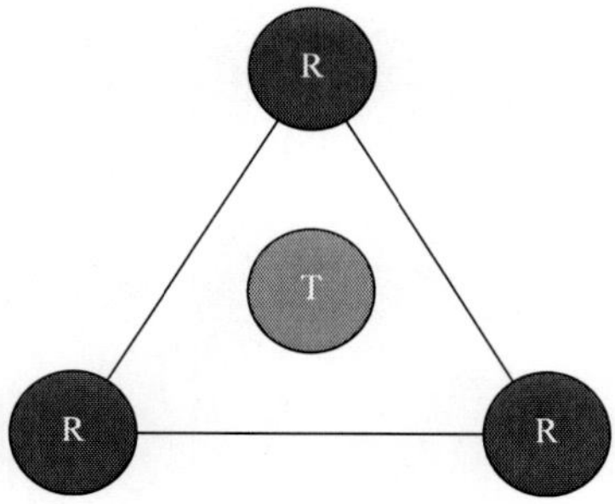

Figure 9.9. Conventional Interferometer Architecture with Three Receive Arrays (R) and One Transmit Array (T)

$$D_2(t_k) = (P_2(t_k) - P_3(t_k)) = 2A^2\rho\left(\cos\left(\frac{4\pi a h_2}{\lambda r(t_k)}\right) - \cos\left(\frac{4\pi a h_3}{\lambda r(t_k)}\right)\right) \quad (9.43)$$

Assuming that $h_1 = h_2 + \Delta h_1$ and $h_2 = h_3 + \Delta h_2$, we can apply a Taylor series approximation for D_1 and D_2.

$$D_1(t_k) = 2A^2\rho\left(\cos\left(\frac{4\pi a h_1}{\lambda r(t_k)}\right) - \cos\left(\frac{4\pi a h_3}{\lambda r(t_k)}\right)\right) = \frac{8\pi A^2 \rho a \Delta h_1}{\lambda r(t_k)} \sin\left(\frac{4\pi a \Delta h_1}{\lambda r(t_k)}\right) \quad (9.44)$$

$$D_2(t_k) = 2A^2\rho\left(\cos\left(\frac{4\pi a h_2}{\lambda r(t_k)}\right) - \cos\left(\frac{4\pi a h_3}{\lambda r(t_k)}\right)\right) = \frac{8\pi A^2 \rho a \Delta h_2}{\lambda r(t_k)} \sin\left(\frac{4\pi a \Delta h_2}{\lambda r(t_k)}\right) \quad (9.45)$$

We now introduce the following metrics:

$$M_1(t_k) = \frac{\lambda \hat{r}(t_k)}{\Delta h_1} D_1(t_k) = 8\pi A^2 \rho a \sin\left(\frac{4\pi a \Delta h_1}{\lambda r(t_k)}\right) \qquad \text{Vertical Multipath Metric} \quad (9.46)$$

$$M_2(t_k) = \frac{\lambda \hat{r}(t_k)}{\Delta h_2} D_2(t_k) = 8\pi A^2 \rho a \sin\left(\frac{4\pi a \Delta h_2}{\lambda r(t_k)}\right) \qquad \text{Horizontal Multipath Metric} \quad (9.47)$$

where $\hat{r}(t_k)$ is an estimate of the radar range to the target at time t_k. The difference operator is chosen in order to reduce effects such as RCS fluctuations. Also, the power difference metrics do not depend on the absolute quantities h_1, h_2, and h_3 and as such are independent of interferometer location with respect to the reflecting plane. The power difference metrics do depend on an estimate of range, which may be corrupted by multipath if it is derived from interferometer measurements.

9.1.3.3 Power Quotient Metrics

The advantage of using quotient metrics is that the effects of signal fluctuations due to RCS fade and glint spikes can be minimized.

$$Q_1(t_k) = \frac{P_1(t_k)}{P_3(t_k)} = \frac{1 + \rho^2 + 2\rho\cos\left(\dfrac{4\pi a h_1}{\lambda r(t_k)}\right)}{1 + \rho^2 + 2\rho\cos\left(\dfrac{4\pi a h_3}{\lambda r(t_k)}\right)} \tag{9.48}$$

$$Q_2(t_k) = \frac{P_2(t_k)}{P_3(t_k)} = \frac{1 + \rho^2 + 2\rho\cos\left(\dfrac{4\pi a h_2}{\lambda r(t_k)}\right)}{1 + \rho^2 + 2\rho\cos\left(\dfrac{4\pi a h_3}{\lambda r(t_k)}\right)} \tag{9.49}$$

As can be seen, the advantage of the quotient metrics is that they do not depend on the return power of signal because power cancels in the quotient formulation. It turns out that each quotient metric is bounded by

$$(1-\rho)^2 \leq Q_i \leq (1+\rho)^2 \tag{9.50}$$

And if $\rho \ll 1$ (low multipath), then the Q_i do not fluctuate, but when ρ is reasonably close to unity (high multipath), there is a significant degree of fluctuation in the quotients. Another value of the quotient metrics is that they can be converted to dB to reflect power differences.

$$Q_1^{dB}(t_k) = P_1^{dB}(t_k) - P_3^{dB}(t_k) \tag{9.51}$$

$$Q_2^{dB}(t_k) = P_2^{dB}(t_k) - P_3^{dB}(t_k) \tag{9.52}$$

Based on the power quotients, we can form two additional metrics:

$$M_3(t_k) = Q_1^{dB}(t_k) \qquad \text{Vertical Multipath Metric} \tag{9.53}$$

$$M_4(t_k) = Q_2^{dB}(t_k) \qquad \text{Horizontal Multipath Metric} \tag{9.54}$$

Note that the power quotient metrics depend on the absolute quantities h_1, h_2, and h_3. Thus the fluctuation statistics can depend on the location of the interferometer antenna with respect to the reflecting plane.

The metrics M_1 and M_3 are measures of the vertical multipath, whereas M_2 and M_4 are measures of the horizontal multipath. In the absence of multipath, all of the power measurements are constant and equal over time, and thus the difference and quotient metrics should all be equal to zero (assuming antenna calibration). In the presence of multipath, the quotients have a power spectrum that reflects periodic signal power. The standard deviation of the metrics would provide a useful measure of the degree of multipath. In the absence of multipath, the standard deviation of all of the metrics is zero, whereas in the presence of multipath, these statistics are nonzero.

As a word of caution, it can be seen in all of the metric equations that, when the range parameter gets large, the sensitivity of these metrics to multipath is less than when range is comparatively small. For example, we have

$$\lim_{r\to\infty} M_1(t_k) = \lim_{r\to\infty}\left(8\pi A^2 \rho a \sin\left(\frac{4\pi a\Delta h_1}{\lambda r(t_k)}\right)\right) = 0 \tag{9.55}$$

$$\lim_{r\to\infty} M_2(t_k) = \pi\left(8\pi A^2 \rho a \sin\left(\frac{4\pi a\Delta h_2}{\lambda r(t_k)}\right)\right) = 0 \tag{9.56}$$

$$\lim_{r\to\infty} M_3(t_k) = \lim_{r\to\infty} Q_1^{dB}(t_k) = 0 \tag{9.57}$$

$$\lim_{r\to\infty} M_4(t_k) = \lim_{r\to\infty} Q_2^{dB}(t_k) = 0 \tag{9.58}$$

Thus it can be seen that, for long range, the standard deviation of these metrics approaches zero. As a result, it only makes sense to apply these metrics over a windowed time interval of interest. Figure 9.10 shows the performance of metrics M_3 and M_4 for a conventional interferometer measuring a low-elevation target out to a maximum range of 1 km. Note that the M_3 metric shows that horizontal multipath has a significant impact on performance for the test environment. Also note that M_3 and M_4 identify the regions where horizontal and vertical multipath are significant.

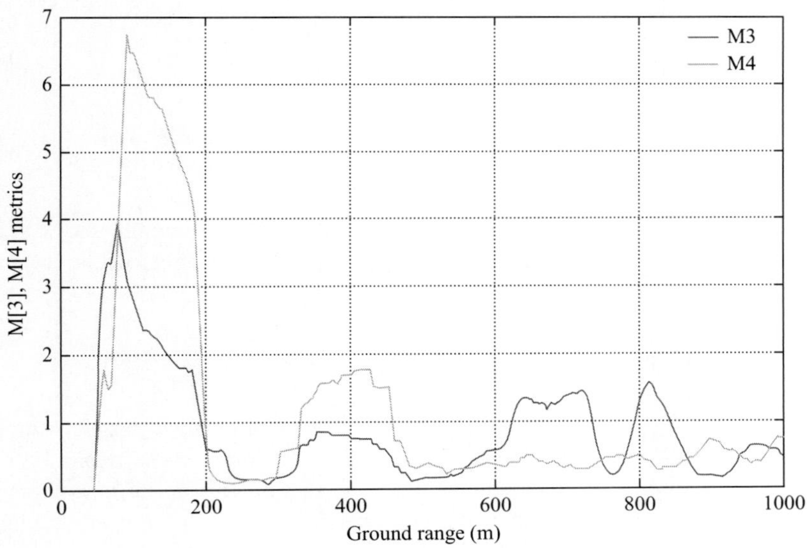

Figure 9.10. Performance of M_3 and M_4 Metrics for a Low-Elevation Target Using a Conventional Interferometer (Courtesy Technovative Applications)

Table 9.1. Summary of Multipath Quantification Metrics Using Interferometry

Multipath Metric	Advantages	Disadvantages
Standard deviation of power difference	Related to the multipath geometry; Related to the interferometer architecture; Independent of interferometer location	Has some dependency on signal fluctuation; Has a dependency on the magnitude of range that restricts the time window for statistics; Requires an estimate of range that may be corrupted by multipath
Standard deviation of power quotient	Related to the multipath geometry; Related to the interferometer architecture; Independent of signal power fluctuations	Has some dependency on interferometer antenna location; Has a dependency on the magnitude of range that restricts the time window for statistics

To summarize, the following four metrics can be used to determine the presence of multipath.

$$std(M_1(t_k)) = \frac{1}{N}\sum_{k=1}^{N}(M_1(t_k) - mean(M_1(t_k)))^2 \tag{9.59}$$

$$std(M_2(t_k)) = \frac{1}{N}\sum_{k=1}^{N}(M_2(t_k) - mean(M_2(t_k)))^2 \tag{9.60}$$

$$std(M_3(t_k)) = \frac{1}{N}\sum_{k=1}^{N}(M_3(t_k) - mean(M_3(t_k)))^2 \tag{9.61}$$

$$std(M_4(t_k)) = \frac{1}{N}\sum_{k=1}^{N}(M_4(t_k) - mean(M_4(t_k)))^2 \tag{9.62}$$

where N is total number of time measurements used to compute the standard deviation. Table 9.1 summarizes the advantages and disadvantages of the proposed metrics.

9.2 Angle Glint

Angle error due to glint [1–4] occurs as a result of unresolved target scatterers interacting with the transmitted waveform to produce a distortion of phase of the received signal at each interferometer antenna. The precise definition of glint is related to the gradient of the constant phase contours. Let $z = f(x,y)$ be a function of two variables. Polar and rectangular coordinates are related by the following change of variables:

$$x = r\cos(\theta) \qquad y = r\sin(\theta) \tag{9.63}$$

and the gradient in polar coordinates is related to the gradient in rectangular coordinates by the following Jacobian transformation.

$$\nabla_P f = \begin{bmatrix} \dfrac{\partial f}{\partial r} \\ \dfrac{1}{r}\dfrac{\partial f}{\partial \theta} \end{bmatrix} = \begin{bmatrix} \dfrac{\partial x}{\partial r} & \dfrac{\partial y}{\partial r} \\ \dfrac{\partial x}{\partial \theta} & \dfrac{\partial y}{\partial \theta} \end{bmatrix} \begin{bmatrix} \dfrac{\partial f}{\partial x} \\ \dfrac{\partial f}{\partial y} \end{bmatrix} = J \cdot \nabla_C f \tag{9.64}$$

Thus

$$J = \begin{bmatrix} \cos(\theta) & \sin(\theta) \\ -\sin(\theta) & \cos(\theta) \end{bmatrix} \tag{9.65}$$

The formal definition of glint is the apparent displacement of the target position relative to the mean target position resulting from phase scintillation due to multiple target scattering interactions. Glint is related to the angular derivative of the phase function, and we now derive the exact expression for glint. For the function f, we begin by computing

$$\frac{1}{r}\frac{\partial f}{\partial \theta} = [-\sin(\theta) \quad \cos(\theta)] \begin{bmatrix} \dfrac{\partial f}{\partial x} \\ \dfrac{\partial f}{\partial y} \end{bmatrix} = \cos(\alpha)\|\nabla_C f\| \tag{9.66}$$

or

$$\frac{\partial f}{\partial \theta} = r\cos(\alpha)\|\nabla_C f\| \tag{9.67}$$

where α and θ are defined in Figure 9.11.

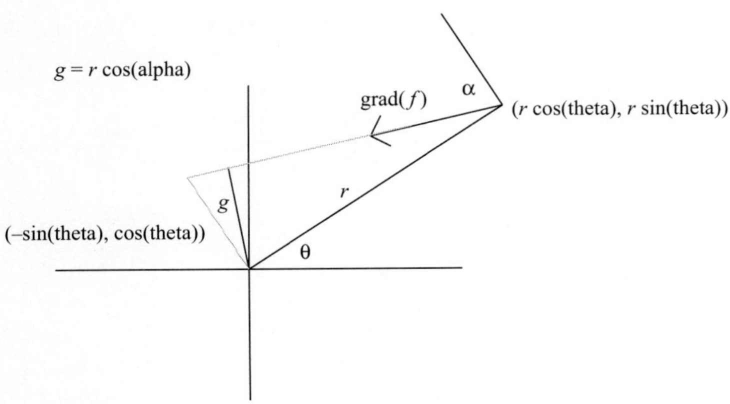

Figure 9.11. Geometry of the Gradient of a Function f(x,y)

Let (x_0, y_0) be the virtual scattering point (defined as the point of rotation) for a target complex. Then the two-way phase at any point (x, y) of the signal scattered from this virtual scattering point is defined as

$$\varphi(x,y) = \frac{4\pi}{\lambda}\sqrt{(x-x_0)^2 + (y-y_0)^2} \tag{9.68}$$

and

$$\nabla_C \varphi = \frac{4\pi}{\lambda}\begin{bmatrix} \dfrac{(x-x_0)}{\sqrt{(x-x_0)^2+(y-y_0)^2}} \\ \dfrac{(y-y_0)}{\sqrt{(x-x_0)^2+(y-y_0)^2}} \end{bmatrix} \tag{9.69}$$

$$\|\nabla_C \varphi\| = \frac{4\pi}{\lambda} \tag{9.70}$$

Therefore,

$$\frac{\partial \varphi}{\partial \theta} = r\cos(\alpha)\|\nabla_C \varphi\| = \frac{4\pi}{\lambda} r\cos(\alpha) \tag{9.71}$$

Because the glint (g) is defined to be the distance from the origin to the point on the line defined by the gradient vector that is closest to the origin (see Figure 9.11), we have,

$$g = \frac{\lambda}{4\pi}\frac{\partial \varphi}{\partial \theta} = r\cos(\alpha) \tag{9.72}$$

Thus, when the gradient vector is pointed directly toward the origin ($\alpha = \pi/2$), glint is zero ($g = 0$). The preceding derivation assumes that the virtual scattering point (x_0, y_0) is fixed; i.e., x_0 and y_0 do not depend on the variables x and y. However, it could be the case that, for a complex target, the virtual scattering point depends on x and y. In that situation, we have the following:

$$\nabla_C \varphi = \frac{4\pi}{\lambda}\begin{bmatrix} \dfrac{(x-x_0)}{\sqrt{(x-x_0)^2+(y-y_0)^2}}\left(1-\dfrac{\partial x_0}{\partial x}\right) - \dfrac{(y-y_0)}{\sqrt{(x-x_0)^2+(y-y_0)^2}}\left(\dfrac{\partial y_0}{\partial x}\right) \\ \dfrac{(y-y_0)}{\sqrt{(x-x_0)^2+(y-y_0)^2}}\left(1-\dfrac{\partial y_0}{\partial y}\right) - \dfrac{(x-x_0)}{\sqrt{(x-x_0)^2+(y-y_0)^2}}\left(\dfrac{\partial x_0}{\partial y}\right) \end{bmatrix} \tag{9.73}$$

and

$$\|\nabla_C \varphi\| \neq \frac{4\pi}{\lambda} \tag{9.74}$$

The correct expression for glint in this case is

$$g = \frac{1}{\|\nabla_C \varphi\|} \frac{\partial \varphi}{\partial \theta} \tag{9.75}$$

Therefore, to be exact, an expression $\|\nabla_C \varphi\|$ must be computed. This involves computing partial derivatives with respect to both r and θ and using the expression for J, which relates the gradient in rectangular coordinates to the gradient in polar coordinates by

$$\nabla_C \varphi = J^{-1} \nabla_P \varphi \tag{9.76}$$

The angle error due to glint is the angle defined by the ratio of glint and range to the target; that is,

$$\alpha_{\text{glint}} = \sin^{-1}\left(\frac{g}{r}\right) \approx \frac{g}{r} = \frac{1}{r\|\nabla_C \phi\|} \frac{\partial \phi}{\partial \theta} \tag{9.77}$$

Because, in general, actual target scatterer locations relative to a radar are unknown, it is appropriate to treat glint as a random parameter. For targets consisting of several distributed scatterers, glint can be estimated using the following expression [1,2]

$$\sigma_{\text{glint}} = \frac{L}{4} \tag{9.78}$$

where L is the projected target cross-range extent relative to the line of sight to the radar. Thus, the angle error due to glint is a function of the range from the target to the radar (r):

$$\sigma_\theta^{\text{glint}} = \frac{L}{4r}. \tag{9.79}$$

As an example of glint, consider a target that is 1 m in length and consists of two scatterers located at each end of the target. When the target is rotated about the midpoint and the interferometer is 3 km from the target (the range is significant only to show the target is in the far field), Figure 9.12 shows the effect of glint when one scatterer has a reflectivity 5 times that of the other (a 7-dB difference). The actual location of the dominant scatterer can be described by 0.5 cos(θ), but the effect of glint demonstrates that the radar can incorrectly determine that the apparent location of the dominant scatterer lies at a point beyond the target's physical extent. Figure 9.13 shows glint noise after the position of the dominant scatterer is removed. Notice that glint noise varies from −0.25 m to 0.25 m, showing that glint can cause the radar to estimate a target position that is ¼ m beyond the target's extent.

Angle glint can be mitigated by implementing frequency hopping among radar dwells. By changing frequency, the relative wavelength separation between scatters is altered for each dwell, and thus the glint error is changed. If multiple frequency changes can be implemented (3 or 4), then the glint error is effectively randomized

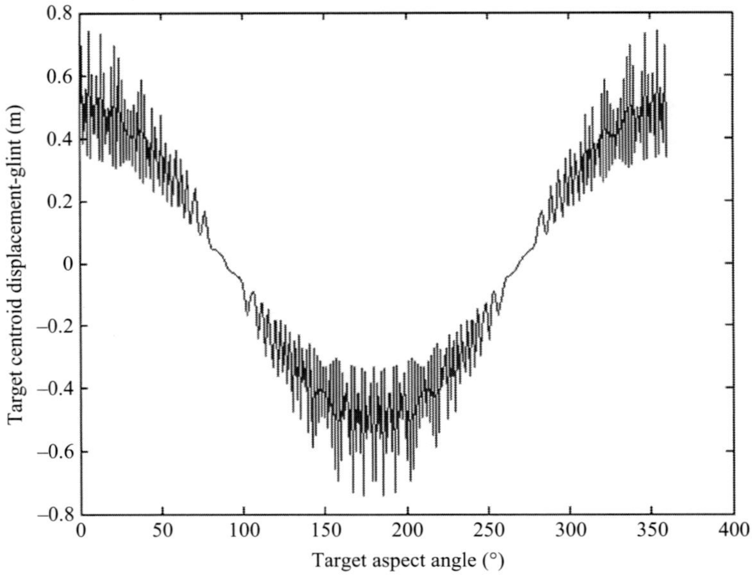

Figure 9.12. Glint Effect for Two-Scatterer Target

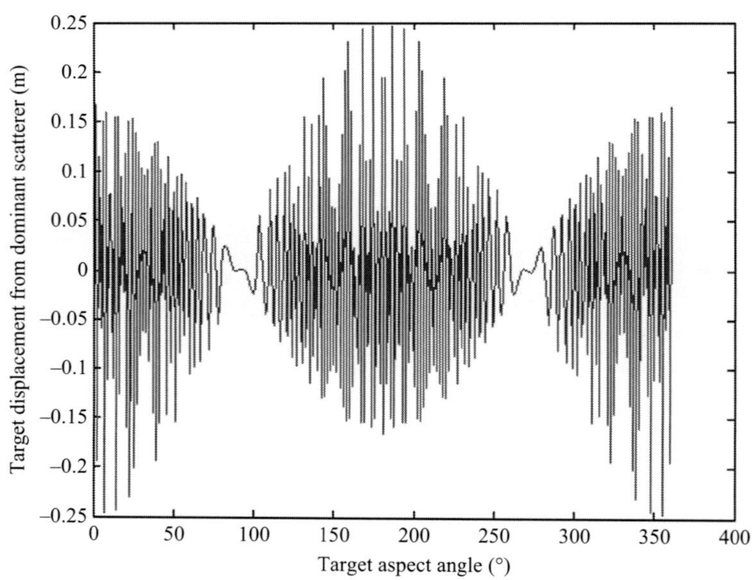

Figure 9.13. Glint Noise Effect After Removing the Position of the Dominant Scatterer

and can be averaged to mitigate the effects of glint. In addition, angle error outliers due to glint can be identified and eliminated using a Kalman filter.

9.3 ADC Timing Jitter

Timing jitter is caused by jitter in the analog-to-digital converter (ADC) clock strobe. In every ADC, the timing of the digitized data is determined by a strobe pulse from a clock. If the timing of this pulse varies, it creates a timing jitter that manifests itself as an error in signal amplitude. Figure 9.14 demonstrates how a timing error results in an amplitude error for a narrowband IF sinusoidal signal.

To estimate the effect of timing jitter on angle-of-arrival, we first consider how an amplitude error can cause a phase error.

$$I = (1+\varepsilon_A)\cos(\phi) \tag{9.80}$$

$$Q = (1+\varepsilon'_A)\sin(\phi) \tag{9.81}$$

$$\hat{\phi} = \tan^{-1}\left(\frac{(1+\varepsilon_A)\cos(\phi)}{(1+\varepsilon'_A)\sin(\phi)}\right) \approx \phi + \frac{\tan(\phi)}{1+\tan^2(\phi)}(\varepsilon_A - \varepsilon'_A) \approx \phi + \frac{\sin(2\phi)}{2}(\varepsilon_A - \varepsilon'_A) \tag{9.82}$$

$$\sigma_\phi = \frac{\sqrt{2}}{2}|\sin(2\phi)|\sqrt{E(\varepsilon_A\varepsilon'_A)} \tag{9.83}$$

To assess the effect of timing, we relate amplitude errors to timing jitter using the following equations:

$$\varepsilon_A = \varepsilon_t\frac{dA}{dt} \tag{9.84}$$

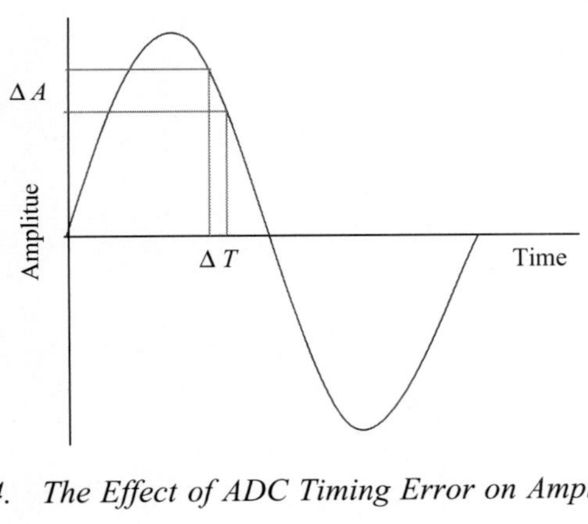

Figure 9.14. The Effect of ADC Timing Error on Amplitude Error

$$E(\varepsilon_A\varepsilon_A') = \sigma_t E\left(\frac{dA}{dt}\frac{dA^*}{dt}\right) \tag{9.85}$$

From [4], the expected value of the conjugate product is

$$E\left(\frac{dA}{dt}\frac{dA^*}{dt}\right) = -\frac{d^2R_A}{d\tau^2}\bigg|_{t=0} \tag{9.86}$$

where R_A is the autocorrelation of the target signal. For a rectangular-shaped spectrum of bandwidth B and signal power S_{target}, we have [6].

$$R_A(t) = S_{\text{target}}\frac{\sin(\pi Bt)}{\pi Bt} = S_{\text{target}}\left(1 - \frac{(\pi Bt)^2}{3!} + \frac{(\pi Bt)^4}{5!} - \ldots\ldots\right) \tag{9.87}$$

Substituting (9.87) into (9.86), we have the following expression that relates the spectrum to the variance amplitude rate:

$$E\left(\frac{dA}{dt}\frac{dA^*}{dt}\right) = S_{\text{target}}\frac{(\pi B)^2}{3} \tag{9.88}$$

$$E\left(\varepsilon_A\varepsilon_A'\right) = S_{\text{target}}\sigma_t^2\frac{(\pi B)^2}{3} \tag{9.89}$$

$$\sigma_\phi^{\text{jitter}} = \frac{\sqrt{2}}{2}\left|\sin(2\phi)\right|\sqrt{S_{\text{target}}}\,\sigma_t\frac{\pi B}{\sqrt{3}} \le \frac{\sqrt{2}}{2}\sqrt{S_{\text{target}}}\,\sigma_t\frac{\pi B}{\sqrt{3}} \tag{9.90}$$

From (9.19) and (9.21), we can relate angle error to phase error for each interferometer architecture. For the digital interferometer,

$$\sigma_\theta^I = \frac{\lambda}{\sqrt{2}\pi D\cos(\theta)}\sigma_\phi^{\text{jitter}} \tag{9.91}$$

and for the monopulse interferometer

$$\sigma_\theta^M = \frac{\lambda}{\pi D\cos(\theta)}\sigma_\phi^{\text{jitter}} \tag{9.92}$$

9.4 I and Q Imbalances

I and Q imbalances are a result of a mismatch between the I and Q channels in a radar receiver. The mismatch can be due to either phase or amplitude imbalances between channels. The result of the imbalances is a loss in signal-to-noise ratio and an error in angle-of-arrival due to phase errors. The simplest model for I and Q imbalance, including thermal noise n_I and n_Q, is the following:

$$I = a_I\cos(2\pi ft + \phi_I) + n_I \tag{9.93}$$

$$Q = a_Q \sin(2\pi ft + \phi_Q) + n_Q \tag{9.94}$$

$$a_Q = a_I + \varepsilon \tag{9.95}$$

$$\phi_Q = \phi_I + \alpha \tag{9.96}$$

$$n_I \propto N\left(0, \frac{1}{\sqrt{2SNR}}\right) \tag{9.97}$$

$$n_Q \propto N\left(0, \frac{1}{\sqrt{2SNR}}\right) \tag{9.98}$$

Figure 9.15 shows the effect of a 10 percent amplitude error and 10° phase error for a 50-kHz signal. Notice that a spurious response appears at 150 kHz and that there is a reduction in signal compression gain of about 2 dB. Because angle error is related to phase error, the average angle error due to I and Q imbalance phase error for a digital interferometer is determined by computing the two components of phase error that result from both amplitude and phase imbalances. We first compute the phase error that results from I and Q phase imbalances.

$$\hat{\phi} = \tan^{-1}\left(\frac{\sin(\phi + \alpha)}{\cos(\phi)}\right) \tag{9.99}$$

$$\hat{\phi} \approx \phi + \cos^2(\phi)\alpha \tag{9.100}$$

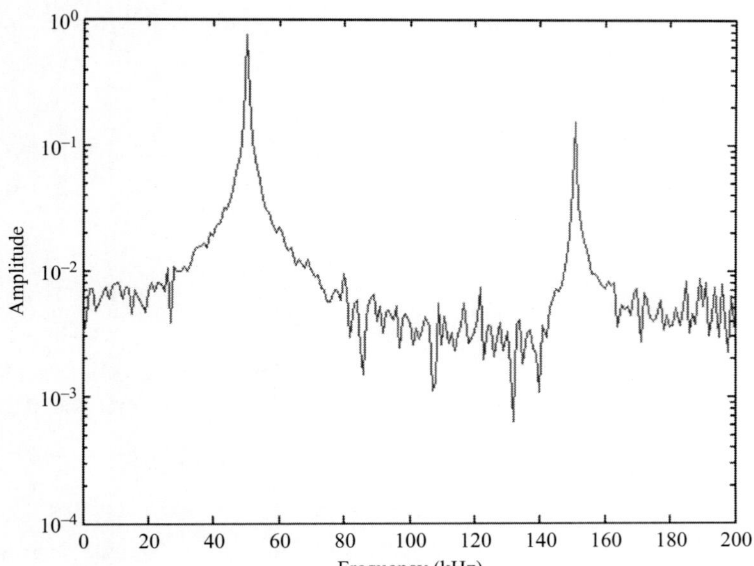

Figure 9.15. The Spectrum for a 50-kHz Signal with 10 Percent Amplitude Error and 10° of Phase I and Q Imbalances

$$\sigma_\phi = \frac{\alpha}{2\pi}\int_0^{2\pi} \cos^2(\varsigma)\,d\varsigma = \frac{\alpha}{2} \tag{9.101}$$

Now compute the component of phase error that results from amplitude I and Q imbalances.

$$\hat{\phi} = \tan^{-1}\left(\frac{(1+\varepsilon)\sin(\phi)}{\cos(\phi)}\right) \tag{9.102}$$

$$\hat{\phi} \approx \phi + \sin(\phi)\cos(\phi)\varepsilon = \frac{\sin(2\phi)}{2}\varepsilon \tag{9.103}$$

$$\sigma_\phi = \frac{\varepsilon}{2\pi}\int_0^{2\pi} \sin(\varsigma)\cos(\varsigma)d\varsigma = \frac{2\varepsilon}{\pi} \tag{9.104}$$

Assuming the two components of I and Q imbalance error are independent, the combined phase error due to phase and amplitude imbalance is

$$\sigma_\phi^{IQ} = \sqrt{\left(\frac{\alpha}{2}\right)^2 + \left(\frac{2\varepsilon}{\pi}\right)^2} \tag{9.105}$$

Again, we can use and relate phase error to angle error for each interferometer concept with the following results:

$$\sigma_\theta^I = \frac{\lambda}{\sqrt{2}\pi D\cos(\theta)}\sigma_\phi^{IQ} \tag{9.106}$$

$$\sigma_\theta^M = \frac{\lambda}{\pi D\cos(\theta)}\sigma_\phi^{IQ} \tag{9.107}$$

9.5 Quantization Effects

Quantization error can lead to errors in interferometry angle estimation due to errors in the measurement of phase. Both the phase shifter and the ADC can introduce phase quantization error.

9.5.1 Phase Shifter Quantization Error

Table 9.2 shows error effects due to phase quantization for a single array with *N* number of elements and *Nbits* number of quantization bits.

The interferometer angle estimate is affected only if the phase error is different in each interferometer channel. Thus, it is necessary that the phase shifter bits be set

Table 9.2. Error Effects Due to Phase Quantization

Number of Bits	**2**	**4**	**6**	**8**
Peak phase error $\left(\frac{\pi}{2^{Nbits}}\right)$	45°	11.25°	2.81°	0.7°
Root mean square (RMS) phase error $\left(\frac{\pi}{2^{Nbits}\sqrt{3}}\right)$	26°	6.5°	1.6°	0.4°
Gain (loss) $1-\frac{\pi^2}{3\cdot 2^{2Nbits}}$ $N \gg 1$	0.99 dB	0.06 dB	______	______
Beam pointing error $\sigma_\theta^{\text{quant}} = \frac{1.15\cdot\theta_3}{N2^{Nbits}}$	$\frac{1.15\cdot\theta_3}{4N}$	$\frac{1.15\cdot\theta_3}{16N}$	$\frac{1.15\cdot\theta_3}{64N}$	$\frac{1.15\cdot\theta_3}{256N}$
$\theta_3 = 3$ dB beamwidth, $N =$ number of array elements				

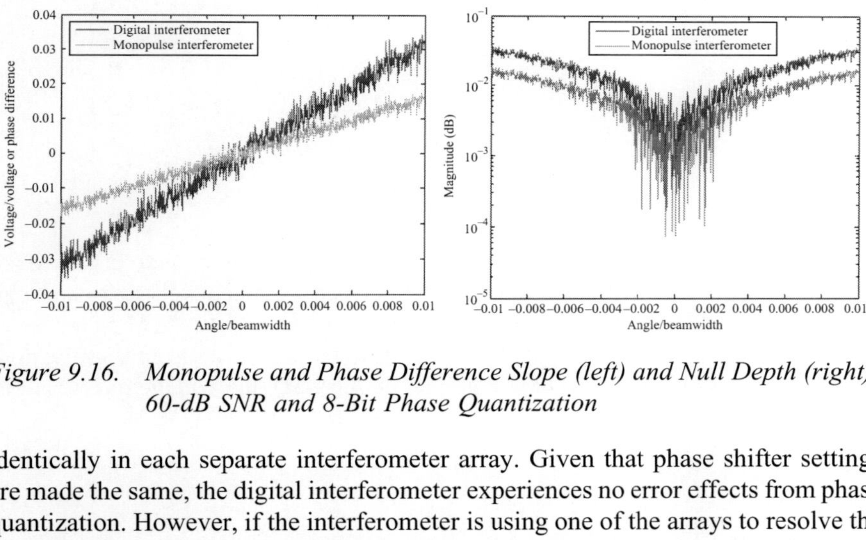

Figure 9.16. Monopulse and Phase Difference Slope (left) and Null Depth (right); 60-dB SNR and 8-Bit Phase Quantization

identically in each separate interferometer array. Given that phase shifter settings are made the same, the digital interferometer experiences no error effects from phase quantization. However, if the interferometer is using one of the arrays to resolve the interferometer phase ambiguity, then beam pointing error must be taken into account to make sure that the error is substantially less than the ambiguity angle.

For a monopulse array, phase shifter quantization has an effect on monopulse error. Figure 9.16 shows the slope and null depth effects for monopulse and digital interferometry due to quantization errors. Notice the increased error for monopulse interferometry, compared to digital interferometry.

9.5.2 ADC Phase Quantization Error

The ADC converts analog voltage into digital voltage states. Figure 9.17 shows the voltage levels for b number of bits in the ADC. The receiver thermal noise is set at

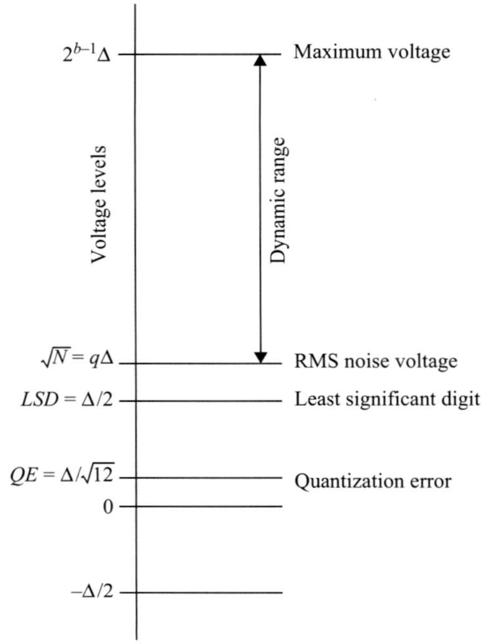

Figure 9.17. ADC Quantization Levels and Error

some number q voltage levels (typically $q = 1$–3). The dynamic range of the ADC is defined by the ratio of the maximum voltage to the voltage level where noise is set. The minimum voltage level Δ is also set and shown in Figure 9.17 as the least significant digit (LSD). Because the ADC quantization error is uniformly distributed over the voltage interval determined by the LSD, the voltage quantization error is given by

$$\sigma_{\text{quant}} = \frac{\Delta}{\sqrt{12}} \tag{9.108}$$

However, phase is determined as the ratio of in-phase and quadrature voltage.

$$\varphi = \tan^{-1}\left(\frac{I}{Q}\right) \tag{9.109}$$

$$\sigma_{\varphi}\sec^2(\varphi) = \frac{1}{Q}\sqrt{\sigma_I^2 + \frac{I^2}{Q^2}\sigma_Q^2} = \frac{\sigma_{\text{quant}}}{Q}\sqrt{1 + \frac{I^2}{Q^2}} \tag{9.110}$$

$$\begin{aligned}\sigma_\varphi &= \sigma_{\text{quant}} \frac{1}{Q\sec^2(\varphi)} \sqrt{1 + \frac{I^2}{Q^2}} \\ &= \frac{\cos(\varphi)}{Q} \sigma_{\text{quant}} \\ &= \frac{\cos(\varphi)}{Q} \frac{\Delta}{\sqrt{12}} \\ &= \frac{\sin(\varphi)}{I} \frac{\Delta}{\sqrt{12}}\end{aligned} \tag{9.111}$$

Using the fact that $\sqrt{\sin^2(\varphi) + \cos^2(\varphi)} = 1$, we have

$$\sigma_\varphi = \frac{1}{\sqrt{I^2 + Q^2}} \frac{\Delta}{\sqrt{12}} \tag{9.112}$$

If we let S denote the signal power, we can rewrite (9.112) as

$$\sigma_\varphi = \frac{\Delta}{\sqrt{12S}} = \frac{q\Delta}{q\sqrt{12S}} = \frac{\sqrt{N}}{q\sqrt{12S}} \tag{9.113}$$

$$\sigma_\varphi = \frac{1}{q\sqrt{12\frac{S}{N}}} \tag{9.114}$$

Equation (9.114) shows that the effect of quantization noise on phase error is inversely proportional to the square root of the *SNR*. The higher the signal power is, the less relatively small perturbation effects due to quantization have on phase error. For an interferometer, angle-of-arrival is computed as the difference of phase measurements, and, using (9.114), we have

$$\varphi_1 - \varphi_2 = \frac{2\pi}{\lambda} D \sin(\theta) \tag{9.115}$$

$$\sqrt{2}\sigma_\varphi = \frac{2\pi}{\lambda} D \cos(\theta) \sigma_\theta \tag{9.116}$$

and

$$\sigma_\theta = \frac{\lambda}{2\pi D \cos(\theta) q \sqrt{6\frac{S}{N}}} \tag{9.117}$$

For the monopulse interferometer, we have

$$\varphi = \tan^{-1}\left(imag\left(\frac{d}{s}\right)\right) = \tan^{-1}\left(imag\left(\frac{(v_1+\eta_1)-(v_2+\eta_2)}{(v_1+\eta_1)+(v_2+\eta_2)}\right)\right) \tag{9.118}$$

$$\begin{aligned} d &= (v_1+\eta_1)-(v_2+\eta_2) \\ s &= (v_1+\eta_1)+(v_2+\eta_2) \end{aligned} \tag{9.119}$$

where d is a difference voltage, s is a sum voltage, v_1 and v_2 are voltages from the two interferometer antennas, and η_1 and η_2 are the errors in the measurement of the voltages.

The angle error resulting from ADC quantization error depends on how the difference and sum voltages are determined. For the monopulse interferometer, voltages from two separated arrays are subtracted and summed to compute d and s.

$$\begin{aligned} \varphi &= \tan^{-1}\left(imag\left(\frac{(v_1-v_2+\eta_1-\eta_2)}{(v_1+v_2+\eta_1+\eta_2)}\right)\right) \\ &= \tan^{-1}\left(imag\left(\frac{(v_1-v_2+\eta_{dif})}{(v_1+v_2+\eta_{sum})}\right)\right) \end{aligned} \tag{9.120}$$

If we assume that η_1 and η_2 are independent and identically distributed, then η_{dif} and η_{sum} are independent. Using the result derived for the digital interferometer, we have

$$\frac{d}{s} \approx \tan(\varphi) + \frac{n_{diff}}{\cos(\varphi)} \tag{9.121}$$

$$\begin{aligned} \sigma_\theta^M &= \frac{\lambda}{\pi D\cos(\theta)}\sigma_{d/s} \\ &= \frac{\lambda}{\pi D\cos(\theta)}\frac{1}{\sqrt{(v_1-v_2)^2+(v_1+v_2)^2}}\frac{\Delta}{\sqrt{12}} \\ &= \frac{\lambda}{\pi D\cos(\theta)}\frac{1}{\sqrt{2(v_1^2+v_2^2)}}\frac{\Delta}{\sqrt{12}} \\ &= \frac{\lambda}{2\pi D\cos(\theta)q\sqrt{6SNR}} \end{aligned} \tag{9.122}$$

Thus angle-of-arrival error due to ADC quantization has the same effect on the monopulse interferometer as on the digital interferometer.

9.6 Wideband Effects

Wideband signals and noise distorts the monopulse pattern due to an effective beam shift over the band of the signal that is due to antenna bandwidth limitations. This effect is significantly pronounced for the monopulse interferometer because the monopulse slope is computed at the center frequency of the band. For the digital interferometer, phase is computed from the peak response of the matched filter, and, as long as the frequency response is not distorted or dispersed, the matched filter yields the correct phase regardless of signal bandwidth. For the monopulse interferometer, the voltage from the difference pattern is in error essentially due to antenna bandwidth limitations. Because antenna bandwidth is not a factor for signals impinging at broadside ($\theta = 0°$), there is no degradation of angle estimation at broadside. Figure 9.18 shows that there is no effect on slope and null performance for a 1-GHz signal impinging at broadside for an array of 64 elements at half wavelength spacing. Figure 9.19 shows that, for a 40° scan angle for the same array, the slope and null performance is degraded due to wideband effects. These wideband effects can be mitigated by introducing time delay units (TDUs) in a

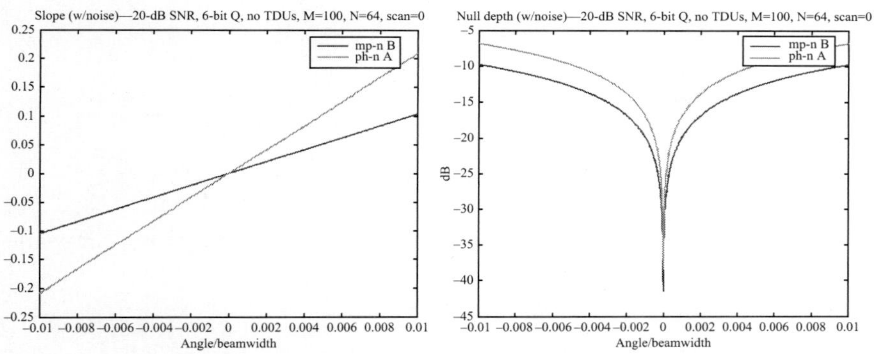

Figure 9.18. Monopulse and Phase Difference Slope (left) and Null Depth (right): 1-GHz Bandwidth; 20-dB SNR; 0° Scan Angle

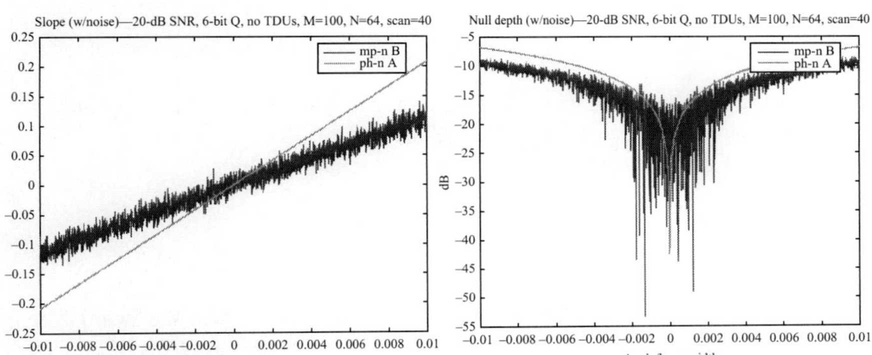

Figure 9.19. Monopulse and Phase Difference Slope (left) and Null Depth (right): 1-GHz Bandwidth; 20-dB SNR; 40° Scan Angle

subarray architecture for the interferometer antennas in order to increase the overall antenna bandwidth. However, angle error requirements may dictate more subarrays for wideband signals in order to reduce antenna bandwidth limitation effects.

9.6.1 Antenna Dispersion Loss

For a wideband waveform, *dispersion losses* occur when the antenna bandwidth is not matched to the waveform bandwidth. Typically narrowband phased-array radars implement a narrowband beam former for beam steering, which is a cost-effective method for array architectures using digital beam forming or analog narrowband phase shifters. However, for large bandwidths, frequency dispersion can degrade the SNR and signal integrity. The magnitude of the degradation due to dispersion is a function of the antenna size, the waveform bandwidth, and the angle-of-arrival. Dispersion losses can particularly affect linear frequency modulation (LFM) waveforms where the swept bandwidth is large.

Consider an LFM waveform with the following parameters:

B = swept bandwidth
T = time duration
TB = time-bandwidth product
B/T = slope of the time-frequency sweep
f_0 = initial frequency

Then the output of the matched filter for each element in a linear phased array after narrowband beam forming can be expressed as

$$s_k(t,\theta) = \overbrace{\exp\left(2\pi j\frac{B}{T}\frac{1}{2}\left((t+t_k)^2 - t_k^2\right) + f_0 t_k\right)}^{\text{LFM waveform}} \overbrace{\exp\left(-2\pi j\frac{B}{T}\frac{1}{2}t^2\right)}^{\text{Matched filter}} \overbrace{\exp(-2\pi j f_0 t_k)}^{\text{Beam former phase}} \tag{9.123}$$

$$t_k = \frac{2R}{c} + \frac{d\cdot k}{c}\sin(\theta) \tag{9.124}$$

where R is the range of the target, d is the element spacing, θ is the target angle-of-arrival, and c is the speed of light (3×10^8 m/s).

The output of the beamformer at each time t is given by

$$S(t,\theta) = \sum_{k=1}^{N} s_k(t,\theta) \tag{9.125}$$

In theory, at each time t and each angle θ, the power from the beam former should be

$$\left\|S(t,\theta)\right\|^2 = N^2 \tag{9.126}$$

Thus, a measure of the dispersion loss is the following ratio:

$$\text{Disp\_Loss}(\theta) = \frac{\frac{1}{T}\int_0^T \|S(t,\theta)\|^2 dt}{N^2} \tag{9.127}$$

Figure 9.20 shows the dispersion loss for a 1-m array antenna for an LFM waveform with $TB = 1{,}000$, $R = 20$ km, $f_0 = 10$ GHz, $d = \lambda/2$ for bandwidths of 100 MHz, 200 MHz, and 400 MHz. To maintain acceptable phase integrity and angle estimation performance, less than 1-dB dispersion loss is acceptable. Note that Figure 9.20 shows that an LFM bandwith of less than 100 MHz provides acceptable performance over a ±60° field of view. If the field of view is restricted to ±45°, then the acceptable bandwidth is 200 MHz.

For a fixed antenna size, dispersion effects can be mitigated by using time delay beam steering, which is practical only at a subarray level or by factoring a wideband waveform into smaller narrowband segments. An LFM waveform naturally lends itself to breaking up the chirp waveform into a sequence of smaller chirped waveforms where the bandwidth of each smaller chirp is on the order of the antenna bandwidth. This sequential method of generating an LFM waveform is termed *frequency jump burst* (FJB). If an FJB sequence is divided into M

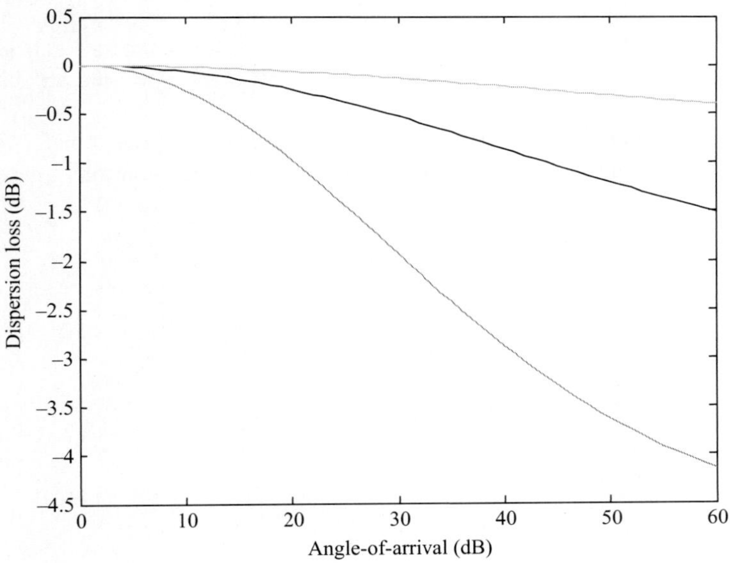

Figure 9.20. LFM Dispersion Loss Versus Angle for a 1-m Array for 100 MHz, 200 MHz, and 400 MHz Bandwidths at a Frequency of 10 GHz

sub-pulses each of smaller bandwidth then the total computations required to compute the FFT for the entire sequence is given by

$$Total\ Operations \approx 5NM \log_2(N/M) \tag{9.128}$$

where N is the order of the FFT. Figure 9.21 shows what such a chirped waveform might look like. The gaps in time shown in Figure 9.21 produce time grating lobes that could degrade target imaging. The location and magnitude of these grating lobes can be controlled by adjusting the gap spacing. The array must be resteered with each smaller waveform to reduce the dispersion loss. Although the resteering operation can be achieved in just a few nanoseconds, the time allotted to this function can be significant if the total bandwidth is divided into several smaller chirps. Also, an issue of eclipsing must be considered, that is, earlier transmitted pulses interfering with later transmitted pulses. Eclipsing also precludes dividing the FJB waveform into a number of arbitrary smaller segments. Thus, for any particular system, a trade-off study is required to assess the performance impact of FBJ on mission applications.

9.6.2 Channel Transfer Function Mismatch

Channel transfer function mismatch (CTFM) occurs due to mismatch in hardware errors between interferometer channels [8] and an error that is difficult to correct due to the nature of its spectral bandwidth. The error is not a bias error that can be corrected in calibration, and it is not similar to broadband noise that can be averaged in time. Typically, CTFM errors occur in an array architecture that incorporates multiple subarrays where the RF hardware components for each subarray include low noise amplifiers, phase shifters, combiner receiver, and an analog-to-digital converter.

To understand channel mismatch error, we assume that the transfer function errors are uncorrelated from channel to channel. In some situations, these errors

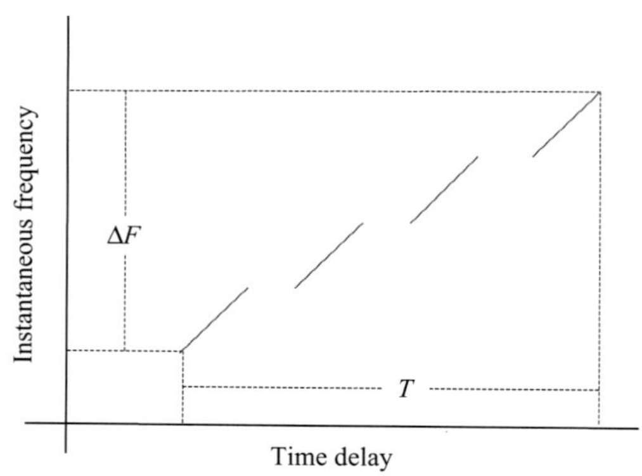

Figure 9.21. Stepped Chirp Waveform to Mitigate Dispersion Effects

may be correlated, such as in thermal antenna gradient effects and time delay unit quantization, and in such cases significant angle bias can result [7]. At first glance, it may seem that channel transfer mismatch errors might be mitigated through the calibration and alignment of the channels. However, the temporal correlation of these errors is sufficiently short that calibration and alignment techniques using estimated error statistics such as space-time adaptive processing do not mitigate channel mismatch errors. Only channel equalization using time taps do so.

As in Figure 6.1, the interferometer produces two voltages at the output of the two channels. Let v_1 and v_2 denote the voltages measured in the two channels, and let h_1 and h_2 be the respective channel impulse response functions. If s_1 and s_2 are the signal functions for each channel, then

$$v_1(t) = s_1(t) \otimes h_1(t) \tag{9.129}$$

$$v_2(t) = s_2(t) \otimes h_2(t) \tag{9.130}$$

Computing the phase difference for the digital interferometer [8],

$$\begin{aligned}
\Delta\phi &= \angle v_1 v_2^* \\
&= \angle(s_1(t) \otimes h_1(t))(s_2(t) \otimes h_2(t))^* \\
&= \angle(s_1(t) \otimes h_1(t))\left(s_2^*(t) \otimes h_2^*(t)\right) \\
&= \angle\int s_1(t+\tau)h_1(\tau)d\tau \int s_2^*(t+\xi)h_2^*(\xi)d\xi \\
&= \angle\iint s_1(t+\tau)s_2^*(t+\xi)h_1(\tau)h_2^*(\xi)d\tau d\xi \\
&= \angle\iint e^{2\pi j\omega(t+\tau)} e^{\frac{2\pi}{\lambda}j\frac{D}{2}\sin(\theta)} e^{-2\pi j\omega(t+\xi)} e^{\frac{2\pi}{\lambda}j\frac{D}{2}\sin(\theta)} h_1(\tau)h_2^*(\xi)d\tau d\xi \\
&= \angle e^{\frac{2\pi}{\lambda}jD\sin(\theta)} \int e^{2\pi j\tau\omega} h_1(\tau)d\tau \int e^{2\pi j\xi\omega} h_2^*(\xi)d\xi \\
&= \angle e^{\frac{2\pi}{\lambda}jD\sin(\theta)} H_1(\omega)H_2^*(\omega)
\end{aligned} \tag{9.131}$$

where H_1 and H_2 are the channel transfer functions with the following form:

$$H_1(\omega) = (1 + \beta_1(\omega))e^{j\alpha_1(\omega)} \tag{9.132}$$

$$H_2(\omega) = (1 + \beta_2(\omega))e^{j\alpha_2(\omega)} \tag{9.133}$$

Thus,

$$\Delta\varphi(\omega) = \frac{2\pi}{\lambda} D\sin(\theta) + \alpha_1(\omega) - \alpha_2(\omega) \tag{9.134}$$

Assuming a narrowband signal and that the α_i are independent,

$$\sigma_{\phi}^{\text{CTFM}} = \sqrt{2}\alpha_{\text{RMS}} \tag{9.135}$$

Angle error due to CTFM can be described as follows [8]:

$$\sigma_{\theta}^{\text{CTFM}} = \frac{\lambda}{2\pi D} \frac{\sqrt{2}\alpha_{\text{RMS}}}{\cos(\theta)} \tag{9.136}$$

For the monopulse interferometer, a similar analysis can be performed.

$$\begin{aligned} \frac{d}{s} &= \text{Im}\left(\frac{e^{-j\varphi}H_2 - e^{j\varphi}H_1}{e^{-j\varphi}H_2 + e^{j\varphi}H_1}\right) = \text{Im}\frac{(1+\beta_2)e^{-j(\varphi-\alpha_2)} - (1+\beta_1)e^{j(\varphi+\alpha_1)}}{(1+\beta_2)e^{-j(\varphi-\alpha_2)} + (1+\beta_1)e^{j(\varphi+\alpha_1)}} \\ &\approx \tan(\varphi) + f(\alpha_1, \alpha_2, \beta_1, \beta_2) \end{aligned} \tag{9.137}$$

Notice that the monopulse ratio is affected by both the channel amplitude and phase errors, whereas the digital interferometer is affected by the channel phase errors only. Now assume that the amplitude errors are zero, and compute the monopulse interferometer error due solely to phase error.

$$\sigma_{\theta}^{M} = \frac{\lambda}{\pi D \cos(\theta)} \sigma_{d/s} \tag{9.138}$$

and

$$\sigma_{d/s} = \sqrt{\left(\left.\frac{\partial f}{\partial \alpha_1}\right|_{\alpha_1=0}\right)^2 + \left(\left.\frac{\partial f}{\partial \alpha_2}\right|_{\alpha_2=0}\right)^2}\,\alpha_{\text{RMS}} = \frac{4|1+\cos(\phi) - \cos(2\phi)|}{|\cos(\phi)|}\alpha_{\text{RMS}} \tag{9.139}$$

Channel transfer function mismatch is more complicated in the way it affects a monopulse interferometer as opposed to a digital interferometer due to both amplitude and phase effects.

Channel Timing Errors

In a digital interferometer, a common clock controls the timing of receive channels. The timing of the signals from the antenna through the receiver to the analog-digital-converter (ADC) must be equalized in order to measure angle-of-arrival accurately. If the electrical paths are not matched, a phase bias is introduced, creating an angle bias. Thus, channel timing errors lead to a phase bias among interferometer channels. The good news is that timing errors can be calibrated with signal injection. By injecting signals with known phase behind the antenna but before the low noise amplifier (LNA), the phase can be measured at the output of the ADC. Any phase differences can be noted and accounted for in order to phase-balance the channels. With proper channel calibration, channel timing errors can be made negligible. It should be noted that phase measurements can be sensitive to environmental effects, especially temperature, and that frequent signal injection should occur to maintain phase calibration.

Table 9.3. Summary of Interferometer Performance in the Presence of Errors

Error Source	Digital Interferometer	Monopulse Interferometer
Thermal noise	$\sigma_\theta^I = \dfrac{\theta_3}{\pi\cos(\theta)\sqrt{S/N}}$	$\sigma_\theta^M = \dfrac{\theta_3}{\frac{\pi}{2}D\cos(\theta)\sqrt{2S/N_\Delta}}$
Glint	$\sigma_\theta^{\text{glint}} = \dfrac{L}{4r}$	$\sigma_\theta^{\text{glint}} = \dfrac{L}{4r}$
Multipath	$\sigma_\theta^{\text{multipath}} = \dfrac{\rho\theta_3}{(8G_{se}(peak))^{1/2}}$	$\sigma_\theta^{\text{multipath}} = \dfrac{\rho\theta_3}{(8G_{se}(peak))^{1/2}}$
ADC timing jitter	$\sigma_\phi = \dfrac{\sqrt{2}}{2}\lvert\sin(2\phi)\rvert\sqrt{S_{\text{target}}}\sigma_t\dfrac{\pi B}{\sqrt{3}}$	$\sigma_\phi = \dfrac{\sqrt{2}}{2}\lvert\sin(2\phi)\rvert\sqrt{S_{\text{target}}}\sigma_t\dfrac{\pi B}{\sqrt{3}}$
I & Q imbalances	$\sigma_\phi^{IQ} = \sqrt{\left(\dfrac{\alpha}{2}\right)^2 + \left(\dfrac{2\varepsilon}{\pi}\right)^2}$	$\sigma_\phi^{IQ} = \sqrt{\left(\dfrac{\alpha}{2}\right)^2 + \left(\dfrac{2\varepsilon}{\pi}\right)^2}$
Phase quantization	Negligible if phase states are balanced among array channels	$\sigma_\theta^{\text{quant}} = \dfrac{9\theta_3}{N2^{Nbits}}$ N = number of phase shifters
ADC quantization	$\sigma_\theta = \dfrac{\lambda}{2\pi D\cos(\theta)\sqrt{6\dfrac{S}{N}}}$	$\sigma_\theta = \dfrac{\lambda}{2\pi D\cos(\theta)\sqrt{6\dfrac{S}{N}}}$
Channel transfer function mismatch	$\sigma_\phi^{\text{CTFM}} = \sqrt{2}\alpha_{\text{RMS}}$	(See text.)
Channel timing errors	Negligible with CAL	Negligible with CAL
Antenna dispersion	Moderate for wideband signals because each interferometer array is summed separately providing wider band array	Significant for wideband signals because all arrays are summed and differenced together, providing narrower band array
Target spectral distortion	Negligible with TDOA	Negligible with TDOA

9.7 Error Summary

The results of the preceding analyses quantify the effects of various radar errors on angle estimation performance using an interferometer. Table 9.3 summarizes the errors that affect digital and monopulse interferometer architectures.

References

1. P. Z. Peebles, *Radar Principles*, John Wiley & Sons, New York, 1998.
2. F. E. Nathanson, *Radar Design Principles*, Scitech Publishing, Mendham, NJ, USA, 1999.

3. M. A. Richards, J. A. Scheer, and W. A. Holm, *Principles of Modern Radar, Basic Principles*, Scitech Publishing, Raleigh, NC, USA, 2010.
4. D. K. Barton, *Modern Radar System Analysis*, Artech House, Norwood, MA, USA, 1988.
5. S. M. Sherman, *Monopulse Principles and Techniques*, Artech House, Norwood, MA, USA, 1984.
6. A. Papoulis, *Probability, Random Variables, and Stochastic Processes*, McGraw-Hill, New York, 1965.
7. J. T. Neesmith, E. J. Holder, L. E. Corey, and R. L. Howard, "The pointing accuracy of phased-array radars with correlated phase errors," *Proc. IEE International Radar Conference*, London, 1987, pp. 87–90.
8. B. Keel and E. J. Holder, "Angle estimation degradation due to channel mismatch errors," Internal GTRI Technical Memorandum, 1994.
9. D. K. Barton, "Multipath fluctuation effects in track-while-scan radar," *IEEE Transactions on Aerospace and Electronic Systems*, vol. AES-15, Nov. 1979.
10. P. Beckman and A. Spizzichino, *The Scattering of Electromagnetic Waves from Rough Surfaces*, Artech House, Norwood, 1987.

Chapter 10

Tropospheric Effects on Angle-of-Arrival

Usually the primary reason to implement an interferometric radar architecture is to enhance angle accuracy, and one source of angle error is the effect of signal propagation through the earth's atmosphere. In particular, all radars experience errors due to propagation phenomena that occur in the troposphere, which is generally considered to encompass the atmosphere below 30 km. Angle bias is degraded by refraction effects in the troposphere, whereas angle precision is degraded by turbulence. In this chapter, we characterize the effects of angle bias due to refraction and angle precision due to turbulence. As a result, angle bias due to refraction can be calibrated and removed, if possible, and any residual error can be quantified in the interferometer angle error budget. Also, the effects of angle precision due to turbulence can be determined and accounted for in the same angle error budget. For turbulence it will be important to understand the spatial correlation of errors that will determine the magnitude of the impact of turbulence effects on angle precision for an interferometer.

Within the troposphere, temperature, humidity, and atmospheric pressure combine to produce both large-scale and small-scale lensing effects that can bend and distort radar signals, in turn, degrading angle-of-arrival measurements. The large-scale effect is due to refraction through the atmosphere caused by gradients in the index of refraction as a function of altitude. Refraction is mostly predictable given accurate measurements of temperature, humidity, and pressure versus altitude. As a result, algorithms have been created by several radar developers to correct for this error, and essentially most of radar refractivity error can be corrected under most refractive conditions. However, for long-range propagation through the troposphere, the remaining uncorrected error can be significant in severe refractive environments. In extreme refractive environments such as ducting or superconducting, the refractive error cannot easily be corrected.

The small-scale error effect is due to turbulence in the atmosphere, which is more of a random error effect with correlation times ranging from 0.5 to 2 s. Kolmogorov [1] was the first to characterize the randomness of turbulence through the introduction of structure functions. In particular, he defined the refractive index structure function (C_n^2) that characterizes the variability of the troposphere as a function of altitude. The magnitude of turbulence effects is smaller than the magnitude of refraction effects, but turbulence effects are nearly impossible to correct due to their inherent randomness. Additionally, the temporal correlation times are sufficiently long that most radars cannot average the effect.

Turbulence is also correlated spatially, and because the interferometer uses phase differencing, some of the correlated turbulence effects are canceled. The interferometer receive antennas are generally closely separated, and the angular difference between the propagation paths from the target to both antennas is small. For coherent phase trilateration where the antennas are separated by large distances (>100 m), the turbulence errors are decorrelated and thus must be accounted for in a system error budget. One particular application where these effects are canceled is using an interferometer to guide an interceptor to engage a target. When the interceptor and target are sufficiently close in angle (depending on the geometry for engagement), the spatially correlated turbulence errors cancel.

10.1 Tropospheric Refraction Effects

Large-scale signal bending results when a radar signal propagates through the troposphere due to the large-scale variation in index of refraction. In the upper troposphere, this large-scale variation is primarily caused by the variable density of the atmosphere with altitude. At lower altitudes, for instance below 6 km, water vapor variation provides an additional substantial contribution to refractive index changes. For low-elevation operation, this bending can contribute to significant estimation error on target locations. At 2° elevation, for instance, the variation in large-scale refraction can contribute as much as 20 km of cross-range error for high-altitude or earth-orbiting targets located above the troposphere.

Most modern long-range surveillance radar systems have some form of correction for errors due to tropospheric refraction [2,3,4]. Residual refraction errors usually amount to about 10–15% of the total refraction error before correction. Inclusion of the tropospheric effects can be achieved through scaling relationships obtained by approximate ray tracing through combined exponential refractivity profiles in the troposphere. In contrast to simpler models, such as the 4/3 earth model of tropospheric bending, this approach correctly accounts for spatial correlation for all elevations.

10.1.1 Geometric Optics (Ray Tracing)

The procedure used to model refraction effects is derived from the equations of motion for a light ray traveling through an inhomogeneous medium [5]. For an inhomogeneous medium, the index of refraction $n(p)$ at a point p in the medium is related to the speed of light in the medium v and the speed of light in vacuum c by the formula $n(p) = \frac{c}{v(p)}$. With this definition, one can compute the travel time τ for an electromagnetic path along an arclength-parameterized path s of length S:

$$T = \frac{1}{c}\int_{S} n(s)ds \tag{10.1}$$

We parameterize the arclength using the independent variable z by $s = (x(z), y(z), z)$. Then we make the following change of variables in (10.1):

$$\begin{aligned} ds &= \sqrt{dx^2 + dy^2 + dz^2} \\ &= dz\sqrt{1 + \left(\frac{dx}{dz}\right)^2 + \left(\frac{dy}{dz}\right)^2} \end{aligned} \tag{10.2}$$

and taking the differential with the change of variables the principle of least time can be formulated as follows:

$$0 = \delta T = \frac{1}{c}\delta \int_{z=A}^{z=B} n(x, y, z)\sqrt{1 + \dot{x}^2 + \dot{y}^2}\, dz \tag{10.3}$$

To derive the solution, we use the calculus of variations where

$$L(x, y, \dot{x}, \dot{y}, z) = n(x, y, z)\sqrt{1 + \dot{x}^2 + \dot{y}^2} \tag{10.4}$$

and the Euler-Lagrange equations that define the solution become

$$\frac{\partial L}{\partial x} - \frac{d}{dz}\frac{\partial L}{\partial \dot{x}} = 0 \tag{10.5}$$

$$\frac{\partial L}{\partial y} - \frac{d}{dz}\frac{\partial L}{\partial \dot{y}} = 0 \tag{10.6}$$

Substituting, we have

$$\frac{\partial n}{\partial x}\sqrt{1 + \dot{x}^2 + \dot{y}^2} - \frac{d}{dz}\frac{n(x, y, z)\dot{x}}{\sqrt{1 + \dot{x}^2 + \dot{y}^2}} = 0 \tag{10.7}$$

$$\frac{\partial n}{\partial y}\sqrt{1 + \dot{x}^2 + \dot{y}^2} - \frac{d}{dz}\frac{n(x, y, z)\dot{y}}{\sqrt{1 + \dot{x}^2 + \dot{y}^2}} = 0 \tag{10.8}$$

Now assume that the propagation path lies in a plane perpendicular to the x-y plane, and, without loss of generality, assume the propagation lies entirely in the x-z plane. Then (10.7) determines the propagation path and reduces to

$$n'\sqrt{1 + \dot{x}^2} - \frac{d}{dz}\frac{n\dot{x}}{\sqrt{1 + \dot{x}^2}} = 0 \tag{10.9}$$

$$n'\left(1 + \dot{x}^2\right)^2 - \left(n'\dot{x}^2 + n\ddot{x}\right)\left(1 + \dot{x}^2\right) + n\dot{x}^2\ddot{x} = 0 \tag{10.10}$$

$$n'\left(1 + \dot{x}^2\right) - n\ddot{x} = 0 \tag{10.11}$$

Continuing, we can arrive at the following relationships:

$$\frac{n'}{n} = \frac{\ddot{x}}{(1 + \dot{x}^2)} \tag{10.12}$$

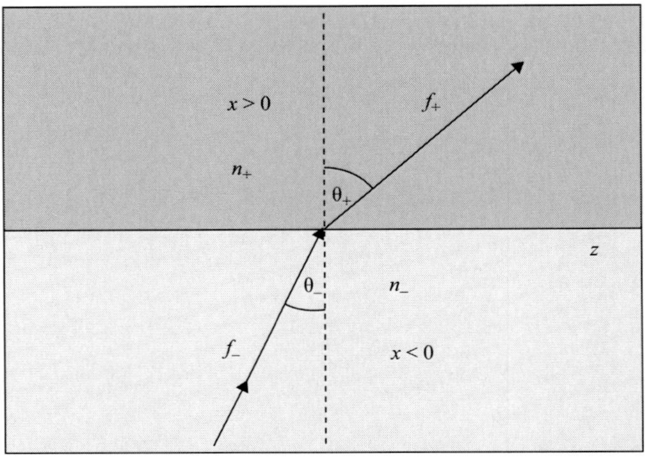

Figure 10.1. Propagation Geometry for a Two-Layered Refractivity Medium

$$\frac{d}{dx}\ln(n) = \frac{1}{2}\frac{d}{dx}\ln\left(1 + \dot{x}^2\right) \tag{10.13}$$

$$\ln(n) - \frac{1}{2}\ln\left(1 + \dot{x}^2\right) = \text{const} \tag{10.14}$$

$$\frac{n}{\sqrt{1 + \dot{x}^2}} = e^{\text{const}} = \alpha \tag{10.15}$$

and

$$n(x) = \alpha\sqrt{1 + \dot{x}^2} \tag{10.16}$$

Now consider a propagation path from one index layer to another as described in Figure 10.1. For convenience, we define a function f to represent the solution of the propagation path where $x = f(z)$. The slope of the path in the negative region is $\phi_- = \pi/2 - \theta_-$, and the slope of the path in the positive region is given by $\phi_+ = \pi/2 - \theta_+$. Thus referring to (10.7) with $\dot{x} = f'(z)$,

$$f'_- = \tan(\phi_-) \quad \text{and} \quad f'_+ = \tan\left(\phi_+\right) \tag{10.17}$$

Substituting in (10.7),

$$\frac{n}{\sqrt{1 + (f')^2}} = \alpha \tag{10.18}$$

we have

$$\frac{n_-}{\sqrt{1 + (f'_-)^2}} = \alpha = \frac{n_+}{\sqrt{1 + (f'_+)^2}} \tag{10.19}$$

$$\frac{n_-}{\sqrt{1+\tan^2(\phi_-)}} = \frac{n_+}{\sqrt{1+\tan^2(\phi_+)}} \tag{10.20}$$

$$\frac{n_-}{\sec(\phi_-)} = \frac{n_+}{\sec(\phi_+)} \tag{10.21}$$

$$n_-\cos(\phi_-) = n_+\cos(\phi_+) \tag{10.22}$$

$$n_-\sin(\theta_-) = n_+\sin(\theta_+) \tag{10.23}$$

Equation (10.23) is a statement of Snell's law, and, in fact, Snell's law can be used to compute the refracted path for electromagnetic signals.

10.1.1.1 Refractive Index

Significant work was accomplished in the 1960s to characterize and standardize the effects of atmospheric refraction. The primary contributor to this effort was R. B. Bean, who made major contributions to the development of standard atmospheric refractive models [6,7,8]. The refractive index of air is very close to 1, typically equaling 1.000310 at sea level. Since it is tedious to use the number of decimals required to capture the fine detail, a new unit, the refractive index N unit, was defined where

$$N = (n-1)\,10^6 \tag{10.24}$$

N is typically around 310–315 at sea level. The value of N can be calculated from this formula:

$$N = 77.6\frac{P}{T} + 3.73\times 10^5\,\frac{e}{T^2} \tag{10.25}$$

where

$$P = \text{dry pressure}, \sim 1000\,\text{mb} \tag{10.26}$$

$$T = \text{temperature}, \sim 300\,\text{K} \tag{10.27}$$

$$e = \text{water vapor partial pressure} \sim 40\,\text{mb} \tag{10.28}$$

The dry term depends only on pressure and temperature, whereas the wet term depends also on the water vapor concentration. The temperature, pressure, and water vapor pressure vary with time and location. Pressure falls exponentially with height where it drops to around 37% of the sea level value at around 8 km altitude. Temperature usually falls by 1°C/100 m in the first few kilometers above sea level.

Water vapor partial pressure is more complicated because it is strongly governed by weather and is limited to the saturated vapor pressure, that is, the amount of moisture the air can hold. Once the temperature drops below 0°C, the excess water vapor condenses to form clouds. The saturated water vapor pressure is around 40 mbar at 300 K (a warm day) and 6 mbar at 273 K (freezing). The height above the ground where the air temperature decreases to 0°C is called the zero degree isotherm. It is typically at a few kilometers in altitude, near the cloud base.

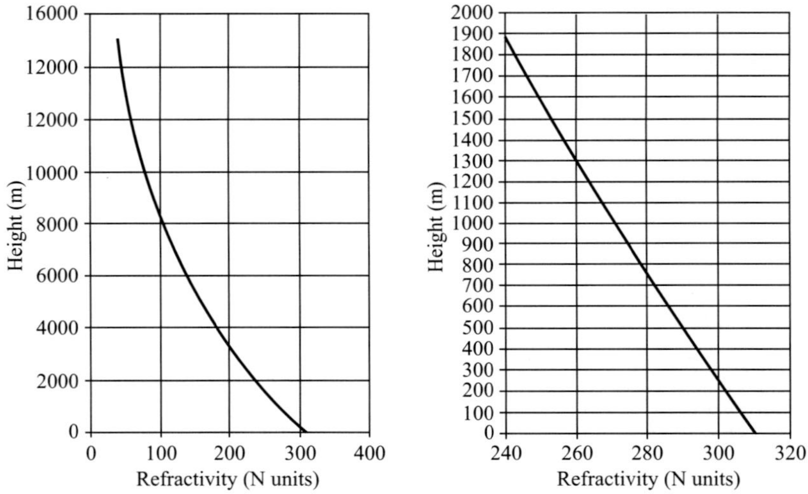

Figure 10.2. Refractive Index Versus Height

As a result, the refractive index normally falls exponentially with height in a standard atmosphere. The scale height of the exponential is ~7.4 km, as is shown in (10.29). Figure 10.2 shows that in the first 1000 m, a straight line with a slope ~−40 N/km provides a reasonably accurate approximation to the refractive index.

$$n(h) = 1 + N \cdot 10^{-6} \cdot e^{-\frac{h}{7350}} \approx 1 + 315 \cdot 10^{-6} \cdot e^{-\frac{h}{7350}} \tag{10.29}$$

10.1.1.2 Geometric Optics in Polar Coordinates

Figure 10.1 shows a propagating signal path with an initial elevation angle denoted by θ_0. For a spherical earth, Snell's law is expressed by [4],

$$n_0 r_0 \cos(\theta_0) = n_1 r_1 \cos(\theta_1) \tag{10.30}$$

where Figure 10.3 shows the definition of the terms in (10.30).

$$r_0 = r_e \tag{10.31}$$

$$r_1 = r_e + h \tag{10.32}$$

The angle α is determined by [4],

$$\alpha = \sin^{-1}\left(\frac{ds \cdot \cos(\theta_0)}{r_1}\right) \tag{10.33}$$

If we integrate over multiple refractive layers, we have the refracted path length *S*,

$$S = \int_S ds = \sum_{i=1}^{N} ds_i \tag{10.34}$$

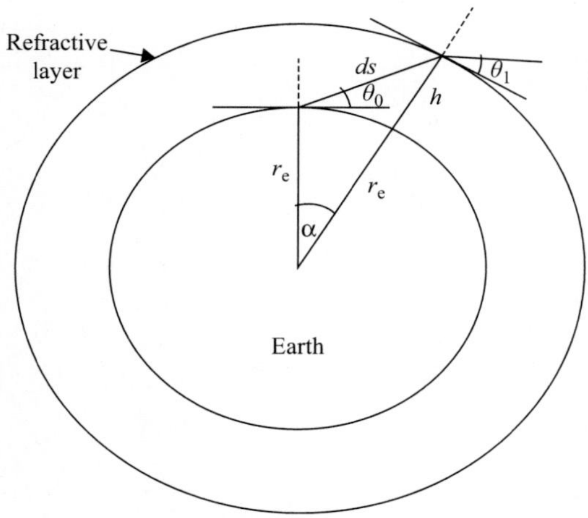

Figure 10.3. Spherical Model for Propagation

and the unrefracted path length L is defined by

$$L = \sqrt{r_{\text{final}}^2 + r_e^2 - 2r_{\text{final}}r_e\cos(\text{A})} \tag{10.35}$$

where

$$\text{A} = \int_{\text{A}} d\alpha = \sum_{i=1}^{N} \alpha_i \tag{10.36}$$

The apparent angle error is determined by integrating the arclength curvature [4].

$$\delta\theta = \int_S \kappa(s)ds = \sum_{i=1}^{N} \frac{d\phi_i}{ds}\Delta s_i \tag{10.37}$$

Or equivalently

$$\delta\theta = \int_S \kappa(s)ds = \sum_{i=1}^{N} \frac{n_0 r_e \cos(\theta_0)}{n_N^2 r_N} \frac{dn_i}{dr}\Delta s_i \tag{10.38}$$

We define the true angle error as the difference between the initial angle θ_0 and the geometric angle between the radar and the target. The true angle error is typically greater than the apparent angle error because the apparent angle error is an average over the path length. Figure 10.4 shows the true error due to refraction for various initial elevation angles and target altitudes using geometric optics.

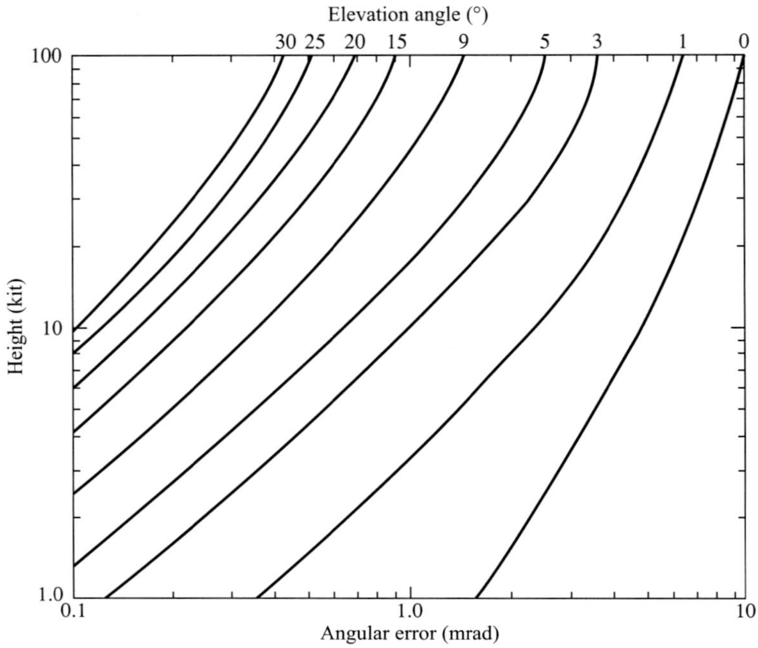

Figure 10.4. Refraction Error for Target Elevation Angles and Altitudes [4]

10.1.1.3 4/3 Earth Model

From Figure 10.5, we can clearly see that a horizontally pointed radar beam propagates farther over the horizon than we would otherwise expect. The curvature of any given ray relative to the earth's surface is given by

$$\frac{d\phi}{ds} = \frac{1}{R} + \frac{dn}{dh} \tag{10.39}$$

where R is the radius of the earth, n is the radar refractive index, h is the height of the ray above the earth's surface at a distance s from the source around the surface, ϕ is the angle at the intersection of the radar beam and a circle with radius $R + h$, which is concentric to the earth (see Figure 10.5).

We can quantitatively calculate the magnitude of the effect by combining the refraction curvature effect and the earth's curvature. This gives what is called the radar effective radius R'.

$$\frac{1}{R'} = \frac{1}{R} + \frac{dn}{dh} \tag{10.40}$$

$$R' = \frac{R}{1 + R\dfrac{dn}{dh}} \tag{10.41}$$

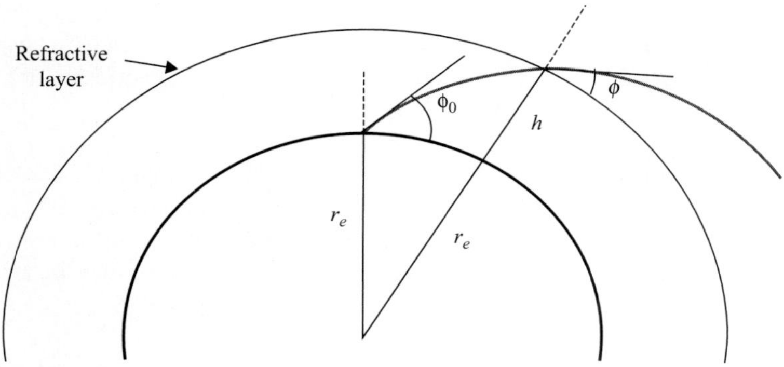

Figure 10.5. Propagation Geometry Through a Refractive Layer Boundary

Table 10.1. Elevation Angle Refraction Error (mrad) for Targets at 3 km, 6 km, and 9 km Altitude for Both Geometric Optics and 4/3 Earth Refraction Models

		3 km Altitude			
	2°	5°	10°	20°	30°
Geometric optics	1.3956	0.6268	0.3173	0.1544	0.0975
4/3 earth	1.4031	0.6485	0.3306	0.1613	0.1018
		6 km Altitude			
	2°	5°	10°	20°	30°
Geometric optics	2.2513	1.0823	0.5580	0.2728	0.1724
4/3 earth	2.4848	1.2545	0.6549	0.3217	0.2034
		9 km Altitude			
	2°	5°	10°	20°	30°
Geometric optics	2.8306	1.4207	0.7419	0.3646	0.2304
4/3 earth	3.3983	1.8255	0.9733	0.4813	0.3046

Using a value of $dn/dh = -4 \times 10^{-8}\text{m}^{-1}$, we find that R' is approximately $4/3R$. The value for dn/dh can vary with meteorological conditions, which affects the approximation of $4/3R$. Replacing R with R' compensates for the bending of the ray due to atmospheric effects (in the standard atmosphere), and simplifies to $d\phi/ds = 1/R'$. In other words, the rays would travel in straight lines with height affected only by the earth's curvature, which simplifies the process that determines the height of the propagated path. As can be seen, the general effect of radar refraction in the standard atmosphere is to make the earth seem bigger and thus flatter than it really is.

Table 10.1 shows a comparison of geometric optics (ray tracing) and the 4/3 earth method for determining the refraction error for targets located at various elevation angles and target altitudes of 3, 6, and 9 km. For a target at 3 km altitude, the 4/3 earth technique is a reasonable approximation for the geometric optics computation of angle error due to refraction. As the target altitude increases, the 4/3 earth approximation is less accurate than for lower-altitude targets. As a result, the 4/3 earth approximation is considered valid for target altitudes less than 3 km. The preferred method for refraction error estimation is geometric optics.

10.1.2 Ray Tracing Adjoint Operator

To develop the observation operator, an approximate fast ray-tracing procedure can be implemented that relies only on straight-line propagation through the refractive atmosphere. This approach is very similar to the approach taken in [9] for refraction computations and makes use of the equation of geometrical optics in arclenth-parameterized coordinates:

$$\ddot{x} = \nabla \ln(n) - \nabla \ln(n)^T \dot{x}\,\dot{x} \tag{10.42}$$

Here x is the position along the ray emanating from the target object with initial vector direction $\dot{x}(0) = u$, $n = n(x)$ is the index of refraction at the point x, and $\nabla \ln(n)$ denotes the gradient of the natural logarithm of the index of refraction evaluated at the position x. Due to the nonlinear nature of the equation, quantitative solution requires a numerical approach, such as a Runge-Kutta numerical integrator. The incorporation of such techniques into adjoint computations are reasonably well understood, although they are often difficult to implement efficiently for the adjoint computation.

The approach taken here for the construction of the observation operator is to use a nonstandard approximation for the bending, taking advantage of the relationship between curvature and path bending. The curvature κ is defined in terms of the rate of bending of an arclength parameterized path. For such a path, the rate of bending is related to the rate of rotation of the tangent vector $\dot{x}(s)$. This term is easy to compute for an arclenth parameterized path because the length of the tangent vector does not change. A classical formula from calculus tells us that, in this case,

$$\ddot{x}(s) = \kappa(s)\,\hat{N}(s) \tag{10.43}$$

where $\hat{N}(s)$ is the inward pointing normal. The relationship between these quantities for a short path segment of constant curvature is shown in Figure 10.6.

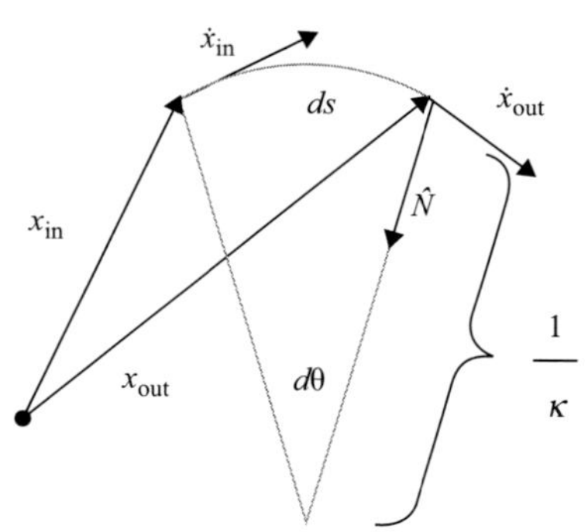

Figure 10.6. Relationship Between Curvature and Bending

From this picture, it is clear that, for an infinitesimal path segment, $ds = \kappa^{-1} d\theta$, or for our purposes, $d\theta = \kappa\, ds$. Integration of both sides leads to the formula for the bending of the path:

$$\theta_{\text{bending}} = \int_0^L \kappa(s)\, ds \tag{10.44}$$

Comparing this formula with (10.42) and (10.43) leads to the following observations:

$$\begin{aligned}\theta_{\text{bending}} &= \int_0^L \kappa(s)\, ds \\ &= \int_0^L \ddot{x}(s) \cdot \hat{N}(s)\, ds \\ &= \int_0^L (\nabla \ln(n(x(s)) - \nabla \ln(n(x(s)) \cdot \dot{x}(s)\, \dot{x}(s)) \cdot \hat{N}(s)\, ds \\ &= \int_0^L \nabla \ln(n(x(s)) \cdot \hat{N}(s)\, ds \end{aligned} \tag{10.45}$$

10.2 Tropospheric Turbulence Effects

10.2.1 Basic Theory for RF Turbulence

Radio frequency turbulence degrades the phase stability of a radar signal by distorting the constant phase surfaces and creating intensity fluctuations in the signal. These effects are linearly related to turbulence strength and roughly proportional to the propagation distance of the signal through the troposphere. This degradation is a result of variations in atmospheric refractivity that exist due to small-scale temperature and humidity fluctuations along the propagation path. These fluctuation regions behave like weak lenses that serve to focus and defocus the radar signal. Although the index-of-refraction fluctuations are small, they impose phase fluctuations on the radar signal that grow nearly linearly with propagation distance. Because the propagation path length through the troposphere scales approximately as the reciprocal of the sine of the elevation angle at low elevations, turbulence effects can dominate both phase-fluctuation and amplitude-fluctuation errors at low elevations where the path lengths can be 100 km or longer.

To understand the origin of the small-scale refractivity fluctuations that lead to RF turbulence, consider first the well-known effect of turbulent flow in a fluid, such as the atmosphere [10,11,12,13,14]. Fluid motion is generally thought of in terms of being either laminar or turbulent. Laminar flow is characterized by smoothly varying velocities across the entire fluid, a state that can be described by a small number of parameters. The Reynolds number *Re* is the bulk flow parameter that determines the flow characteristics of the fluid in regions well separated from the flow boundary, $Re = l_0 v_0 \nu^{-1}$, and is defined in terms of a characteristic system length l_0, multiplied by a characteristic system velocity v_0 and the reciprocal of the kinematic viscosity of the fluid ν. Laminar fluids are characterized by a low Reynolds number, a condition indicating that small variations in the fluid velocity will equalize out to the average fluid velocity. A turbulent fluid is characterized by a high Reynolds number, implying that small fluctuations will be amplified. Amplification of these fluctuations results in an energy cascade mechanism through which the energy of simple bulk fluid motion is transferred to smaller scales through the formation of vortices or eddies. These smaller-scale eddies, when considered as a system, have characteristic Reynolds numbers smaller than that of the bulk. Due to the viscosity effects that resist the formation of large velocity gradients, the effective Reynolds number of the eddies decrease with size. For small enough eddy size, the kinetic energy is dissipated as heat rather than passed on to smaller length scales.

The atmosphere generally has a very large Reynolds number for eddies down to centimeter length scales and thus behaves as a turbulent fluid on most scales of interest in radar propagation. This turbulent motion convects and mixes air with different temperature and humidity properties, and this inhomogeneous mixture persists due to the low thermal conductivity of air and the relatively low diffusivity of humidity. Because the index of refraction of the atmosphere depends strongly on temperature and humidity, this mixture process results in a spatially varying mixture of air with differing indices of refraction across a wide range of length scales.

10.2.2 Turbulence-Induced Radar Effects

Atmospheric turbulence degrades both the phase and amplitude of a return signal. The phase degradation increases the error in angle-of-arrival estimation by as much as 100–200 μrad in bad turbulent conditions. This corresponds to approximately 500 m in cross-range error for a target located 2000 km from the radar. In these conditions, turbulence error is the dominant contributor to angle-of-arrival error. Amplitude degradation leads to scintillation that can cause variations in SNR by as much as 4 dB, thereby degrading search, track, and target classification performance.

Temporal and spatial averaging serve to mitigate turbulence-induced phase and amplitude errors. In this section, analysis quantifies the temporal and spatial decorrelation intervals that permit effective mitigation of errors due to turbulence.

Turbulence errors are caused by eddies set up in the troposphere that act as refractive and diffractive lenses. The sizes of these eddies are bounded by the inner-scale (l_0) and outer-scale (L_0) of turbulence. The inner scale is usually assumed to be on the order of a few centimeters, and the outer scale is on the order of a few tens or hundreds of meters; the power spectral density of the turbulence is a function of these parameters. Kolmogorov was the first to estimate the form of the power spectral density of velocity fluctuations in turbulence, and much work has followed to extend and refine these estimates. In particular, Von Karman's methods are generally used in practice because these methods have proven to be reliable in estimating the effects of turbulence due to a more complete form of the power spectral density. However, we will use the Kolmogorov power spectral density in the presentation that follows due to its relative simplicity compared with these other forms.

10.2.3 Turbulence-Induced Radar Scintillation

When a microwave signal propagates in the turbulent atmosphere, small variations in the index of refraction in space and time caused by microscopic changes in the temperature and humidity result in signal amplitude fluctuations and random changes to angle-of-arrival (AOA) radar measurements. The amplitude variation, called scintillation, can cause severe fades of a radar signal and thus degrade radar performance. AOA fluctuations result in a tracking error. As the elevation angle decreases, the thickness of the atmosphere between radar and target increases, magnifying the atmospheric degradation effects. For this reason, a quantitative understanding of the effects of small-scale atmospheric turbulence on radar performance becomes increasingly important for low elevation angles.

A notional radar with aperture diameter $B = 6.75$ m, wavelength $\lambda = 3$ cm, and target range $L = 2000$ km is examined to illustrate the impact of turbulence effects on tracking error. Two cases are considered, corresponding to moderate and severe turbulence. It is shown that radar track accuracy in severe turbulent conditions is significantly deteriorated by a small-scale turbulence.

10.2.4 Radar Beam Fluctuation at the Target

For the analysis of radar-beam propagation in a turbulent atmosphere, we use the Rytov approximation [12,13,14] for the propagation model and the turbulent power spectral density function derived by Kolmogorov, $\Phi_n(\kappa) = 0.033\, C_n^2\, \kappa^{-11/3}$. The Kolmogorov spectrum is valid only over the inertial subrange, $2\pi/L_0 < \kappa < 2\pi/l_0$.

In the Rytov method [9], an electromagnetic field is presented in the form $U(r) = U_0(r)e^{i\Psi(r)}$, where U_0 is the field in free space.

$$\psi(r) = \chi + iS_1 = \ln(A/A_0) + iS_1 \tag{10.46}$$

is the complex phase perturbation, $\chi = \ln(A/A_0)$ is the logarithm of the amplitude, A and A_0 are the amplitudes in the turbulent atmosphere and free space, respectively, and S_1 is the difference between the perturbed and unperturbed phase. The Rytov method permits us to calculate statistical moments of χ and S when a statistical model of the turbulent medium is known. Using the Kolmogorov turbulence spectrum [1], the variances of the log amplitude and AOA fluctuations of a spherical wave on the uplink propagation path are given by

$$\sigma_\chi^2 = 0.56(2\pi/\lambda)^{7/6}\int_0^L C_n^2(h(\eta))(\eta/L)^{5/6}(\eta - L)^{5/6}d\eta \tag{10.47}$$

$$\sigma_\theta^2 = 0.264\frac{\pi^2 k^2}{B^2}\int_0^L\int_{\frac{2\pi}{L_0}}^{\frac{2\pi}{l_0}} C_n^2(h(\eta))\kappa^{-8/3}\cos^2\left(\frac{\eta\cdot(L-\eta)}{2kL}\kappa^2\right)\left(1 - J_0\left(\kappa B\frac{\eta}{L}\right)\right)d\kappa d\eta \tag{10.48}$$

Here $C_n^2(h)$ is the vertical profile of the refractive index structure characteristic, k is the wave number equal to $2\pi/\lambda$, and $h(\eta)$ is the altitude of the propagation path above the ground at the distance η from the radar along the path.

10.2.5 Space-to-Ground Turbulence Analysis

To characterize the strength of turbulence along the path, we use vertical profiles of $C_n^2(h)$ measured with the Millstone Hill UHF radar in Westford, Massachusetts. The two representative sets of data corresponding to moderate and severe turbulent conditions are shown in Figure 10.7. The estimated values of the log amplitude and AOA variance of a radar signal at the target calculated from the equations for σ_x^2 and σ_θ^2 are given in Tables 10.2 and 10.3.

Under the conditions considered here, the RMS AOA fluctuation along a typical low-elevation propagation path with severe turbulence is approximately 150 μrad. It is seen that for severe turbulent conditions and an elevation angle of 0.5°, the variance of log amplitude fluctuations at the target is $\sigma_x^2 = 0.19$. This means that for severe turbulent conditions, the turbulence-induced random target brightness modulation is 56%. For moderate turbulent conditions and the same elevation angle, the target brightness modulation is 14%.

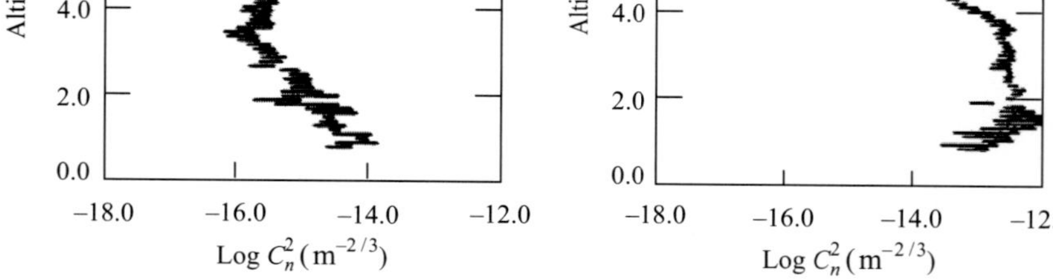

Figure 10.7. Vertical Profiles of C_n^2 for Moderate (left) and Bad (right) Turbulent Conditions from the Radar Measurements [13]

Table 10.2. Severe Turbulent Conditions, Log-Amplitude and Angle-of-Arrival Statistics

Elevation	**0.5°**	**1°**	**2°**	**5°**
σ_χ	0.44	0.36	0.2	0.14
σ_ϕ (μrad)	172	144	119	60

Table 10.3. Moderate Turbulent Conditions, Log-Amplitude and Angle-of-Arrival Statistics

Elevation	**0.5°**	**1°**	**2°**	**5°**
σ_χ	0.13	0.11	0.08	0.04
σ_ϕ (μrad)	54	46	36	19

In the case of a homogeneous and isotropic random medium, the correlation function for log amplitude between two targets can be written:

$$B_\chi(L,\rho) = 0.033k^2 4\pi^2 \int_0^L C_n{}^2(\eta)d\eta \int_{\frac{2\pi}{L_0}}^{\frac{2\pi}{\ell_0}} d\kappa J_0\left(\kappa\eta\frac{\rho}{L}\right)\kappa^{-8/3}\sin^2\left(\frac{\left(\eta - \frac{\eta^2}{L}\right)}{2k}\kappa^2\right) \tag{10.49}$$

For phase differences, the structure function, which is very similar to (10.49), is used to calculate correlations:

$$D_S(L,\rho) = 0.033k^2 8\pi^2 \int_0^L C_n{}^2(\eta)d\eta \int_{\frac{2\pi}{L_0}}^{\frac{2\pi}{\ell_0}} d\kappa \left[1 - J_0\left(\kappa\eta\frac{\rho}{L}\right)\right]\kappa^{-8/3}\cos^2\left(\frac{\left(\eta - \frac{\eta^2}{L}\right)}{2k}\kappa^2\right) \tag{10.50}$$

In both cases, the η integral is a path integral along the line between the radar and the first target (distance L away), ρ is a distance between two points in the plane perpendicular to the η integration path, k is the wave number of the signal ($= 2\pi/\lambda$), and $C_n{}^2$ quantifies variations in the refractive index.

These two equations are central to the simulation of turbulence and are first used to calculate statistical variables for single targets. The error on an angular measurement is related to (10.50) by

$$\sigma_\theta = \frac{\lambda}{2\pi\rho}\sqrt{D_S(L,\rho)} \tag{10.51}$$

where in the case of a phased-array radar, ρ refers to the half-width of the array. In the case of infrared sensors, we take it to be the diameter of the optics in front of the focal plane array. The error on the log amplitude is

$$\sigma_\chi = \sqrt{B_\chi(L,0)} \tag{10.52}$$

This quantity can then be used to calculate the error on the radar cross section (in decibels):

$$\sigma_{RCS} = \sqrt{10\log_{10}(e^{4\sigma_\chi^2}(e^{4\sigma_\chi^2}-1))} \tag{10.53}$$

In the case of two targets, (10.49) and (10.50) are used to calculate correlation matrices for log amplitude and phase. The log amplitude calculation is straightforward, using (10.49) and setting ρ equal to the distance between the targets. The phase correlations are more complicated because one must account for the distance across the radar face as well as for the distance between the two targets. Equations (10.49) and (10.50) can be numerically computed for various radar frequencies.

10.2.5.1 Spatially Correlated Turbulence

In addition to considering the turbulence correlations using one radar to observe two targets, we can also consider using two radars or distributed arrays to view one target, as in an interferometer. In this case, (10.49) can be used to examine the amplitude fluctuation correlations seen at the two radars. This is plotted in Figure 10.8 for the moderate and severe turbulence profiles given previously for a target at 1° elevation. The results for B_χ have been normalized to $\rho = 0$, effectively showing a correlation coefficient. The correlation drops off as separation increases but overshoots zero and

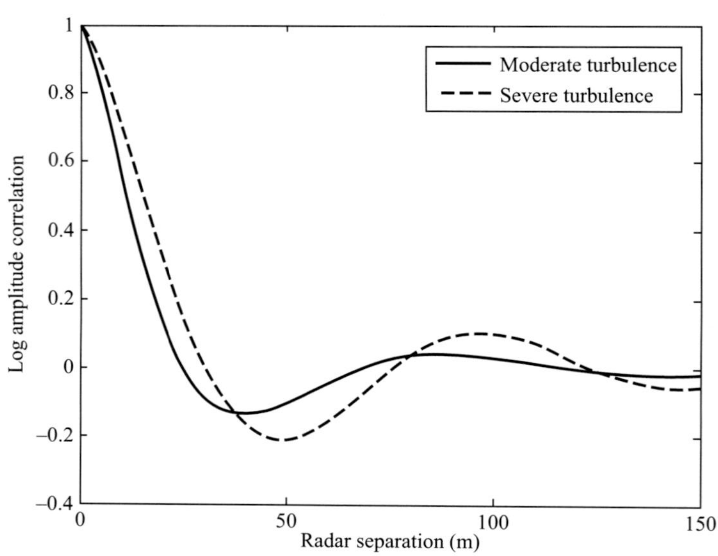

Figure 10.8. Log Amplitude Correlation Versus Radar Separation

then oscillates around it. This is, of course, similar to any two-wave interference pattern in which waves constructively and destructively add. Note that the strength of the turbulence has some effect on the location of the minima and maxima.

Applying the results in Figure 10.8 to interferometry, we see that for small antenna separations (less than 10 m), the angle-of-arrival error is highly spatially correlated, and thus for angle estimation using phase differencing, the turbulence error is negligible. However, for antenna separations in excess of 10 m, the decorrelation of the spatial turbulence effects can contribute to an additional error source for interferometry. This effect is most noticeable at relatively low RF frequencies where large baselines are needed to achieve the required angle accuracies.

References

1. A. N. Kolmogorov, “The local structure of turbulence in an incompressible viscous fluid for very large Reynolds numbers,” *C. R. (Doki) Acad. Sci., U.S.S.R.*, vol. 30, pp. 301–305, 1941.
2. M. I. Skolnik, *Radar Handbook*, 2nd Edition, McGraw Hill, New York, 1990.
3. P. Z. Peebles, *Radar Principles*, John Wiley & Sons, New York, 1998.
4. M. E. Lopez and W. W. Vickers, “Refraction correction of rocket tracking radar inputs in near Real time,” *J. Atmos. Sci.*, vol. 29, pp. 893–899, Jul. 1972.
5. M. Born and E. Wolf, *Principles of Optics: Electromagnetic Theory of Propagation, Interference, and Diffraction of Light*, 7th Edition, Cambridge University Press, Cambridge, UK, 2002.
6. B. R. Bean and G. D. Thayer, *CRPL Exponential Reference Atmosphere*, National Bureau of Standards Monograph 4, Washington, D.C., USA, 1959.
7. W. B. Sweezy and B. R. Bean, “Correction of atmospheric refraction errors in radio height finding,” *Journal of Research of the NBS, Radio Propagation*, vol. 67D, no. 2, pp. 139–142, March–April 1963.
8. B. R. Bean and E. J. Dutton, *Radio Meteorology*, Dover Publications, New York, 1966.
9. I. M. Gelfand and S. V. Fomin, *Calculus of Variations*, Prentice-Hall, Englewood Cliff, NJ, USA, 1963.
10. V. I. Tatarskii, *Wave Propagation in a Turbulent Medium*, McGraw-Hill, New York, 1961.
11. A. Ishimaru, *Wave Propagation and Scattering in Random Media*, Vol. 2, Academic, New York, 1978.
12. Rytov, S. M., Y. A. Kravtsov, and V. I. Tatarskii, *Principles of Statistical Radiophysics, Vol. 4: Wave Propagation Through Random Media*, Springer, Berlin, Germany, 1989.
13. S. A. Cohen, “Radar measurements of turbulent eddy dissipation rate in the troposphere: A comparison of techniques,” *J. Atm. Ocean Technology*, vol. 12, 85–95, 1995.
14. L. C. Andrews and R. L. Phillips, *Laser Beam Propagation in Random Media*, SPIE Optical Engineering Press, Bellingham, WA, USA, 1998.

Appendix A

Discrete Fourier Transform

The *discrete Fourier transform* (DFT) is the algorithm of choice to implement a sampled version of the Fourier transform. The DFT of a time signal $x[n]$ is given by

$$X[k] = \sum_{n=0}^{N-1} x[n] e^{-j(2\pi/N)kn} \tag{A.1}$$

The DFT can be interpreted as a sampled version of the discrete time Fourier transform (DTFT) defined by

$$X(f) = \sum_{n=0}^{N-1} x[n] e^{-j2\pi nf} \tag{A.2}$$

Note that

$$X[k] = X(f)|_{f=k/N} \tag{A.3}$$

The DFT is normally implemented in practice using the famous fast Fourier transform (FFT) algorithm. The FFT is an efficient way to implement the DFT and is not a transform in and of itself. Furthermore, the FFT is actually a large and growing class of algorithms optimized for different transform sizes, real versus complex data, and so forth. The choice of FFT algorithm is important when considering computational speed and quantization noise. In general, the number of operations required to implement the FFT is on the order of $N\log_2(N)$, where N is the length of the discrete time sequence that is being transformed. The actual number of total arithmetic operations is

$$\text{Total operations} = 5N\log_2(N) \tag{A.4}$$

The apparent loss of amplitude in the DFT when the input signal frequency does not match the DFT sample frequency is called straddle loss. Straddle loss can be mitigated by (1) zero padding the FFT algorithm to increase the sample frequency and (2) applying a weighting function to the data to degrade the resolution. The average straddle loss is 1.54 dB when no weighting is used and 0.65 dB when a Hamming window is used. Of course, Hamming weighting imparts a 1.34-dB mismatch loss at the outset, but the amplitude deviation is less than applying no window due to the loss in resolution.

Of course, zero padding increases the number of computations required to compute the DFT. If P is the number of zeros added to the input sequence, then the number of computations required to implement the zero padded DFT is

$$\text{Total operations} = 5(N + P)\log_2(N + P) \tag{A.5}$$

Appendix B

The Matched Filter

In this appendix, we derive the expression for the matched filter that optimizes the signal-to-noise ratio for a given signal [1,2]. We show that the matched filter is directly proportional to the signal shifted in time divided by the noise power and thus is matched to the signal and the noise.

Let $s(t)$ represent a real signal and $\psi(t)$ represent the analytic signal associated with $s(t)$. Thus

$$\psi(t) = s(t) + j\hat{s}(t) \tag{B.1}$$

where $\hat{s}(t)$ is the Hilbert transform of $s(t)$ defined as follows:

$$\hat{s}(t) = \frac{1}{\pi}\int_{-\infty}^{\infty}\frac{s(\alpha)}{t-\alpha}d\alpha \tag{B.2}$$

Let $h(t)$ represent the real impulse response for the matched filter with transfer function $H(\omega)$. Then the analytic signal representing $\psi_o(t)$, the output of the filtered real signal $s_o(t)$, is given by

$$\psi_o(t) = \frac{1}{2}\psi(t) * h(t) = \frac{1}{2\pi}\int_{-\infty}^{\infty}\frac{1}{2}\Psi_o(\omega)e^{j\omega t}d\omega = \frac{1}{2\pi}\int_{-\infty}^{\infty}\frac{1}{2}\Psi(\omega)H(\omega)e^{j\omega t}d\omega \tag{B.3}$$

where $\Psi_o(\omega)$ and $\Psi(\omega)$ are the Fourier transforms of $\psi_o(t)$ and $\psi(t)$, respectively. The factor of ½ is used so that the real part of the analytic signal ψ corresponds to the real part of the filter output.

Let t_o be defined as the time that the optimal filter response occurs; then at time t_o the peak output signal power is

$$\hat{S}_o = |\psi_o(t_o)|^2 = \left|\frac{1}{2\pi}\int_{-\infty}^{\infty}\frac{1}{2}\Psi(\omega)H(\omega)e^{j\omega t_0}d\omega\right|^2 \tag{B.4}$$

Let the real noise $n(t)$ be modeled as a sampled function of a continuous random noise process $N(t)$ with power spectral density $\mathcal{N}$. The power spectral density of the real noise output is given by $\mathcal{N}(\omega)|H(\omega)|^2$. The average noise power output N_o is expressed as follows:

$$N_o = \frac{1}{2}\frac{1}{2\pi}\int_{-\infty}^{\infty} \mathcal{N}(\omega)|H(\omega)|^2 d\omega \tag{B.5}$$

The factor of ½ is due to the fact that, since the noise $n(t)$ is real, the power of the analytic noise spectrum is twice that of the real noise spectrum. We can now write peak signal power to average noise power as

$$\frac{\hat{S}_o}{N_o} = \frac{\left|\frac{1}{2\pi}\int_{-\infty}^{\infty} \Psi(\omega)H(\omega)e^{j\omega t_0} d\omega\right|^2}{2\frac{1}{2\pi}\int_{-\infty}^{\infty} \mathcal{N}(\omega)|H(\omega)|^2 d\omega} \tag{B.6}$$

Determining the optimal filter $H(\omega)$ from above would appear to be a difficult chore; however by using Schwartz's inequality, the filter can be determined with little difficulty. Schwartz's inequality can be expressed in the following form:

$$\left|\int_{-\infty}^{\infty} A(\omega)B(\omega)d\omega\right|^2 \leq \int_{-\infty}^{\infty} |A(\omega)|^2 d\omega \int_{-\infty}^{\infty} |B(\omega)|^2 d\omega \tag{B.7}$$

An equality occurs if and only if

$$A(\omega) = CB^*(\omega) \tag{B.8}$$

where C is an arbitrary nonzero real constant. Here we choose

$$A(\omega) = \frac{1}{\sqrt{2\pi}}\sqrt{\mathcal{N}}H(\omega) \tag{B.9}$$

and

$$B(\omega) = \frac{1}{\sqrt{2\pi}}\frac{\Psi(\omega)e^{j\omega t_o}}{\sqrt{\mathcal{N}}} \tag{B.10}$$

which yields

$$\left|\frac{1}{2\pi}\int_{-\infty}^{\infty} \Psi(\omega)H(\omega)e^{j\omega t_0} d\omega\right|^2 \leq \frac{1}{2\pi}\int_{-\infty}^{\infty} \mathcal{N}(\omega)|H(\omega)|^2 d\omega \cdot \frac{1}{2\pi}\int_{-\infty}^{\infty} \frac{|\Psi(\omega)|^2}{\mathcal{N}(\omega)} d\omega \tag{B.11}$$

,

and equality occurs when

$$H(\omega) = \frac{C\Psi^*(\omega)e^{-j\omega t_o}}{\mathcal{N}} \tag{B.12}$$

Substituting (B.11) into (B.6) we have,

$$\frac{\hat{S}_o}{N_o} \leq \frac{1}{4\pi} \int_{-\infty}^{\infty} \frac{|\Psi(\omega)|^2}{\mathcal{N}(\omega)} d\omega = \left(\frac{\hat{S}_o}{N_o}\right)_{\max} \tag{B.13}$$

and thus when (B.12) is satisfied, we have equality in (B.13). Thus (B.12) is the expression for the matched filter that optimizes the signal-to-noise ratio. For white noise, we can represent noise as constant over all frequencies

$$\mathcal{N} = \frac{N_o}{2} \tag{B.14}$$

Transforming (B.12), we have

$$h(t) = \frac{2C\psi^*(t - t_o)}{N_o} \tag{B.15}$$

It also follows that the real filter response is

$$h_r(t) = \frac{2Cs^*(t - t_o)}{N_o} \tag{B.16}$$

Since the constant C is arbitrary, we can choose $C = 1/2$ and

$$h_r(t) = \frac{s^*(t - t_o)}{N_o} \tag{B.17}$$

References

1. P. Z. Peebles, *Radar Principles*, John Wiley & Sons, New York, 1998.
2. N. L. Levanon and E. Mozeson, *Radar Signals*, IEEE Press Wiley-Interscience, Hoboken, NJ, USA, 2004.

Appendix C
The Principle of Stationary Phase

The principle of stationary phase states that the major contribution to the energy spectral density comes from when the phase is stationary. In other words, the spectral density energy of frequency is relatively large at a specific time when the rate of change of frequency at that time is relatively small [1]. We illustrate this with an example. Assume that the phase function $f(t)$ has the following form:

$$f(t) \approx f(c) + \frac{f''(t)}{2}(t-c)^2 \tag{C.1}$$

$$f'(c) = 0 \tag{C.2}$$

Now compute the following generalized Fourier integral:

$$U(f) = \int_{-\infty}^{\infty} g(t)e^{ixf(t)}dt \qquad x > 0 \tag{C.3}$$

$$U(f) = \int_{-\infty}^{c-\varepsilon} g(t)e^{ixf(t)}dt + \int_{c-\varepsilon}^{c+\varepsilon} g(t)e^{ixf(t)}dt + \int_{c+\varepsilon}^{\infty} g(t)e^{ixf(t)}dt \tag{C.4}$$

Riemann-Lebesgue lemma: If g is Lebesgue integrable over the interval $[a\ b]$ and if f is continuously differentiable and not constant on any subinterval of $[a\ b]$, then

$$\lim_{x\to\infty} \int_a^b g(t)e^{ixf(t)}dt = 0 \tag{C.5}$$

Proof: We apply integration by parts to the integral in (C.5).

$$\begin{aligned}
\int_a^b g(t)e^{ixf(t)}dt &= \int_a^b g(t)f'(t)e^{ixf(t)}\frac{1}{f'(t)}dt \\
&= \left[\frac{1}{ix}e^{ixf(t)}\left[\frac{g(t)}{f'(t)}\right]\right]_a^b - \int_a^b \frac{1}{ix}e^{ixf(t)}\frac{d}{dt}\left[\frac{g(t)}{f'(t)}\right]dt \\
&\leq \left|\frac{1}{x}\right|\left(\left|\left[e^{ixf(t)}\left[\frac{g(t)}{f'(t)}\right]\right]_a^b\right| + \left|\int_a^b e^{ixf(t)}\frac{d}{dt}\left[\frac{g(t)}{f'(t)}\right]dt\right|\right) \\
&= O\left(\frac{1}{x}\right)
\end{aligned} \tag{C.6}$$

Thus the integral approaches 0 as x approaches ∞, which proves the lemma.

Since f defined by (C.1) and (C.2) is nearly constant over the the interval $[c-\varepsilon \quad c+\varepsilon]$ for small ε but is not constant outside the interval, we apply the Riemann-Lebesgue lemma to (C.4), and for large x we have

$$
\begin{aligned}
U(f) &\approx \int_{c-\varepsilon}^{c+\varepsilon} g(t)e^{ixf(t)}dt \\
&\approx \int_{c-\varepsilon}^{c+\varepsilon} g(c)e^{ix\left(f(c)+f'(c)(t-c)+\frac{f''(c)}{2}(t-c)^2\right)}dt \\
&= \int_{c-\varepsilon}^{c+\varepsilon} g(c)e^{ix\left(f(c)+\frac{f''(c)}{2}(t-c)^2\right)}dt \qquad f'(c)=0 \\
&= g(c)e^{ixf(c)}\int_{c-\varepsilon}^{c+\varepsilon} e^{ix\frac{f''(c)}{2}(t-c)^2}dt \\
&= g(c)e^{ixf(c)}\int_{-\infty}^{\infty} e^{ix\frac{f''(c)}{2}(t-c)^2}dt \\
&= g(c)e^{ixf(c)}\int_{-\infty}^{\infty} e^{-\frac{(t-c)^2}{2\left(\sqrt{\frac{i}{xf''(c)}}\right)^2}}dt \\
&= g(c)e^{ixf(c)}\sqrt{2\pi\left(\sqrt{\frac{i}{xf''(c)}}\right)^2} \qquad \frac{1}{\sqrt{2\pi\sigma^2}}\int_{-\infty}^{\infty} e^{-\frac{z^2}{2\sigma^2}}dz = 1 \text{ (Gaussian)} \\
&= g(c)\sqrt{i}e^{ixf(c)}\sqrt{\frac{2\pi}{xf''(c)}} \\
&= g(c)e^{ixf(c)+i\frac{\pi}{4}}\sqrt{\frac{2\pi}{xf''(c)}}
\end{aligned}
\tag{C.7}
$$

Then the principle of stationary phase is expressed as follows:

$$
|U(f)|^2 \approx \frac{2\pi g^2(c)}{x|f''(c)|} \tag{C.8}
$$

When the second derivative of frequency has a large value, then less energy is contributed, and when the second derivative is near zero, there is considerable contribution to energy. However, since phase is the derivative of frequency, the expression can be stated that signal components that have stationary phase contribute significantly to the signal energy.

Reference

1. N. Levanon and E. Mozeson, *Radar Signals*, IEEE Press, John Wiley & Sons, Hoboken, NJ, USA, 2002.

Appendix D
The Fundamental Theory of Binary Code

We will begin with a review of the theory of binary codes [1]. A binary shift register is defined by the following:

$$x' = Tx \pmod 2 \tag{D.1}$$

where

$$T = \begin{bmatrix} 0 & 1 & 0 & 0 & \cdot & \cdot & \cdot & 0 & 0 \\ 0 & 0 & 1 & 0 & \cdot & \cdot & \cdot & 0 & 0 \\ 0 & 0 & 0 & 1 & \cdot & \cdot & \cdot & 0 & 0 \\ & & \cdot & & & & & \cdot & \cdot \\ & & \cdot & & & & & \cdot & \cdot \\ & & \cdot & & & & & \cdot & \cdot \\ 0 & 0 & 0 & 0 & \cdot & \cdot & \cdot & 0 & 1 \\ a_0 & a_1 & a_2 & a_3 & \cdot & \cdot & \cdot & a_{m-2} & a_{m-1} \end{bmatrix} \tag{D.2}$$

T is the characteristic matrix of the shift register and $\det(T) = a_0$.

Fact: There exists N such that $x_0 = T^N x_0, \; x_0 \neq 0$ and the set $[x_0, \; Tx_0, \; T^2 x_0, \; \ldots, T^N x_0]$ is called the cyclic set.

The matrix A defined by

$$A = [x \mid Tx \mid T^2 x \mid \ldots \mid T^{N-1} x] \tag{D.3}$$

is denoted the group code matrix and the resulting code is cyclic.

Definition: The characteristic polynomial of T is the polynomial defined by

$$\varphi(\lambda) = |T - \lambda I| = \lambda^N + a_{N-1}\lambda^{N-1} + a_{N-2}\lambda^{N-2} + \cdots + a_0 \tag{D.4}$$

Definition: The minimal polynomial of T is the monic polynomial $f(\lambda)$ of lowest degree such that $f(T) = 0$.

For a general matrix, $f(\lambda)$ divides $\varphi(\lambda)$, but in the case of binary shift registers, we have the following:

Fact: If a matrix has the form T, its minimal and characteristic polynomial are the same.

$$f(\lambda) = \lambda^N + a_{N-1}\lambda^{N-1} + a_{N-2}\lambda^{N-2} + \cdots + a_0 \tag{D.5}$$

Let Z be the set of all 2^q matric polynomials in T of degree $\leq q-1$ where q is the degree of the minimal polynomial $f(\lambda)$ of T. Thus we have:

1. Z is an Abelian (commutative) group under addition.
2. The left and right distributive law of multiplication holds.
3. The set Z_0 of all nonzero elements of Z is an Abelian group under multiplication.

As a result, we can say that Z is a Galois field of order 2^q.

Definition: The period r of T is such that $T^r = I$.

Fact: If T is $m \times m$ and nonsingular and $f(\lambda)$ is irreducible, then $r \leq 2^m - 1$ and r divides $2^m - 1$ where r is the period of T.

Definition: If the period of T is $2^m - 1$, then T is said to have maximal period.

Theorem: T has maximal period if and only if $f(\lambda)$ is irreducible, and $f(\lambda)$ does not divide $\lambda^k - 1$ for any $k < 2^m - 1$.

We have shown that any feedback shift register generates a cyclic code. The converse is also true.

Theorem: Let S be any cyclic code; then there is a feedback shift register that generates S.

Waveform Autocorrelation Property

Let c be a maximal length binary code with length $N = 2^q - 1$ and let $s = \exp(j\pi c)$. If M is the circular shift matrix defined as

$$M = \begin{bmatrix} 0 & 1 & 0 & 0 & \cdot & \cdot & \cdot & 0 & 0 \\ 0 & 0 & 1 & 0 & \cdot & \cdot & \cdot & 0 & 0 \\ 0 & 0 & 0 & 1 & \cdot & \cdot & \cdot & 0 & 0 \\ & & \cdot & & & & & \cdot & \cdot \\ & & \cdot & & & & & \cdot & \cdot \\ & & \cdot & & & & & \cdot & \cdot \\ 0 & 0 & 0 & 0 & \cdot & \cdot & \cdot & 0 & 1 \\ 1 & 0 & 0 & 0 & \cdot & \cdot & \cdot & 0 & 0 \end{bmatrix} \tag{D.6}$$

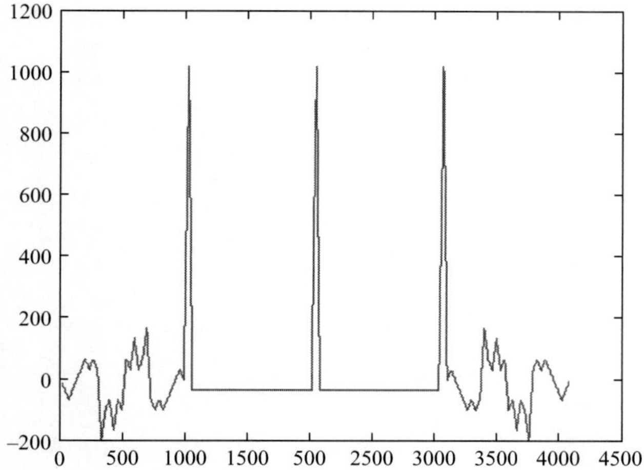

Figure D.1. Circular Autocorrelation of a 1023 Maximal Length Code

then the *i,k* circular autocorrelation of *s* is defined by

$$\begin{aligned} A_{i,k} &= corr_{i,k}(s,s) \\ &= \langle M^i s, M^k s\rangle \\ &= \langle \exp(j\pi M^i c), \exp(j\pi M^k c)\rangle \\ &= \sum \exp\left(j\pi\left(M^i c \oplus_2 M^k c\right)\right) \end{aligned} \tag{D.7}$$

where $\oplus_2$ indicates binary addition. From the group property of maximal length codes, it turns out that $\exp\left(j\pi\left(M^i c \oplus_2 M^k c\right)\right)$ is also a maximal length binary code consisting of 1s and -1s. Since the total number of 1s and -1s differs only by 1, the sum of all the elements of $A_{i,k}$ must equal ± 1 and we have,

$$A_{i,k} = \pm 1 \qquad \text{for all } \ i,k \ \ i \neq k \tag{D.8}$$

And the autocorrelation sidelobes are all ± 1, and, without loss of generality, we choose $A_{i,j} = -1$. Figure D.1 shows the circular autocorrelation of a maximal length code with length $1023 = 2^{10} - 1$.

Reference

1. R. B. Ash, *Information Theory*, Dover Publications Inc., New York, 1965.

Appendix E
Theoretical Development of Kasami Codes

We present the mathematical theory that leads to Kasami codes using the notation developed in Appendix D. Let u be a maximal length code of length $2^N - 1$ where $N = 2K$ is even. For finite Abelian groups, we have the following result

$$\begin{aligned} &2^{2K} - 1 = (2^K - 1)(2^K + 1) \\ &Z_{2^K - 1} \subset Z_{2^{2K} - 1} \end{aligned} \tag{E.1}$$

Now form the sequence v by

$$v(i) = u(mod(i(2^K + 1), 2^{2K} - 1) + 1) \quad \text{for } i = 1 : 2^K - 1 \tag{E.2}$$

then $v \in Z_{2^K - 1}$ and v is maximal length. Now form $w = [v\ v \ldots v]$ where w is $2^K + 1$ copies of v. It turns out that due to the fact that $2^K - 1$ divides $2^{2K} - 1$ and the subgroup property

$$w = \underbrace{[v \quad . \quad . \quad . \quad v]}_{2^k+1} = u(mod(i(2^K + 1), 2^{2K} - 1) + 1) \quad \text{for } i = 1 : 2^K - 1 \tag{E.3}$$

The cross-correlation between u and w is Kasami-like. That is, all the values are between $-(2^K - 1)$ and $+(2^K + 1)$. Furthermore, define

$$\begin{aligned} &ks(1, :) = u \\ &ks(i, :) = u \oplus_2 \left[w(i : 2^{2K} + 1) \quad w(1 : i - 1) \right] \quad \text{for } i = 1 : 2^{2K} \end{aligned} \tag{E.4}$$

Then the matrix ks is a group code with Kasami-like cross-correlation.

Note that each Kasami code is the sum of u with a cycle shift of w. Each cycle shift of w consists of $2^K + 1$ copies of cycle shifts of v. Since v is maximal length, all cycle shifts of v are also maximal length codes with i,k autocorrelation values equal to ± 1 when i not equal to k.

Let $u = [b_1 \quad b_2 \quad . \quad . \quad . \quad b_{2^N - 1}]$, then

$$w = [v \quad v \quad . \quad . \quad . \quad v] \tag{E.5}$$

where

$$v = \begin{bmatrix} b_{m1} & b_{m2} & . & . & . & b_{mp} \end{bmatrix}$$
$$mi = mod(i(2^K + 1),\ 2^N - 1) + 1 \qquad \text{for } i = 1{:}2^K - 1 \tag{E.6}$$

Since $2^K + 1$ divides $2^N - 1$, there are $2^K - 1$ number of $2^K + 1$ order subgroups of the $2^N - 1$ order cyclic group. Now the cross-correlation is computed as

$$corr_k(u, w) = \langle u, M^k w \rangle = \langle u, \begin{bmatrix} M^k v & M^k v & . & . & . & M^k v \end{bmatrix} \rangle \tag{E.7}$$

Each entry of $M^k v$ is multiplied by $2^K + 1$ entries in u that form a cyclic subgroup u. Let $M^k v_i$ denote the ith entry of $M^k v$ and u_S^i the associated ith subgroup of u, which is made up of $2^K + 1$ entries of u. Then continuing,

$$\begin{aligned} corr_k(u, w) &= \langle u, M^k w \rangle = \langle u, \begin{bmatrix} M^k v & M^k v & . & . & . & M^k v \end{bmatrix} \rangle \\ &= \sum_{i=1}^{2^K-1} M^K v_i \sum_{k=1}^{2^K+1} u_S^i(k) \\ &= \sum_{i=1}^{2^K-1} M^K v_i S_i \qquad \text{where } S_i = \sum_{k=1}^{2^K+1} u_S^i(k) \end{aligned} \tag{E.8}$$

Since each u_S^i is made up of $2^K + 1$ entries that have value either $+1$ or -1, S_i can take on only odd integer values over the interval $[-2^K + 1\ \ 2^K + 1]$. Since u is a maximal length code,

$$\sum_{i=1}^{2^K-1} S_i = -1 \tag{E.9}$$

and, in order for u to have autocorrelation

$$A_{i,k} = -1, \qquad \text{for all } i, k \ \ i \neq k \tag{E.10}$$

It will be necessary for each u_S^i to satisfy

$$Q_k = \sum_{i=1}^{2^K-1} \langle M^k u_S^i, u_S^i \rangle = -1, \qquad \text{for all } k \neq i \tag{E.11}$$

To satisfy all the conditions, S_i must be defined as follows:

$$S_i = \begin{cases} 1 & \text{for all } i \text{ except one} \\ -(2^K - 1) & \text{for one value of } i \end{cases} \tag{E.12}$$

Thus,

$$\sum_{i=1}^{2^K-1} S_i = \sum_{i=1}^{2^K-2} S_i - (2^K - 1) = 2^K - 2 - (2^K - 1) = -1 \tag{E.13}$$

Since each $M^k v$ has values $+1$ or -1 and is a maximal length code,

$$\sum_{i=1}^{2^K-1} M^k v_i = -1 \tag{E.14}$$

The sum of any subset of $2^K - 2$ values of $M^k v_i$ will have values either 0 or -2. Thus the cross-correlation has only two values:

$$corr_k(u, w) = \begin{cases} (2^K - 1) + 2 = 2^K + 1 \\ -(2^K - 1) + 0 = -(2^K - 1) \end{cases} \tag{E.15}$$

For Kasami codes to have minimum correlation, the exponent N needs to be even so that $2^N - 1$ factors into two numbers that are reasonably close in magnitude. When N is even, the factorization $2^N - 1 = (2^{N/2} + 1)(2^{N/2} - 1)$ yields values close to $2^{N/2}$. This factorization allows for two cyclic subgroups of the $2^N - 1$ order cyclic group that is defined by a maximal length code u. The subgroup factorization and the properties of maximal length codes (autocorrelation) lead to a maximal length subcode v embedded in u. The cross-correlation properties of u and v are defined by the order of the subgroups, and the factorization provides reasonably low cross-correlation values. For odd values of N, there may exist factorizations that yield two subgroups (e.g., for $N = 11$, $2^{11/2} - 1 = 23*89$) that lead to cross-correlations with two values based on the factorization. However, these factorizations tend to yield higher cross-correlation values because at least one of the factors is large relative to $2^{N/2}$.

Appendix F
Relationship of the Continuous Power Spectrum and Discrete Variance

We show the relationship between the continuous noise power spectrum and the sample variance for a discrete process.

Let $n(t)$ represent a continuous noise process where $n(t)$ is assumed to be mean zero and Gaussian. Since $n(t)$ is band-limited with bandwidth W, the power spectrum is defined by

$$\mathcal{P}(\omega) = \frac{\mathcal{N}}{2} rect\left(\frac{\omega}{2W}\right) \tag{F.1}$$

and autocorrelation function

$$R(\tau) = \frac{\mathcal{N} W}{2\pi} \operatorname{sinc}(W\tau) = \frac{\mathcal{N}}{2\pi} \frac{\sin(W\tau)}{\tau} \tag{F.2}$$

But $0 = R\left(\frac{\pi}{W}\right)$, so the uncorrelated sample interval is defined by $\Delta t = \frac{\pi}{W}$, and the variance is defined by

$$\sigma^2 = R(0) = \frac{\mathcal{N} W}{2\pi} = \frac{\mathcal{N}}{2\Delta t} \tag{F.3}$$

Appendix G
Time-of-Arrival CRLB (Alternative Approach)

We take a less rigorous approach to establish the CLRB in order to provide an alternative derivation of the CRLB for time-of-arrival angle estimation in Section 6.9. We begin with the convolution operator:

$$c(t_d) = |(a * a)(t_d)| = \left| \int_{-\infty}^{\infty} a(t)a^*(t - t_d)dt \right| \tag{G.1}$$

For simplicity, write $c(t_d)$ as though it were real valued and form the derivative

$$c(t_d) = (a * a)(t_d) = \int_{-\infty}^{\infty} a(t)a(t - t_d)dt \tag{G.2}$$

$$c'(t_d) = \frac{dc}{dt_d} = \int_{-\infty}^{\infty} a(t)\frac{d}{dt_d}a(t - t_d)dt \tag{G.3}$$

Then the estimator for the true time delay is the solution of the following:

$$c'(t_d, \varphi) = 0 \tag{G.4}$$

where we have introduced an implicit dependence on phase. Taking differentials, we have

$$\frac{\partial c'}{\partial t_d}dt_d + \frac{\partial c'}{\partial \varphi}d\varphi = 0 \tag{G.5}$$

and

$$\sigma_{t_d}^2 = E(dt_d^2) = \frac{(\partial c'/\partial\varphi)^2}{(\partial c'/\partial t_d)^2} E(d\varphi^2) = \frac{(\partial c'/\partial\varphi)^2}{(\partial c'/\partial t_d)^2} \sigma_\varphi^2$$

$$= \sigma_\varphi^2 \frac{\left(\int_{-\infty}^{\infty} \frac{\partial}{\partial\varphi} a(t) \frac{\partial}{\partial t_d} a(t-t_d)dt + \int_{-\infty}^{\infty} a(t) \frac{\partial^2}{\partial\varphi\partial t_d} a(t-t_d)dt \right)^2}{\left(\int_{-\infty}^{\infty} a(t) \frac{\partial^2}{\partial {t_d}^2} a(t-t_d)dt \right)^2}$$

$$= \sigma_\varphi^2 \frac{\left(\int_{-\infty}^{\infty} \frac{\partial}{\partial\varphi} a(t) \frac{\partial}{\partial t} a(t-t_d)dt + \int_{-\infty}^{\infty} a(t) \frac{\partial^2}{\partial\varphi\partial t} a(t-t_d)dt \right)^2}{\left(\int_{-\infty}^{\infty} \frac{\partial}{\partial t} a(t) \frac{\partial}{\partial t} a(t-t_d)dt \right)^2} \tag{G.6}$$

Since the preceding expression is valid for all t_d, we compute the CRLB for the case when $t_d = 0$.

$$\sigma_{t_d}^2 = \sigma_\varphi^2 \frac{\left(\int_{-\infty}^{\infty} \frac{\partial}{\partial\varphi} a(t) \frac{\partial}{\partial t} a(t)dt + \int_{-\infty}^{\infty} a(t) \frac{\partial^2}{\partial\varphi\partial t} a(t)dt \right)^2}{\left(\int_{-\infty}^{\infty} \frac{\partial}{\partial t} a(t) \frac{\partial}{\partial t} a(t)dt \right)^2} \tag{G.7}$$

We illustrate with the following example:

$$a(t) = a\cos(f(t) + \varphi) \qquad 0 \le t \le T \tag{G.8}$$

$$f(0) = f'(0) = 0 \tag{G.9}$$

where $f'(t)$ is assumed to be monotonic. This example is representative of both linear and nonlinear FM waveforms. Now the preceding integrals become

$$\sigma_{t_d}^2 = \sigma_\varphi^2 \frac{\left(\int\limits_{-\infty}^{\infty} \frac{\partial}{\partial \varphi} a(t) \frac{\partial}{\partial t} a(t) dt + \int\limits_{-\infty}^{\infty} a(t) \frac{\partial^2}{\partial \varphi \partial t} a(t) dt \right)^2}{\left(\int\limits_{-\infty}^{\infty} \frac{\partial}{\partial t} a(t) \frac{\partial}{\partial t} a(t) dt \right)^2}$$

$$= \sigma_\varphi^2 \frac{\left(\int\limits_{-\infty}^{\infty} f'(t) a^2 \sin^2(f(t)+\varphi) dt + \int\limits_{-\infty}^{\infty} f'(t) a^2 \cos^2(f(t)+\varphi) dt \right)^2}{\left(\int\limits_{-\infty}^{\infty} (f'(t))^2 a^2 \sin^2(f(t)+\varphi) dt \right)^2}$$

$$= \sigma_\varphi^2 \frac{\left(a^2 \int\limits_0^T f'(t) dt \right)^2}{\left(a^2 \int\limits_0^T (f'(t))^2 a^2 \sin^2(f(t)+\varphi) dt \right)^2} \geq \sigma_\varphi^2 \frac{\left(\int\limits_0^T f'(t) dt \right)^2}{\left(\int\limits_0^T (f'(t))^2 dt \right)^2}$$

$$= \sigma_\varphi^2 \frac{(f(T)-f(0))^2}{\left(\int\limits_0^T (f'(t))^2 dt \right)^2} = \frac{\sigma_\varphi^2}{\frac{1}{(f(T))^2} \left(\int\limits_0^T (f'(t))^2 dt \right)^2}$$

$$= \frac{\sigma_\varphi^2}{\overline{W}^2} \geq \frac{1}{2W^2 SNR} \tag{G.10}$$

where, for purposes of this example, we have defined the mean bandwidth as follows:

$$\overline{W}^2 = \frac{1}{(f(T))^2} \left(\int\limits_0^T (f'(t))^2 dt \right)^2 \tag{G.11}$$

Note that, for an LFM waveform where $f(t) = bT^2/2$, we have

$$\overline{W}^2 = \frac{1}{\left(\frac{b}{2} T^2 \right)^2} \left(\int\limits_0^T (bT)^2 dt \right) = \frac{4}{9} b^2 T^2 < b^2 T^2 = W^2 \tag{G.12}$$

and

$$\sigma_{t_d}^2 \geq \frac{1}{2W^2SNR} \tag{G.13}$$

The CLRB for range accuracy follows as before.

$$\sigma_{Range}^2 = c^2\sigma_{t_d}^2 \geq \frac{c^2}{2W^2SNR} \tag{G.14}$$

Appendix H
Two-Dimensional Trilateration Using CPT and RGS Ranging Methods—MATLAB® Code

```
%%%%%%%%%%
%% This program computes the location of a target in 2-D using
%% trilateration. The trilateration solution uses phase differencing
%% with four (4) receivers where the trialteration can be either
%% CPT or RGS. The distinction is in which error model for range error is
%% chosen to be used in the code (choose line 15 or line 16)

clear
close all
freq=2e9;          %% radar operating center frequency
bw=2e6;            %% radar signal bandwidth
snr=20;            %% SNR at the receivers
c=3e8;             %% speed of light
lambda=c/freq;  %% wavelength

%%choose error model
%%rang=lambda/(2*pi*10^(snr/20));  %% CPT range error
rang=c/(2*bw*10^(snr/20));    %% RGS range error

%% distance between receivers (size of the area bounded by the receivers)
rrr=100;

%% Location of the radar receivers
r1=rrr*[1  0]/2;
r2=rrr*[0  1]/2;
r3=rrr*[-1  0]/2;
r4=rrr*[0  -1]/2;

%%Location of the target
r=rrr*[.1  .1]/2;
```

```
%%range estimates
R1=norm(r1-r);
R2=norm(r2-r);
R3=norm(r3-r);
R4=norm(r4-r);

%%phase difference estimates
phi12=(R1-R2);
phi23=(R2-R3);
phi34=(R3-R4);
phi41=(R4-R1);

%%vector of phase differences
PD=[phi12 phi23 phi34 phi41];
plot(r1(1),r1(2))
hold on
plot(r2(1),r2(2));
plot(r3(1),r3(2));
plot(r4(1),r4(2));

for kk=1:1000
        err=rang*randn(1,4); %% range error based on error model
        PDE=PD+err;          %% erred phase differences

        rr=[0 0]';
        for k=1:20      %% 20 iterations for Newton-Raphson to converge
                RR1=norm(r1-rr');
                RR2=norm(r2-rr');
                RR3=norm(r3-rr');
                RR4=norm(r4-rr');

                phid12=(RR1-RR2);
                phid23=(RR2-RR3);
                phid34=(RR3-RR4);
                phid41=(RR4-RR1);

                PPD=[phid12 phid23 phid34 phid41];
                dr1=r1-rr';
                dr2=r2-rr';
                dr3=r3-rr';
                dr4=r4-rr';
```

```
            %%compute gradient
            M=[dr2(1)/RR2-dr1(1)/RR1  dr2(2)/RR2-dr1(2)/RR1
                dr3(1)/RR3-dr2(1)/RR2  dr3(2)/RR3-dr2(2)/RR2
                dr4(1)/RR4-dr3(1)/RR3  dr4(2)/RR4-dr3(2)/RR3
                dr1(1)/RR1-dr4(1)/RR4  dr1(2)/RR1-dr4(2)/RR4];

            %%Newton-Raphson algorithm
            rr=rr-inv(M'*M)*M'*(PPD'-PDE');

      end

      xx(kk,:)=rr;

      %%plots the distribution of position estimates
      plot(rr(1),rr(2),'.r')
      xlabel('Cross-Range (m)')
      ylabel('Range (m)')

end
hold off
```

Appendix I
Angle-of-Arrival Determination Using a Rotated Antenna Configuration

Assume that interferometer antennas are configured in a triangle configuration and that the origin is chosen to be equidistant from each antenna. Also assume that the three interferometer antennas are located in the *x-y* plane with *y* being up and that *z* is perpendicular to the array plane. Let *d* define the distance from any antenna to the origin and α_1 and α_2 be the rotation angle of the two left antennas with respect to the *x*-axis and α_3 be the rotation of the right antenna with respect to the *x*-axis. Furthermore, we assume that the *x-y* axes are rotated such that $\alpha_1 = \alpha_2 = \alpha$; then the antenna coordinates are as follows:

$$\vec{a}_1 = [-d\cos(\alpha) \quad -d\sin(\alpha) \quad 0] \tag{I.1}$$

$$\vec{a}_2 = [-d\cos(\alpha) \quad d\sin(\alpha) \quad 0] \tag{I.2}$$

$$\vec{a}_3 = [d\cos(\alpha_3) \quad d\sin(\alpha_3) \quad 0] \tag{I.3}$$

Let $\vec{r}_t$ be the target position vector located at azimuth angle Az, elevation angle El, and range r.

$$\vec{r}_t = [r\sin(Az)\cos(El) \quad r\sin(El) \quad r\cos(Az)\cos(El)] \tag{I.4}$$

Figure I.1 illustrates the antenna coordinate frame.

The interferometer makes a phase measurement at each antenna.

$$\varphi_i = k\|\vec{r}_t - \vec{a}_i\| \tag{I.5}$$

where $k = \frac{2\pi}{\lambda}$.

First order:

$$\Delta\varphi_{12} = \varphi_2 - \varphi_1 = 2kd\sin(\alpha)\sin(El) \tag{I.6}$$

$$\begin{aligned}\Delta\varphi_{13} &= \varphi_3 - \varphi_1 \\ &= kd(\cos(\alpha)+\cos(\alpha_3))\sin(Az)\cos(El) + kd(\sin(\alpha)+\sin(\alpha_3))\sin(El)\end{aligned} \tag{I.7}$$

$$\begin{aligned}\Delta\varphi_{23} &= \varphi_3 - \varphi_2 \\ &= kd(\cos(\alpha)+\cos(\alpha_3))\sin(Az)\cos(El) - kd(\sin(\alpha)-\sin(\alpha_3))\sin(El)\end{aligned} \tag{I.8}$$

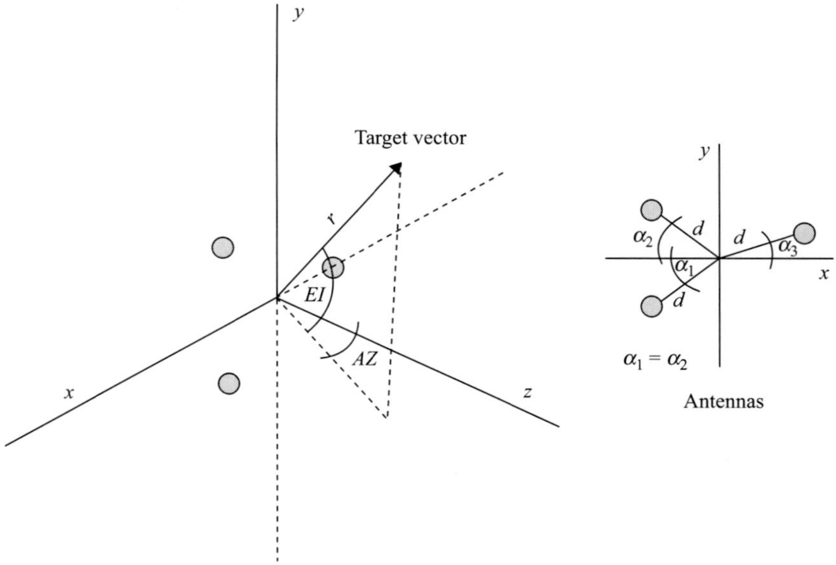

Figure I.1. Coordinate Frame for Measuring Target Angle

and angle is determined as follows:

$$\sin(El) = \frac{\Delta\varphi_{12}}{2kd\,\sin(\alpha)} \tag{I.9}$$

$$\cos(El) = \sqrt{1 - \sin^2(El)} \tag{I.10}$$

$$\sin(Az) = \frac{\Delta\varphi_{13} + \Delta\varphi_{23} - 2kd\,\sin(\alpha_3)\sin(El)}{2kd(\cos(\alpha) + \cos(\alpha_3))\cos(El)} \tag{I.11}$$

Second order: O(1/*r*)

$$\Delta\varphi_{12} = \varphi_1 - \varphi_2 = 2dk\,\sin(\alpha)\sin(El) + \frac{2d^2k}{r}\sin(Az)\cos(El)\sin(El)\sin(\alpha)\cos(\alpha) \tag{I.12}$$

$$\begin{aligned}\Delta\varphi_{13} &= \varphi_1 - \varphi_3 \\ &= kd(\cos(\alpha) + \cos(\alpha_3))\sin(Az)\cos(El) + kd(\sin(\alpha) + \sin(\alpha_3))\sin(El) \\ &\quad - \frac{d^2k}{2r}\sin^2(Az)\cos^2(El)\cos^2(\alpha) - \frac{d^2k}{2r}\sin^2(El)\sin^2(\alpha) \\ &\quad - \frac{dk}{r}\sin(Az)\cos(El)\sin(El)\sin(\alpha)\cos(\alpha) + \frac{d^2k}{2r}\sin^2(Az)\cos^2(El)\cos^2(\alpha_3) \\ &\quad + \frac{dk}{r}\sin(Az)\cos(El)\sin(El)\sin(\alpha_3)\cos(\alpha_3) + \frac{d^2k}{2r}\sin^2(El)\sin^2(\alpha_3)\end{aligned} \tag{I.13}$$

$$\begin{aligned}\Delta\varphi_{23} &= \varphi_2 - \varphi_3 \\ &= kd(\cos(\alpha)+\cos(\alpha_3))\sin(Az)\cos(El) - kd(\sin(\alpha)-\sin(\alpha_3))\sin(El) \\ &\quad -\frac{d^2k}{2r}\sin^2(Az)\cos^2(El)\cos^2(\alpha) - \frac{d^2k}{2r}\sin^2(El)\sin^2(\alpha) \\ &\quad +\frac{dk}{r}\sin(Az)\cos(El)\sin(El)\sin(\alpha)\cos(\alpha) + \frac{d^2k}{2r}\sin^2(Az)\cos^2(El)\cos^2(\alpha_3) \\ &\quad +\frac{dk}{r}\sin(Az)\cos(El)\sin(El)\sin(\alpha_3)\cos(\alpha_3) + \frac{d^2k}{2r}\sin^2(El)\sin^2(\alpha_3)\end{aligned} \tag{I.14}$$

We have the following:

$$\begin{aligned}\Delta\varphi_{13}+\Delta\varphi_{23} &= 2dk(\cos(\alpha_3)+\sin(\alpha))\sin(El) - \frac{d^2k}{r}\sin^2(Az)\cos^2(\alpha) \\ &\quad +\frac{d^2k}{r}\sin^2(Az)\sin^2(El)\cos^2(\alpha) + \frac{d^2k}{r}\sin^2(El)\cos^2(\alpha) \\ &\quad +\frac{d^2k}{r}\sin^2(Az)\sin^2(\alpha_3) - \frac{d^2k}{r}\sin^2(Az)\sin^2(El)\sin^2(\alpha_3) \\ &\quad +\frac{2dk}{r}\sin(Az)\cos(El)\sin(El)\sin(\alpha_3)\cos(\alpha_3) - \frac{d^2k}{r}\sin^2(El)\sin^2(\alpha_3)\end{aligned} \tag{I.15}$$

and angle is determined by the following iterative algorithm:

Step 1: Initialize.

$$\sin(El_1) = \frac{\Delta\varphi_{12}}{2kd\sin(\alpha)} \tag{I.16}$$

$$\cos(El_1) = \sqrt{1-\sin^2(El_1)} \tag{I.17}$$

$$\sin(Az_1) = \frac{\Delta\varphi_{13}+\Delta\varphi_{23} - 2kd\sin(\alpha_3)\sin(El_1)}{2kd(\cos(\alpha)+\cos(\alpha_3))\sin(Az)\cos(El_1)} \tag{I.18}$$

Step 2: Compute El_n for $n=2$.

$$\sin(El_n) = \frac{\Delta\varphi_{12}}{2kd\sin(\alpha) + \frac{2d^2k}{r}\sin(Az_{n-1})\cos(El_{n-1})\sin(\alpha)\cos(\alpha)} \tag{I.19}$$

Step 3: Compute Az_n.

$$\sin(Az_n) = \frac{-B + \sqrt{B^2 - 4AC}}{2A} \tag{I.20}$$

$$A = \frac{d^2k}{r}\left(\cos^2(\alpha_3) - \cos^2(\alpha_3)\right)\cos(El_n) \tag{I.21}$$

$$B = 2dk(\cos(\alpha_3) + \cos(\alpha))\cos(El_n) + \frac{2dk}{r}\sin(El_n)\cos(El_n)\sin(\alpha_3)\cos(\alpha_3) \tag{I.22}$$

$$C = \frac{d^2k}{r}\sin^2(El_n)\left(\sin^2(\alpha_3) - \sin^2(\alpha)\right) + 2kd\,\sin(\alpha_3)\sin(El_n) - \Delta\varphi_{13} - \Delta\varphi_{23} \tag{I.23}$$

$$\cos(Az_n) = \sqrt{1 - \sin^2(Az_n)} \tag{I.24}$$

Step 4: Return to Step 2, and stop after two iterations.

Appendix J

First- and Second-Order Interferometer Angle Measurements—MATLAB® Code

```
%%%% Antennas lie in the x-y plane %%%%
a1=[-d*cos(alpha) -d*sin(alpha) 0]; %% Antenna 1
a2=[-d*cos(alpha) d*sin(alpha) 0];  %% Antenna 2
a3=[d*cos(alpha3) d*sin(alpha3) 0]; %% Antenna 3

rt=r*[sin(Az)*cos(El) sin(El) cos(Az)*cos(El)]; %% Target Position

k=2*pi/lambda;

phi1=k*norm(rt-a1);
phi2=k*norm(rt-a2);
phi3=k*norm(rt-a3);

delphi12=phi1-phi2;
delphi13=phi1-phi3;
delphi23=phi2-phi3;

%%%% First Order Angle Estimation

sinel=delphi12/(2*k*d*sin(alpha));
cosel=sqrt(1-sinel^2);
sinaz=(delphi13+delphi23-
2*k*d*sin(alpha3)*sinel)/(2*k*d*(cos(alpha3)+cos(alpha))*cosel);

Elest1=asin(sinel)*180/pi
Azest1=asin(sinaz)*180/pi

%%%% Second Order Angle Estimation %%%%
```

```
for kk=1:2
    sinel=delphi12/((2*k*d*sin(alpha))- ...
        2*d^2*k*sinaz*cosel*sin(alpha)*cos(alpha)/r);
    cosel=sqrt(1-sinel^2);
    A=d^2*k*(cos(alpha3)^2-cos(alpha)^2)*cosel/r;
    B=2*d*k*(cos(alpha3)+cos(alpha))*cosel+ ...
        2*d^2*k*sinel*cosel*sin(alpha3)*cos(alpha3)/r;
    C=2*k*d*sin(alpha3)*sinel+d^2*k*sinel^2*(sin(alpha3)^2- ...
        sin(alpha)^2)/r-delphi13-delphi23;
    sinaz=(-B+sqrt(B^2-4*A*C))/(2*A);
    cosaz=sqrt(1-sinaz^2);
end

Elest2=asin(sinel)*180/pi
Azest2=asin(sinaz)*180/pi
```

Appendix K

Interferometer Angle Measurements for Distributed Transmit/Receive Antennas—MATLAB® Code

```
lambda=.03;
d=(lambda/8)*1/sqrt(3);  %%Nyquist spacing: d=1/sqrt(3) is the
                         %%correct spacing but introduces 4 times as
                         %%many ambiguities as a conventional
                         %%interferometer
az=45*pi/180;
el=45*pi/180;
range=250;

r=range*[sin(az)*cos(el) sin(el) cos(az)*cos(el)];

ant1=d*[1 0 0];
ant2=d*[-1/2 sqrt(3)/2 0];
ant3=d*[-1/2 -sqrt(3)/2 0];

k=2*pi/lambda;
%%%%%%Clockwise Array
sig13=exp(k*j*(norm(r-ant1)+norm(r+ant3)));
sig21=exp(k*j*(norm(r-ant2)+norm(r+ant1)));
sig32=exp(k*j*(norm(r-ant3)+norm(r+ant2)));
%%%%%%Counter-clockwise Array
sig12=exp(k*j*(norm(r-ant1)+norm(r+ant2)));
sig23=exp(k*j*(norm(r-ant2)+norm(r+ant3)));
sig31=exp(k*j*(norm(r-ant3)+norm(r+ant1)));
%%%%%%Antipodal Array
sig11=exp(k*j*(norm(r-ant1)+norm(r+ant1)));
sig22=exp(k*j*(norm(r-ant2)+norm(r+ant2)));
sig33=exp(k*j*(norm(r-ant3)+norm(r+ant3)));

%%%%%%First Order Estimation: Clockwise Array
```

```
%%Clockwise Array
delph=angle(sig13*sig21');
%%Uncorrelated Mixed Clockwise and Counter-clockwise
deldelph=angle(sig13*sig32'*(sig23'*sig12)');

el_est1=-asin((deldelph)/(3*k*d*sqrt(3)))
az_est1=-asin((delph)/(3*k*d*cos(el_est1)))

%%%%%%%First Order Estimation: Counter-clockwise Array


%%Counter-clockwise Array
delph=angle(sig12*sig31');
%%Uncorrelated Mixed Clockwise and Counter-clockwise
deldelph=angle(sig31*sig23'*(sig32'*sig21)');

el_est2=asin((deldelph)/(3*k*d*sqrt(3)))
az_est2=-asin((delph)/(3*k*d*cos(el_est2)))

%%%%%%%Average Estimates Using Both Arrays
el_est=(el_est1+el_est2)/2
az_est=(az_est1+az_est2)/2

el_est0=el_est;
az_est0=az_est;

%%%%%%%Second Order Estimate Using Antipodal Array
del12=angle(sig11*sig22');
del13=angle(sig11*sig33');
del23=angle(sig22*sig33');
alpha=3*d^2*k/(2*range);
beta=sqrt(3)*d^2*k/range;
A=alpha*beta^2;
B=-(del12+del13)*beta^2;
C=-alpha*del23^2;
sin2el=(-B+sqrt(B^2-4*A*C))/(2*A);
el_est3=sign(el_est0)*abs(asin(sqrt(sin2el)))
sinaz_est=del23/(beta*sin(el_est)*cos(el_est));
az_est3=asin(sinaz_est)
```

Index